Molecular Targeted Drug Discovery

Molecular Targeted Drug Discovery

A Guide to How Modern Medicines are Created

Robert A. Copeland, Ph.D.
Venture Partner, Atlas Venture
Advisor, Google Ventures
Adjunct Professor, Perelman School of Medicine
The University of Pennsylvania
Philadelphia, PA, USA

WILEY

Published by John Wiley & Sons, Inc., Hoboken, New Jersey.

Library of Congress Cataloging-in-Publication Data Applied for:

Hardback ISBN: 9781394341061

Cover Design: Wiley
Cover images created by and courtesy of Robert A. Copeland
Author Photo: Courtesy of Robert A. Copeland

Set in 9.5/12.5pt STIXTwoText by Straive, Chennai, India

To Nancy, Lindsey, Amanda, and Kieran.

And to the memory of two dear friends and treasured mentors: Howard S. Tager and Christopher T. Walsh.

Contents

Foreword

For over 25 years, Bob Copeland and I have worked together to discover and develop novel drugs. If I close my eyes, the first image of Bob that comes to mind is of him consumed in thought. He is a deep analytical thinker prone to describing complex concepts with mathematical equations that the rest of us just nod and pretend to understand. Bob also happens to be a uniquely talented drug discoverer whose work has directly led to five approved drugs to date. And finally, he is a fervent teacher at heart as evidenced by his extensive publication history of 242 (and counting) scientific publications and six textbooks with this one being his seventh.

The discipline of drug discovery has evolved and grown over the past 40 years that I have participated as a practitioner. In its early years, it was an obscure, poorly regarded domain of a small group of researchers working exclusively within pharmaceutical R&D campuses. A university professor once dismissively informed me that "a trained monkey could do drug discovery." But today, the creation of novel medicines is conducted by a broad range of scientists, physicians, and researchers working across a wide range of industrial, government, not-for-profit, and yes, even university settings. Most of us early devotees honed our skills in the craft of drug discovery by "seat-of-the-pants" learning from a limited number of scientific mentors. Academic institutions routinely educate students in foundational sciences, such as chemistry, biology, pharmacology, and clinical medicine, but rarely are these elements successfully taught as the integrated, multifunctional discipline that is drug discovery. That is why Bob's thoughtful effort to assemble his lifetime of experience into this textbook is so relevant. This book is a valuable resource not only for the uninitiated who are entering the field of pharmaceutical R&D but also for experienced functional experts in industry or academia who will benefit from the breadth of teachings and perspectives presented on the pages to follow.

Finding or designing a new drug that ultimately makes it to an FDA or other regulatory agency approval is a long and arduous task in which the odds are stacked against you. There are so many ways to fail that to detail each would be a reference text unto itself. Objective data-driven scientific evidence indeed serves as the bedrock for successful drug discovery. One of my favorite, among many, "Bob-isms" that I have shamelessly pirated is "We don't practice advocacy science." That is Bob's way to constantly remind younger scientists that experimental results should be interpreted without bias and regardless of whether they confirm or conflict with our preconceptions. However, successful drug discovery requires not only a deep knowledge of multiple sciences, but also insight, creativity, and persistence that not everyone possesses. The phrase, "It's not a job, it's an adventure," is apt because drug discovery is not only a profession but a personal mission that one chooses to take on. Many discovery scientists do good, even great, research over their careers, yet never work on a candidate molecule that becomes an approved medicine. Others reach the promised land with one drug, but never again. There also exists an element of subjective feel when it comes to selecting a drug candidate that has the legs to go all the way. A few discovery scientists have this instinct, but most do not. Many in the biopharma industry use the moniker "drug hunter" to describe those exceedingly rare individuals with the scientific acumen, practical experience, and the *je ne sais quoi* intuition who discover several approved drugs over the course of their lives. Bob is one of these exceptional drug hunters. Except that in real life, he is an avid fisherman and not a hunter. So, it would be more appropriate to confer Bob the title of "drug angler."

It has been my pleasure and privilege to work with Bob over such a long time and in many different venues. Our collaborative pursuit of finding and developing novel medicines has been occasionally frustrating, frequently fun, and always intellectually stimulating. But most importantly, it has resulted in improved outcomes and better treatment choices for patients. I feel fortunate to call Bob not just a friend, but a trusted partner in crime, even if I do not fish. Hopefully, this book will provide you with additional drug discovery knowledge and tools to catch your big one.

Peter Ho, M.D., Ph.D.
Center for Biotechnology Education
The Johns Hopkins University
Baltimore, MD, USA

Preface

Over the course of the past 35+ years, I have devoted my professional life and scientific acumen to the discovery and development of novel therapeutic agents to address unmet medical needs of patients. During my career, I have also endeavored to use my knowledge and experience to educate others through teaching, invited lectures, and published works in the form of textbooks and journal articles. My previous textbooks have been focused on protein science and enzymology, two specialized areas of science, each with great relevance to molecular targeted drug discovery.

Throughout my career, however, I have witnessed countless newcomers to drug discovery struggle to assimilate the vastness of the enterprise, the multidisciplinary nature of the discovery process, and more generally the key concepts required for the successful translation of basic science to concrete drug discovery. At the same time, I have also heard what amounts to mythology regarding where and how new medicines are created – or will in the future be created – from multiple sources: news media, politicians, business leaders (often with their own agendas), and learned academic colleagues outside the field of true drug discovery. Hence, I felt that there was a clear and current need for a book that would provide an approachable overview of the drug discovery process, as it applies to modern, molecular targeted therapeutics, for a broad audience of scientists, physicians, and business professionals that are planning to embark on careers in drug discovery and allied fields. Whereas drug discovery was once largely the purview of the biotechnology and pharmaceutical industry, today the art is practiced across a range of organizations from academic institutions to nonprofit organizations, for-profit companies, and even at the intersection of medicine and technology; this broad expansion of drug discovery further amplifies the need for a new book like this.

The book that follows is not intended to be a "how to" manual, filled with detailed protocols and processes. Such protocols and processes are readily available to the interested reader through the primary scientific literature and other

Acknowledgments

I have been greatly aided in the development of this book by Jennifer Castro and Brian Sparling, who have been kind enough to read and provide feedback on all the chapters and appendices of this book. I thank them for their tremendous help and insightful suggestions.

I have also benefitted from fruitful discussions and feedback on different components of the text from Peter T. C. Ho, Dashyant Dhanak, Peter Tummino, Peter Tonge, Peter Meinke, Stuart Ince, Serena Silver, Maureen Lynes, Matthew Daniels, Shane Buker, Nigel Waters, Scott A. Armstrong, Allen Oliff, P. Roy Vagelos, Lloyd Brotman, and Kim Stickland.

My experience and understanding of molecular targeted drug discovery has been honed and enriched over the years by detailed discussions with many talented colleagues and friends, especially Sean and Layla Sullivan, James A. (Jim) Wells, Paul S. Anderson, Robert Gould, Ron Evans, Tyler Jacks, H. Robert (Bob) Horvitz, William (Bill) Kaelin, Jr., George Daley, Paul Workman, Nathanael Gray, Howard Chang, Larry Lasky, P. Ann Boriack-Sjodin, Kenneth Duncan, Scott Ribich, David Schenkein, Jason Rhodes, and many others with whom I have interacted at Merck, Bristol Myers Squibb, GlaxoSmithKline, Epizyme, Inc., Accent Therapeutics, Atlas Venture, Google Ventures, and other institutions.

Finally, as always, I have been supported throughout the writing of this book by my loving family: my wife, Nancy, our two daughters, Lindsey and Amanda, and most recently our grandson, Kieran. I wish also to thank my son-in-law, Kyle Foster, for introducing me to the joys of great coffee – the pharmacodynamics of which fueled my writing. I am eternally grateful to all of them for their love, support, humor, and uplifting spirit, without which this work could not have been completed.

Epigraph

"The book of Nature is written in the language of mathematics."

—Galileo Galilei

"Great things are not done by impulse, but by a series of small things brought together."

—Vincent Van Gogh

"Whosoever saves a single life saves the world entire."

—Jewish Talmud

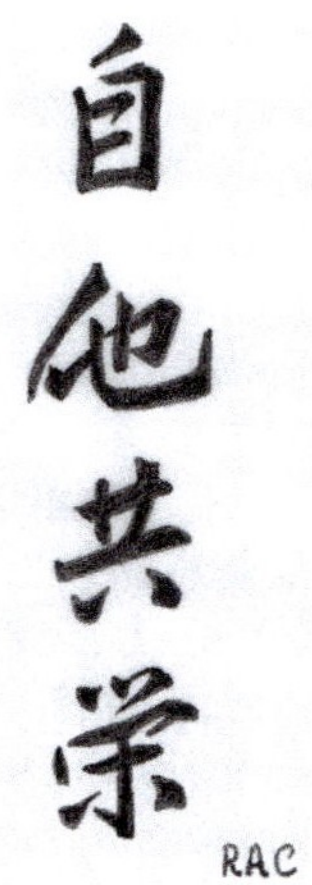

"Mutual Benefit and Prosperity"

—Jigoro Kano, Founder of Judo

specific drug–target interactions and the relationships between compound structure and pharmacological activity. The historic perspective set by this chapter will provide a solid foundation for subsequent chapters, in which we shall explore the sequential stages of molecular targeted drug discovery and the integrated panoply of scientific disciplines that must be brought to bear for successful discovery and development of modern medicines.

1.1 Disease and Treatments in Antiquity

Disease has been a perennial bane of existence for as long as there has been advanced life on earth. Indeed, the fossil record is replete with evidence of inflammation, infections, arthritis, cancers, and other diseases throughout the history of *Homo sapiens* and our hominid forebearers. For example, in 1932, the famed paleontologist Louis Leaky discovered a fossilized jawbone fragment (known as the Kamanmandibular fragment) of an ancestral primate (either *Homo erectus* or *Homo australopithecus*) with clear features of Burkitt's lymphoma (Retsas, 1986; Halperin, 2004).

In kind, the human urge to intervene in disease also has a long and storied history.

We know, for example, that early humans, and potentially Neanderthals, practiced a crude form of surgery in which a hole was made in the skull, presumably for some medical or metaphysical purpose; remarkably, many of these skulls display signs of in-life healing, suggesting that the individuals survived after such "surgery". The ancient Egyptians were among the first to document their attempts at medical intervention in disease. For example, seven papyrus records from the period 3000 to 1500 BCE describe cases of cancer and its treatment. In one of these documents, known as The Edwin Smith Papyrus, eight cases of breast tumors are described that were treated by surgical removal or by cauterization with an instrument referred to as "the fire drill" (Olson, 2005). The disease name, cancer, itself can be traced back to ancient Greece. Hippocrates (460–370 BCE, often referred to as The Father of Medicine) recorded his examination of tumors in patients. He noted that the hard core of the tumors and their spiny projections were reminiscent of the marine arthropod, crabs. He therefore referred to the tumors by the Greek word for crabs, karkinos or karkinoma, from which we derive the modern term, carcinoma. The ancient Romans subsequently translated the term into the Latin word for crabs, cancer (Olson, 2005; Mukerjee, 2010).

1.2 Natural Products as Medicines

Early humans were hunter-gatherers, learning by trial and error (sometimes with life-threatening consequences) what plants, fruits, berries, and other natural materials were available as foods. Over millennia, humans also learned – again by random trial and error – that ingestion or topical application of certain natural materials conveyed specific pharmacological effects. For example, there is clear archaeological evidence that indigenous people of Peru chewed coca leaves to release cocaine as early as 7,000 years ago. Likewise, archaeological evidence suggests that opium poppy was cultivated in Italy and the Middle East some 7,000–8,000 years ago. There is also some evidence that ancient Egyptians applied a poultice of moldy bread to wounds to stave off infections, foreshadowing the discovery of penicillin by thousands of years (see later). Finally, the bark of the white willow (*Salix alba*) has been chewed or brewed into tea as a treatment for pain and fever since the times of the ancient Egyptian, Sumerian, and Assyrian cultures. This was the humble start of the practice of medicine.

As hunter-gathering gave way to more sedentary farming and herding, societal specialization ensued, and specific individuals were tasked with the practice of healing. These healers also became the repositories of the collective knowledge of the natural pharmacopeia, which in the early development of society was verbally passed down from one generation to the next. As written languages developed, accounts of natural medicines were written to preserve knowledge of their collection, preparation, uses, and applications. Some of the earliest examples of medical compendia come from ancient Mesopotamia in the form of cuneiform clay tablets; a collection of over 600 such medical tablets from the library of Ashurbanipal (ca. 2400 BCE) are housed in the British Museum (Thompson, 1923). Some of these provide detailed descriptions of the preparation of natural product–based treatments for various ailments. There is also clear evidence of the use of plant, mineral, and animal materials as remedies for illnesses by the ancient Egyptians. For example, a papyrus scroll from ancient Egypt (ca. 1550 BCE), referred to as Ebers Papyrus, describes over 700 formulas and remedies for various ailments; among these is the heating of a mixture of herbs on a brick, so that a patient could inhale the fumes as a treatment for asthma.

The process of trial-and-error–based refinement of natural materials as therapeutics continued for centuries. Greek and Roman scholars, such as Hippocrates, Galen of Pergamum, and Pliny the Elder, made attempts to collect, codify, and record their societies' understanding of the medical uses of natural products in books that became main treaties on medicine well into the Middle Ages and Renaissance (Cartwright and Armstrong, 2021). At the same time, Persian and Islamic scholars were likewise refining, cataloging, and recording their understandings of the medicinal uses of natural products. Plant-, mineral-, and animal-derived natural products were used in isolation or compounded with others to form treatments. In some cases, the products were ingested or applied themselves, and in other cases they were mixed with water, wine, or vinegar to produce teas, fortified drinks, plasters, and the like.

Medicinal natural products were commonly provided by traders of spices and peppers during the 12th–16th centuries; indeed, the term apothecary originally referred to a place where wine, spices, and herbs were stored and sold. By the middle of the 16th century, however, apothecaries became a separate and more specialized trade in medicinal products, which continued to evolve in practice to the modern pharmacies that we know today.

1.3 Isolation of Active Ingredients and Chemical Synthesis

If one uses a knife to score the seed pod of an opium poppy plant (*Papaver somniferum*), a milky latex will ooze out from the cut. When this is collected and dried, it results in a dark resin known as opium (the word opium is derived from the Ancient Greek, meaning juice of a plant). The medicinal uses of opium (dulling the senses, relieving pain, and inducing sleep) were known since early Sumerian culture, and opium use was widely distributed across Europe, the Middle East, and Asia as early as the Bronze Age. At the beginning of the 19th century, an apothecary apprentice, named Fredrich Wilhelm Serturner, began experiments to extract pure compounds from opium and test these for pharmacological effects in dogs, mice, and sometimes humans. Through these studies, he isolated crystals of an insoluble alkaloid (an organic compound containing at least one nitrogen atom), which, when dissolved in alcohol, produced the narcotic effects of opium. Hence, Serturner designated this purified compound *principium somniferum* (i.e., the principal or basis of sleep) of opium. He later named the compound morphium (referring to the Greek god of sleep and dreams, Morpheus), which was subsequently renamed morphine (Figure 1.1). Morphine was soon being produced and sold commercially, using the purification methods developed by Serturner. Sadly, because of his self-experimentation, Serturner became addicted to the morphine that he first identified.

The isolation of morphine opened a floodgate of efforts to isolate the pharmacologically active compound(s) from various medicinal plants. As articulated by Francois Magendie, pure substances were thought to be required to make systematic observations of the pharmacological effects on animals and their tissues and to understand their mechanisms of action. Over the remainder of the 19th century, active ingredients were isolated and identified for many well-established medicinal natural products. This included caffeine as the active ingredient of coffee; nicotine from tobacco; codeine and thebaine, in addition to morphine, from opium; digitalis (digoxin) from foxglove; and atropine from nightshade (Figure 1.1). This remarkably productive period can hence be viewed as the beginning of the modern science of pharmacology.

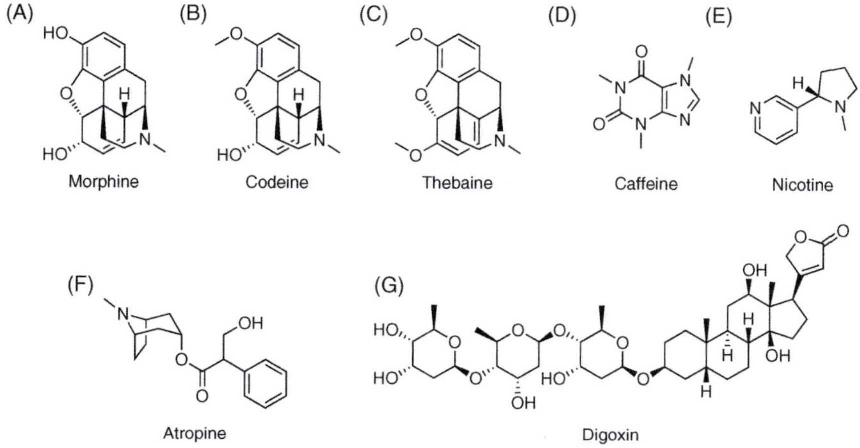

Figure 1.1 Examples of pharmacologically active ingredients isolated from medicinal plants during the 19th century. (A) Morphine, (B) codeine, (C) thebaine, all three isolated from opium, (D) caffeine isolated from coffee beans, (E) nicotine isolated from tobacco leaf, (F) atropine isolated from nightshade, and (G) digoxin (digitalis) isolated from foxglove.

An additional early and important 19th-century example of the development of a purified, pharmacologically active ingredient from a natural source is the development of acetylsalicylic acid (Aspirin) starting from willow bark. As stated earlier, the ability of willow bark to relieve pain and fever was widely known for over a millennium. The active ingredient of willow bark was first isolated by Henri Leroux in 1829, and named salicin, after the genus name for the willow tree, *Salix*. The chemical structure of salicin is illustrated in Figure 1.2. The compound consists of a glucosyl sugar attached to an aryl alcohol (salicylic alcohol). In 1838, Raffaele Piria successfully cleaved salicin into its component sugar and salicylaldehyde (Figure 1.2). He then oxidized the salicylic aldehyde (salicylaldehyde) to produce salicylic acid (Figure 1.2), which became a standard treatment for pain, fever, and rheumatism, albeit with severe renal and gastrointestinal side effects associated with high-dose, chronic use.

In parallel with these successes in isolating active compounds from natural materials, the 19th century also saw great progress in the field of synthetic organic chemistry, in large part due to the great demand for plentiful and inexpensive fabric dyes as alternatives to high-priced, scarce botanical sources. The commercial value of fabric dyes created significant industrial investment in organic chemistry, with large, successful companies becoming the critical patrons of this scientific discipline. One such company was Fredrich Bayer & Company, which was founded in 1863 as a dyestuff producer, but which had the foresight

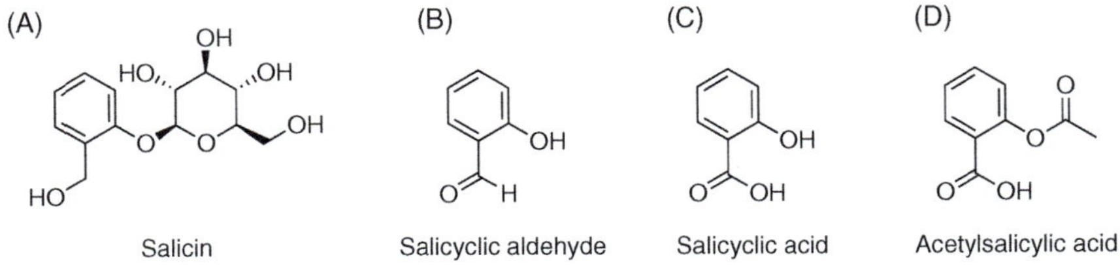

Figure 1.2 Chemical structures of salicin isolated from white willow bark and its derivatives. (A) Salicin, (B) salicylic aldehyde, (C) salicylic acid, and (D) acetylsalicylic acid (Aspirin).

to expand into other areas that might benefit from the versatility and emerging power of organic chemistry. In 1894, Bayer hired Felix Hoffman, a young chemist whose father suffered from rheumatoid arthritis and was becoming intolerant of the bitter taste and gastrointestinal side effects of chronic use of salicylic acid. Following the earlier synthesis of acetylsalicylic acid (Figure 1.2) by Charles Gerhardt in 1853 and Hugo von Gilm in 1859, Hoffman perfected a method for producing acetylsalicylic acid on an industrial scale. Bayer commercialize this under the tradename Aspirin, as a more well-tolerated treatment for pain, fever, and inflammation (although Aspirin attenuates the side effects of salicylic acid, it does not eliminate them, and long-term use of Aspirin remains problematic for patients); Aspirin became a major product for the Bayer company. Hoffman also developed a synthetic route to a morphine derivative, diamorphine. Bayer commercialize diamorphine under the tradename Heroin, ironically as a nonaddictive alternative to morphine.

There is considerable controversy over Hoffman's claim to have invented Aspirin (Sneader, 2000), and clearly Heroin had significant unintended societal consequences. Nevertheless, the investments made in organic chemistry, initially for the development of fabric dyes, were a boon to science and to the development of pharmaceuticals. As we shall see next, the development of organic dyes also played an important role in the development of a mechanistic theory of drug action.

The isolation of active ingredients from natural materials continued to be an important part of pharmacology well into the 20th century. Two critical examples of this are the discovery and development of penicillin as the first antibiotic therapy, and the discovery of the statins as cholesterol-lowering therapeutics.

The story of the discovery of penicillin by Alexander Fleming has been presented many times. Briefly, Fleming return from vacation in early September 1928 to find that a Petri dish containing colonies of *Staphylococcus* bacteria had been contaminated by mold and that the zone surrounding the mold was cleared of bacteria. This observation suggested to Fleming that the mold had secreted some chemical(s) that blocked bacterial growth. He identified the mold and successfully identified the active ingredient of the mold secretion as penicillin. However, the limited amount of penicillin secreted by the mold and the limited stability of the isolated compound precluded any clinical testing. Howard Florey, Ernst Chain, and colleagues at Oxford University took up the challenge of producing penicillin at a scale that would allow for its clinical use. These efforts allowed Florey to demonstrate that penicillin could protect mice against fatal *Streptococci* infections – the first demonstration of *in vivo* antibiotic activity. Following this critical demonstration, on February 12, 1941, penicillin produced by the Oxford group was first administered to a human patient. Albert Alexander, a 43-year-old police officer, had gotten a minor injury while gardening that became infected and quickly spread to affect his eyes, face, and lungs. He was injected with penicillin and made a rapid and dramatic recovery. Sadly, supplies of the drug were limited and so he eventually succumbed to the infection once the penicillin supply was exhausted. Nevertheless, the remarkable activity demonstrated in this one patient stimulated interest in producing penicillin on an industrial scale; interest was further heightened by the onset of World War II and the prospect of British troops suffering and dying from infections on the battlefield. Realizing that scaling and manufacturing sufficient penicillin for the war effort was beyond the capabilities of the wartime United Kingdom, help was enlisted from the United States Department of Agriculture and ultimately from the US pharmaceutical industry (in particular, Merck, Pfizer, Squibb, and Lederle). These efforts allowed sufficient supplies of penicillin to be procured in time for the D-Day invasion of Normandy and ushered in a new era of antibiotic research, discovery, and clinical use in fighting bacterial infections that continues to the present day. An excellent review of the development of penicillin can be found in the American Chemical Society International Historic Chemical Landmarks, Discovery and Development of Penicillin (http://www.acs.org/content/acs/en/education/whatischemistry/landmarks/flemingpenicillin.html).

By 1960, it was clear that elevated levels of cholesterol in systemic circulation were a major risk factor for coronary heart disease. Hence, the medical community became focused on identifying agents that could reduce systemic cholesterol levels, in the hope that this would in turn reduce the risk of heart attack and stroke in patients. Serum cholesterol comes from two major sources: about 33% from food sources, and about 67% from *de novo* biosynthesis, through a 20-step enzymatic pathway that starts with acetyl-coenzyme A.

In the mid-1970s, Akira Endo, a biochemist at the Sankyo Company in Japan, isolated a compound from the *Penicillium citrinum* fungus that was found to be an inhibitor of the enzyme 3-**h**ydroxy-3-**m**ethyl-**g**lutaryl-coenzyme A reductase (HMG-CoA reductase). HMG-CoA reductase converts HMG-CoA to mevalonic acid and is the rate-limiting enzyme in the metabolic pathway of cholesterol biosynthesis. This compound, known as either compactin or mevastatin (Figure 1.3), was shown to lower systemic levels of cholesterol in

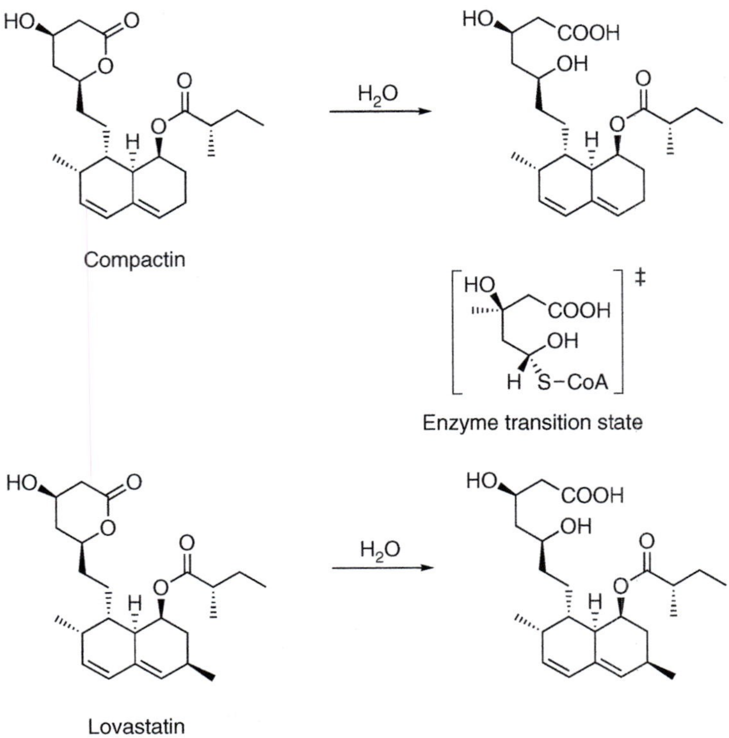

Figure 1.3 Chemical structures of compactin (top) and lovastatin (bottom) and their pharmacologically active hydrolysis products compared to the proposed transition state of the enzyme–substrate complex (middle). Adapted from Copeland and Anderson (2002).

multiple, preclinical animal species and soon thereafter clinical studies were started in Japan.

Around this same time, P. Roy Vagelos, a renowned biochemist who had done seminal work on the enzymology of long-chain fatty acid biosynthesis, was recruited to head Merck Research Laboratories. Merck had already made inroads into cholesterol biosynthesis when Vagelos and his colleague, Alfred (Al) Alberts, arrived at the company. Vagelos asked Alberts to lead a program to discover inhibitors of HMG-CoA reductase. Relying on his biochemistry expertise, Alberts designed and implemented a screening assay of HMG-CoA reductase enzyme activity and began screening synthetic compounds and isolated natural products for inhibitory activity. Soon the team had identified a natural product isolated from fermentation broth of the fungus *Aspergillus terreus*, that was a potent inhibitor of the enzyme, lovastatin (Alberts et al., 1980; Alberts, 1988). Lovastatin (also known by the brand name Mevacor) was very similar in chemical structure to compactin, and the two compounds both inhibit HMG-CoA reductase by a common mechanism. The compounds are not themselves potent inhibitors of the enzyme but undergo hydrolytic ring opening to species that resemble an intermediate state of the enzyme's reaction with substrate, known as the transition state (Figure 1.3).

After completing animal safety studies, Mevacor entered clinical trials in the United States in April 1980. However, in September 1980, the Merck team learned that the Sankyo Company had abruptly halted its clinical trials of compactin. It was rumored that this was because the company had found that the compound induced tumors in a preclinical animal species. Out of an abundance of caution and concern for patient safety, Vagelos decided to voluntarily pause Merck's clinical trial of lovastatin. In 1983, after extensive preclinical safety studies and at the urging of the scientific and medical community, Merck reinitiated the clinical development of Mevacor, which in 1987 became the first statin drug to be approved by the FDA for the treatment of hypercholesterolemia. Lovastatin and several other statin drugs continue to be used today to help patients maintain healthy levels of circulating cholesterol.

I cannot end this section without adding an important sidenote. The influence of P. Roy Vagelos on the development of molecular targeted drug discovery cannot be overstated. His training in biochemistry and enzymology convinced him to radically change the way Merck searched for new drug starting points. He almost single-handedly shifted the company away from animal and cellular screening to the concept of defining a specific enzyme as a target for disease intervention, developing a robust assay for the activity of that enzyme, and screening chemical libraries in search of modulators of enzyme activity, as a starting point for drug discovery. This was a seismic sea change in how drugs are discovered and was soon adopted by every major pharmaceutical and biotechnology company. Much of what we shall discuss in the subsequent chapters of this book is founded on the molecular target concept that Vagelos brought into focus for drug discovery. For an interesting and readable account of his life and work in science, medicine, and business, the reader is enthusiastically referred to his autobiography (Vagelos and Galambos, 2004).

1.4 The Receptor Theory of Drug Action

As the 19th century rolled on, scientists continued to isolate and identify the active pharmaceutical ingredients from various medicinal natural sources (Bynum, 1970). With knowledge of the chemical structures of these active compounds accumulating, a vague understanding emerged that related chemical structures could elicit common physiological effects (what is known today as a structure–activity relationship or SAR). However, what was lacking was a unifying, mechanistic model of drug action. This deficit was largely addressed in the early 20th century by a small group of scientists that conceptualized and experimentally and mathematically verified a unifying mechanism of drug action, the receptor model. Two scientists independently led the way in conceptualizing this model: Paul Ehrlich of Berlin, Germany and John Newport Langley of Cambridge, the United Kingdom.

As a medical student, Paul Ehrlich became fascinated with the ability of certain aniline fabric dyes to selectively stain certain animal tissues and cells. His studies of the theory and practice of cell staining by these dyes formed the basis of his doctoral thesis. He followed up these studies as a junior professor, showing that dyes could be classified as basic, acidic, or neutral, and that specific dyes could be used to stain granules of white blood cells. His initial observations of the staining of bacteria ultimately led to the Gram method of bacterial staining that is still used today. These studies made clear to Ehrlich that certain dyes stained specific cells, but not others, and to wonder what the basis of this specificity might be. Together with his subsequent studies of antibody–antigen interactions, these observations led him to propose the existence of receptive substances (later referred to as receptors) in or on cells to which specific dyes and other chemicals can bind to form a bimolecular complex of dye and receptor (more generally, compounds that bind to a receptor are referred to today as ligands). Ehrlich extended this hypothesis to suggest that the physiological action of drugs is predicated on their first binding to a specific cellular receptor. This prescient concept is captured in Ehrlich's famous aphorism: "*Corpora non agunt nisi fixata* – A substance will not work unless it is bound" (Ehrlich, 1913, 1959–1960). The reader should note that in the vernacular of modern pharmacology, the term "receptor" is usually reserved for membrane-bound proteins that bind specific ligands and thereby initiate some type of cellular response. This is but one of many classes of proteins that can be targeted by small-molecule drugs, as we shall see further in Chapter 2. In Ehrlich's use, and for our present purpose, however, we shall use the term "receptor" to more generically refer to any target to which a ligand – including a drug molecule – may bind. Using this receptor concept as the *sine qua non* of drug action, Ehrlich, together with his associate Sahachiro Hata, was the first to systematically screen a large library of compounds for pharmacological activity, as potential treatments for syphilis. Hata had previously shown that rabbits could be infected with the bacterium that causes syphilis. Ehrlich and Hata then tested hundreds of compounds for anti-syphilis activity by injecting them into the infected rabbits. The 606th compound tested, arsphenamine, proved effective in treating syphilis and was subsequently marketed as Salvarsan to treat syphilis in humans. Ehrlich deemed Salvarsan a "magic bullet" – a compound that could selectively kill an infectious organism without harming the host. He also was the first to coin the term "chemotherapy", referring to the use of chemically defined, pure compounds as drugs (Ehrlich and Hata, 1910).

John Newport Langley's research was focused on physiology and the phenotypic impact of specific alkaloids. In the late 19th century, he studied the effects of two alkaloids, jaborandi and pilocarpine on the salivary gland of cat and found that the two compounds acted in a mutually antagonistic fashion. This led him to assume that the two drugs acted by forming complexes with a common "receptive substance" of the cell. He hypothesized that which of the two compounds prevails was determined by their chemical affinity for the cell and the concentration of compound at the site of action. For the rest of the 19th century, Langley focused his experimental work on the effects of nicotine on the autonomous nervous system. In 1905, he examined the ability of nicotine to induce tonic contractions in the leg muscle of fowl (Langley, 1905). He found that when the leg muscle was denervated, nicotine was still able to induce contractions, suggesting that the compound acted directly on muscle cells. He also found that curari (a plant-derived muscle paralyzer, today known as curare) antagonized the ability of nicotine to induce contractions. These data again supported his hypothesis of a receptive substance of the cells with which the compounds nicotine and curari each bound, in a mutually exclusive manner, to form a complex.

The receptor concept(s) initially articulated by both Ehrlich and Langley were conceptual in nature. Langley recognized that the concept needed to be further examined from a more quantitative perspective. He therefore made the decision to have an undergraduate student, who first majored in mathematics and then switched to physiology, take up these studies. That student was Archibald Vivian Hill.

Hill examined the effects of nicotine and curari (today referred to as curare) on excised abdominal muscle from the frog *Rana temporaria*, using an instrument designed and built in Langley's laboratory (Hill, 1909). For these studies, the muscle was attached to a lever that allowed measurement of the timing and amplitude of muscle contraction by the displacement of the lever. Hill allowed the excised muscle to relax in Ringer's solution before being immersed in a dilute solution of nicotine to induce contraction. He found that the onset of contractions could be well described by the following kinetic equation (note that here we are using more modern terms within the equation and not the original terminology of Hill).

$$E = E_{eq}\left(1 - e^{-k_1 t}\right) \tag{1.1}$$

where E (for *Effect*) represents the amplitude of the displacement of the lever, t is time, k_1 is the observed rate constant, and E_{eq} is a constant equal to the maximum or full amplitude possible for a contraction as the system approaches equilibrium (Figure 1.4A).

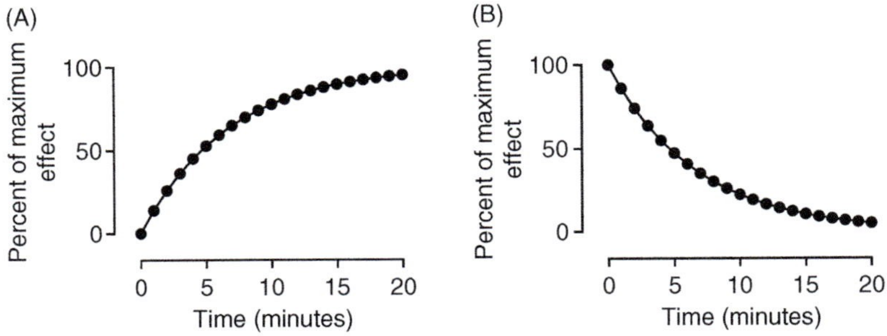

Figure 1.4 (A) Idealized plot of the onset of muscle contractions induced by nicotine. (B) Idealized plot of the relaxation of nicotine-induced muscle contractions after washing and incubation in Ringer's solution. Data in (A) are fit to Equation 1.1 and that in (B) are fit to Equation 1.2.

Next, the muscle that had contracted in dilute nicotine was washed and left to relax in Ringer's solution. Hill found that the rate of relaxation could also be described by a simple kinetic equation (Figure 1.4B):

$$E = E_{eq}e^{-k_2t} \tag{1.2}$$

In later chapters, we will encounter several pharmacological events that conform to equations very similar to Equations 1.1 and 1.2 and yield graphs such as those shown in Figure 1.4. Appendix 1 provides a fuller description of these equations and their applications to various aspects of drug discovery.

These mathematical relationships may be explained by either of two hypotheses: a physical hypothesis related to the rate of compound diffusion through the muscle tissue and a chemical hypothesis related to the rate of association and dissociation of nicotine with a receptor of the muscle cells. To distinguish between these possibilities, Hill derived equations for the amplitude (y) as a function of nicotine concentration (N) and used these to compare the two hypotheses by two methods: (1) comparison of the total amplitudes and (2) comparison of the velocities of contractions as a function of N, at constant temperature. Both methods were only consistent with the chemical hypothesis. Thus, consistent with the receptor hypothesis, Hill concluded that the action of nicotine could be described by the following equilibrium:

$$N + A \rightleftharpoons NA$$

where N is nicotine and A is the receptor molecule. A similar equilibrium can be derived for curari combining with the receptor.

Hill then combined Equations 1.1 and 1.2 to derive an equation relating the amplitude of contraction (y) at equilibrium to concentration of nicotine (N):

$$y = \frac{N}{k' + kN} - M \tag{1.3}$$

where k' and k are experimental constants and M is a correction term for background signal, which Hill equated with the minimum concentration of nicotine necessary to cause any contraction. Equation 1.3 was a critical contribution to pharmacology, as it described the degree or amplitude of a biological effect as a function of the concentration of ligand at equilibrium. The form of Equation 1.3 describes a square hyperbola when effect is plotted as a function of ligand concentration. Over the first half of the 20th century, similar hyperbolic relationships between biological effect and ligand concentration were derived for drug action on cells by Alfred J. Clark (1933; see also Maele et al., 2002), for enzyme activity as a function of substrate concentration by Victor Henri (1903) and Leonor Michaelis and Maud Menten (1913), and for gas molecules adsorption on to metal surfaces by Irving Langmuir (1918).

The equilibrium described by Hill for nicotine binding to the muscle cell receptor can be generalized for any ligand (L) and any receptor (R) as

$$L + R \rightleftharpoons RL$$

We can likewise recast Equation 1.3 into a more contemporary and generalized form for any biological effect (E) caused by the combination of a ligand with a specific receptor (i.e., target), as a function of ligand concentration ($[L]$). For this purpose, we replace y in Hill's original equation with E_{eq} (the amplitude of effect at equilibrium), replace N with $[L]$, and we explicitly define Hill's experimental constants k' and k. The constant k' is equated with EC_{50}/E_{max}, where EC_{50} is the concentration of ligand that produces a half-maximal amplitude of effect and E_{max} is the maximum amplitude of effect that would be expected at infinite ligand concentration. The constant k is equated with $1/E_{max}$. Finally, the correction factor M of Hill's original equation is ignored. With these substitutions, Equation 1.3 becomes

$$E_{eq} = \frac{[L]}{\dfrac{EC_{50}}{E_{max}} + \dfrac{[L]}{E_{max}}} \tag{1.4}$$

which can be rearranged to

$$E_{eq} = \frac{E_{max}}{1 + \dfrac{EC_{50}}{[L]}} \tag{1.5}$$

Equation 1.5 is a central equation of pharmacology and biochemistry, as it describes the common concentration–response relationship between ligand concentration and biological effect (response); the equation is commonly referred to as the Hill–Langmuir equation, in honor of Archibald V. Hill and Irving Langmuir. As we shall see, this equation describes many biochemical, biophysical, and biological processes that are mediated by ligand binding to a target receptor. We shall encounter this equation in various forms throughout the subsequent

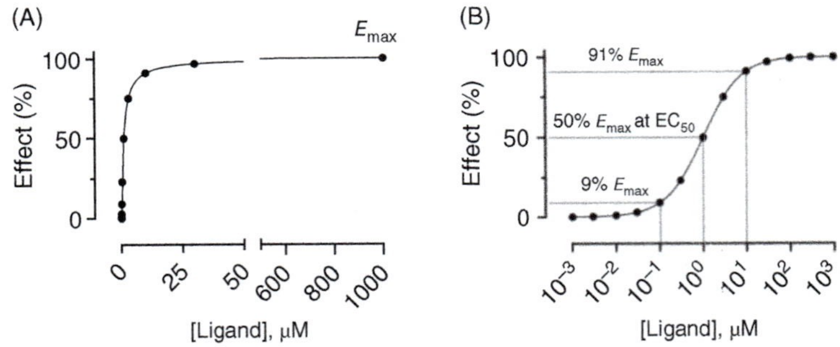

Figure 1.5 (A) A typical concentration–response plot for a biological effect related to specific target occupancy as a function of ligand concentration. Data fit to Equation 1.5. (B) As in (A) but plotted with the ligand concentration (*x*-axis) on a $\log_{10}$ scale.

chapters of this book, and the reader will likewise encounter forms of Equation 1.5 in books and primary literature in the fields of biochemistry, biophysics, pharmacology, medicinal chemistry, and the like (see also Appendix 1). As stated earlier, the form of Equation 1.5 is that of a square hyperbolic function of effect amplitude as a function of $[L]$, as illustrated in Figure 1.5A. As the concentration of ligand is increased, the curve asymptotically approaches the value of E_{max}.

If the *x*-axis is plotted on a logarithmic scale, rather than a linear scale (a semilog plot), the resulting plot presents an S-shaped curve. This curve is characterized by a low-amplitude plateau at low ligand concentrations (approaching zero amplitude at zero ligand concentration, assuming no background signal), a high-amplitude plateau at high ligand concentrations (asymptotically approaching E_{max}), and a semi-linear segment spanning a range of amplitudes between approximately 25% and 75% of E_{max}. The midpoint of this semi-linear segment occurs at an *x*-axis value of EC_{50} (Figure 1.5B). Plots of this form are referred to as concentration–response plots and provide a straightforward means of estimating the EC_{50} for a ligand–receptor interaction from visual inspection of the graph, and more precisely through nonlinear curve fitting methods (Motulsky and Christopoulos, 2004). The curve illustrated in Figure 1.5A and B is for a simple case of one molecule of ligand binding to one molecule of receptor. In this case, the steepness of the curve is such that at a ligand concentration one decade below the EC_{50}, one observes only 9% of the maximal amplitude of effect (E_{max}) and at one decade above the EC_{50}, one observes about 91% of E_{max} (see Copeland, 2023, for more complex situations in which the interactions of ligand and receptor are other than one-to-one). As with Equation 1.5, semilog scale concentration–response plots are of broad utility across many pharmacological and other applications.

The EC_{50} is thus an experimental measure of ligand affinity for a specific receptor or of the effectiveness of a ligand in inducing a specific cellular or organismal response. The more effective a ligand is in eliciting a biological effect, the lower the concentration of that ligand required to reach the EC_{50}. Hence, the EC_{50} is inversely related to the strength of interaction (i.e., affinity) between the ligand and its target receptor; that is, the lower the value of EC_{50}, the greater the strength of interaction. The term EC_{50} is generic for any biochemical, biophysical, or biological effect caused by a ligand molecule. In specific cases, this term is modified to reflect a more specific effect that is being observed (see Appendix 1). For example, if the ligand results in an inhibition of target activity, then the term EC_{50} is replaced by IC_{50}, where the *I* indicates inhibition. Likewise, if the effect of ligand is to augment the activity of a target, then the relevant term may be referred to as an AC_{50}, where the *A* indicates activation. Again, we will encounter the use of the EC_{50} (or IC_{50} or AC_{50}) value as a relative measure of ligand effectiveness in multiple applications throughout the remainder of this book.

In summary, the collective works of Ehrlich, Langley, Hill, Clark, and others in the early years of the 20th century have firmly established the receptor theory of drug action as a central tenet of modern pharmacology. Although we have here used the term "receptor" to generically refer to any macromolecular target of drug action in the development of receptor theory, in subsequent chapters, we shall more explicitly define different types of macromolecular targets of drug action. Thus, we can state the central tenet of modern pharmacology as follows.

> Binding of drug to a specific biological target (most commonly a protein) modulates the physiological activity of that target. This physiological modulation of target activity results in specific pharmacodynamics associated with the target function, and in turn to disease-modifying phenotypic effects within the cell or organism.

As we shall see throughout the rest of this book, this central tenet forms both the strategic and tactical basis for the practice of molecular targeted drug discovery.

1.5 Early Chemotherapy: Attempts to Mimic Nature

With the receptor theory of drug action gaining acceptance, the middle years of the 20th century saw a flurry of attempts to antagonize the physiological action of small-molecular-weight metabolites by synthetic compounds. The synthetic compounds were designed to retain key structural features of the natural compounds, to thereby ensure binding to the relevant receptor, but with sufficient structural excursions to block the normal activity of the receptors; this strategy became known as *rational drug design*. For many of these early efforts, the identity of the target receptors was discovered only subsequently to the design and testing of the synthetic analogues.

In the late 1940s, Sidney Farber made note of earlier studies demonstrating that administration of folic acid to leukemia patients worsened the course of their disease. Conversely, a diet deficient in folic acid could produce some benefit for such patients. These data stimulated Farber and his team to investigate a structural analogue of folic acid, aminopterin as a treatment for children with acute leukemia (Farber et al., 1948). His team treated 16 children with acute leukemia, 10 of whom achieved temporary remission (62.5%). The treatment was, however, accompanied by significant side effects. He later found that another folic acid analogue, amethopterin, also known as methotrexate, was better tolerated and therefore subsequent studies focused on methotrexate as a treatment for acute leukemia. The chemical structures of folic acid, aminopterin and methotrexate are shown in Figure 1.6. One can clearly see from this figure that there is considerable structural similarity among these three compounds. A decade after Farber reported his results, biochemical studies identified the target

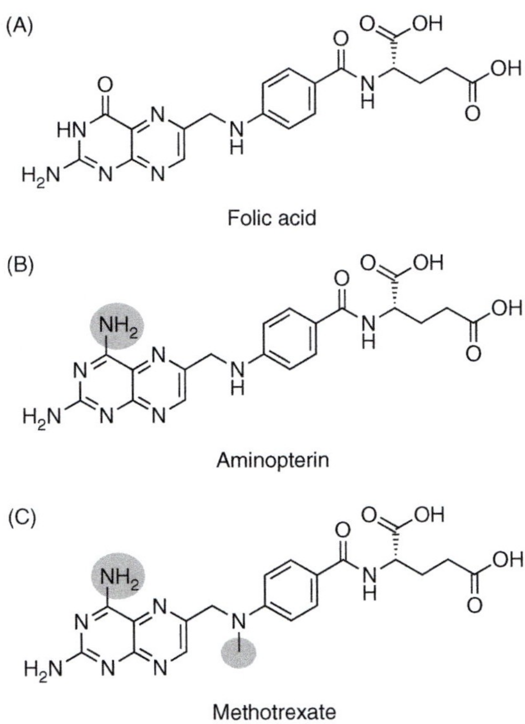

Figure 1.6 Chemical structures of folic acid and its analogues. (A) Folic acid, (B) aminopterin, and (C) methotrexate (amethopterin). The gray circles in (B) and (C) are meant to highlight the structural changes made relative to the founder compound, folic acid.

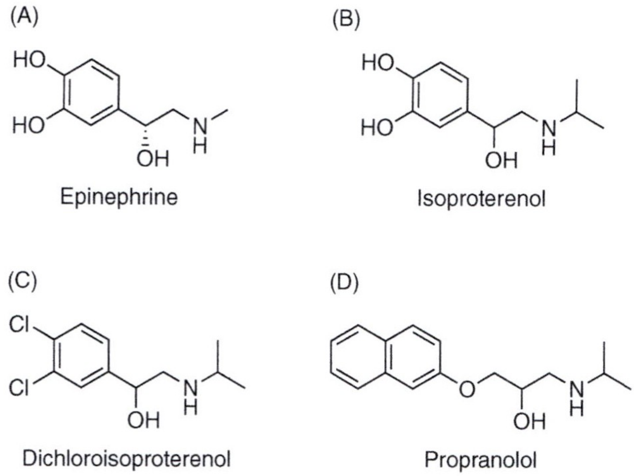

Figure 1.7 Structure of epinephrine (adrenaline) and its pharmacological analogues. (A) Epinephrine, (B) isoproterenol, (C) dichloroisoproterenol (DCI), and (D) propranolol.

of aminopterin and methotrexate as the enzyme dihydrofolate reductase (Osborn et al., 1958; Futterman, 1957; see also the review by Huennekens. 1996).

In the late 1950s, Sir James Black set his scientific focus on blocking the action of epinephrine (adrenaline) on the heart. A structural analogue of epinephrine, isoproterenol (Figure 1.7), was known to more selectively stimulate the heart; hence, Black used isoproterenol as a starting point for his work. It was known that the action of isoproterenol was terminated by metabolic substitution of one of the two phenyl hydroxide moieties by a methyl group. Based on this observation, chemists at Lilly substituted both phenyl hydroxide moieties with chlorine atoms to produce dichloroisoproterenol (DCI). When Black's team tested DCI, they indeed found that this antagonized the action of isoproterenol. A chemist on Black's team suggested that the size of the dichlorophenyl group of DCI could be mimicked by a fused biaryl ring system; this replacement resulted in propranolol, a potent antagonist of isoproterenol which became the first commercial beta-blocker for treatment of hypertension, irregular heart rate, and other related conditions. Figure 1.7 illustrates the close structural relationships among epinephrine, isoproterenol, DCI, and propranolol. It is now known that all these compounds act as ligands for the beta-adrenergic receptor protein.

A final example of rational drug design from the mid-20th century is provided by the work of Gertrude B. Elion and George H. Hitchings on analogues of purines (Figure 1.8). Adenine and guanine are the two purine bases that, along with the pyrimidine bases cytosine and thymine, form the nucleotide constituents of deoxyribose nucleic acid (DNA). Elion and Hitchings believed that purine analogues could have antiproliferative activities that would form the basis of

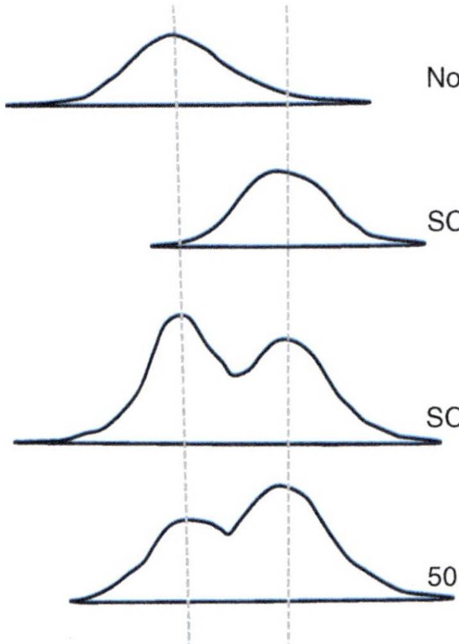

Figure 1.9 Longsworth scanning diagrams (Longsworth, 1942) representing the electrophoretic mobility of Hb–CO from normal, sickle cell anemia (SCA), sickle cell trait (SCT) erythrocytes, and a 50:50 mixture of normal and SCA Hb–CO. The dashed vertical lines are merely to aid the eye of the reader and represent the peak positions for normal and SCA Hb–CO electrophoretic mobility. *Source:* Adapted from Pauling et al. (1949).

Hb. Normal Hb and Hb–CO had isoelectric points of 6.68 and 6.87, respectively, whereas the SCA Hb and Hb–CO had values of 6.91 and 7.09, respectively. In Figure 1.9, I have made redrawings of the electrophoretic patterns observed by Pauling and coworkers (Pauling et al., 1949) and illustrated as Longsworth scanning diagrams (Longsworth, 1942), for the Hb–CO forms of normal, SCA, SCT, and a 50:50 mixture of normal and SCA Hb–CO. Several points are abundantly clear from examination of Figure 1.9. First, it is obvious that the peak positions for normal and SCA Hb–CO are quite distinct. Second, the samples from SCT patients display a bimodal pattern with one peak coincident with that of normal Hb–CO and the other peak coincident with that of SCA Hb–CO. This pattern strongly suggests that patients with SCT have both forms of Hb within their erythrocytes. However, the relative amplitudes of the two peaks in these samples differ from that of a simple, equal mixture of the two Hb–CO species (normal plus SCA). Pauling and team quantified the area under the curve for each of the two peaks from five SCT patients and found that the mean values indicated that SCT patients contain 39% of the SCA Hb form and 61% of the normal Hb form.

Pauling et al. went on to speculate on the cause of these differences in electrophoretic mobility. They inferred that the most plausible explanation is a difference in the number of ionizable groups in the two Hb forms. They suggested that in the pH region of their isoelectric points, the net number of positive charges (i.e., total number of cationic groups minus the number of anionic groups) is greater in SCA Hb than in normal Hb. Based on the difference in isoelectric points between normal and SCA Hb–CO (0.22), they suggested that the difference corresponds to an additional two to four positive charges per molecule of SCA Hb (note that Hb is a tetramer composed of two alpha chains and two beta chains, see later). They further inferred from these data that normal patients are homozygous for a normal Hb gene, that SCA patients are homozygous for a mutant Hb gene, and that SCT patients are heterozygous, reflecting one normal and one mutant allele.

The work reported by Pauling et al. (1949) revealed, for the first time, that an allelic change in a single gene results in a change in composition of the encoded protein that dramatically affects the function (i.e., aggregation potential in the absence of gaseous ligands) of that protein, in a manner that is disease-causing (i.e., pathogenic). Thus, this report represents the first example of a molecular basis for human disease: that the disease phenotype can be causally related to a specific genetic alteration that confers a pathogenic modulation of protein function. The epiphanic nature of these findings relates directly to a key tenet of molecular targeted drug discovery today: *Drug-mediated modulation of aberrant (pathogenic) target function, is the requisite predicate of disease (phenotypic) modification* (Copeland and Boriack-Sjodin, 2018).

About a decade after Pauling et al. published these results, Ingram (1957) reported that the beta chain of SCA Hb differs from that of normal Hb by a single amino acid: a glutamic acid (net charge −1) at position 6 of the beta chain of normal Hb is replaced by a valine (net charge 0) in SCA Hb. As Hb exists as a tetramer of two alpha and two beta chains, the glutamic acid to valine change of the beta chain 6 position would confer a net +2 change in charge, going from normal to SCA Hb. This difference is in quantitative agreement with the prediction made by Pauling et al. (1949), based on the difference in isoelectric points between normal and SCA Hb–CO.

1.7 ACE Inhibitors: An Early Example of Targeted Therapeutics

By the late-1960s, the concept of targeting a specific biological "receptor" with selective, synthetic compounds as ligands, was beginning to take hold as a rational approach to pharmacology. Concurrent advances in synthetic organic chemistry, protein chemistry (e.g., peptide synthesis and amino acid–sequencing methods), structural biology, enzymology, and methods for purifying or semi-purifying proteins with retention of specific biochemical activity set the stage for more rational and focused approaches to modulating a specific biological activity for pharmacological effect, by targeting a specific protein. Hence, the 1970s and 1980s are replete with successful examples of this general approach. Perhaps one of the most illustrative examples from this period is the design of small-molecule inhibitors of the enzyme angiotensin-converting enzyme (ACE) as treatments for hypertension (high blood pressure).

The carboxyl-terminal two amino acids (His-Leu; see Copeland, 1994, 2023, for the three-letter abbreviations of the amino acid constituents of proteins) of the decapeptide angiotensin I must be cleaved off to form the octapeptide angiotensin II (Figure 1.10). Angiotensin II increases blood pressure in mammals as a vasoconstrictor itself, and by stimulating release of aldosterone, a hormone that in turn stimulates increased potassium excretion from cells and concomitant sodium and water retention. Hydrolysis of the carboxyl-terminal dipeptide of angiotensin I is performed by the enzyme ACE. The enzymatic activity of ACE also contributes to increasing blood pressure by hydrolyzing, and thereby inactivating, the vasodilating nonapeptide bradykinin in two sequential dipeptidase steps, as illustrated in Figure 1.10. In the 1960s, it was known that the venom of certain snakes could, among other effects, lower blood pressure in their bite victims by stabilizing angiotensin I and bradykinin. Bakhle (1968) developed a method for semi-purifying ACE activity from cell-free extracts of dog lung and demonstrated that a mixture of peptides from the venom of the Brazilian viper (*Bothrops jararaca*) could inhibit the enzymatic angiotensin I to angiotensin II

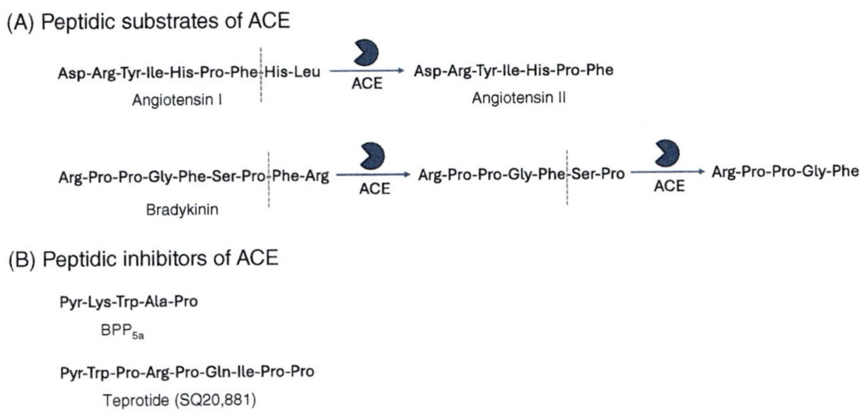

Figure 1.10 (A) Hydrolysis of angiotensin I to form angiotensin II and the two-step hydrolysis of bradykinin, both catalyzed by angiotensin-converting enzyme (ACE). The dashed, vertical lines indicate the peptide bond that is hydrolyzed by the enzyme. (B) Examples of peptidic inhibitors of ACE isolated from the venom of the Brazilian viper (Ferreira et al., 1970). In these sequences, Pyr is the abbreviation for pyroglutamic acid, which results from the cyclization of an amino-terminal glutamic acid residue to form a lactam; this modified amino acid is also sometimes referred to as 5-oxoproline.

converting activity of the lung extract; he coined the term bradykinin-potentiating factor (BPF) to refer to this inhibitory peptide mixture. Two years later, Ferreira et al. (1970) isolated and sequenced nine BPF peptides from this venom, the smallest of which was a pentapeptide, referred to as bradykinin potentiating peptide 5a (BPP_{5a}: Figure 1.10), that showed good potency in inhibiting ACE, but had short-lived effects on *in vivo* blood pressure due to its susceptibility to enzymatic hydrolysis in systemic circulation. At the same time, Miguel A. Ondetti, a peptide chemist, and David W. Cushman, a biochemist, both working at the Squibb Institute for Medical Research, started research into ACE inhibitors as a potential means of treating hypertension. They noted that a nonapeptide BPP, also isolated by Ferreira et al. (1970), contained multiple proline residues which might make this peptide more stable *in vivo*. They synthesized this peptide, referred to as teprotide or SQ20,881 (Figure 1.10), confirmed its inhibitory activity against ACE, and showed that the peptide was effective *in vivo* as an antihypertensive drug. However, the cost of manufacturing and the lack of oral bioavailability of SQ20,881 precluded further development of this compound. Nevertheless, their studies of BPP_{5a}, SQ20,881, and related peptide analogues cemented some important features of the SAR for ACE inhibition. For example, these studies suggested that binding affinity for ACE would be optimized by having Ala-Pro as the carboxyl-terminal dipeptide of the inhibitor. The cumulative data also suggested to Ondetti and Cushman that ACE was in fact a zinc metalloprotein and that a zinc-coordinating moiety would also be an important component of a small-molecule inhibitor.

About this time, Byers and Wolfenden (1973) published a paper demonstrating that L-benzylsuccinic acid (Figure 1.11) was a potent inhibitor of the zinc exopeptidase carboxypeptidase A. Reading this work, Ondetti and Cushman made the inference that the carboxylate of benzylsuccinic acid was acting to coordinate the active site zinc ion, and that a similar strategy might work for the related exopeptidase, ACE. However, they hypothesized that a key difference in active site architecture between the two enzymes is the need for ACE to accommodate a dipeptide leaving group, whereas carboxypeptidase A cleaves a single amino acid from the carboxyl-terminus of its substrates. Hence, they reasoned that a small molecule containing a carboxylic acid, and an Ala-Pro mimetic might prove to be a selective ACE inhibitor. They keyed in on D-2-methylsuccinyl-L-proline (SQ13,297), but first decided to synthesize the related, but simpler compound succinyl-L-proline (Figure 1.11). When tested, succinyl-L-proline proved to be a modest inhibitor of ACE but displayed the selectivity of activity across several assays of excised guinea pig ileum: it displayed inhibition of angiotensin I and augmentation of bradykinin activities, both as expected of an ACE inhibitor, but no effects on the activities of angiotensin II or acetylcholine (Table 1.1). Based on these data, the D-2-methylsuccinyl-L-proline was synthesized and

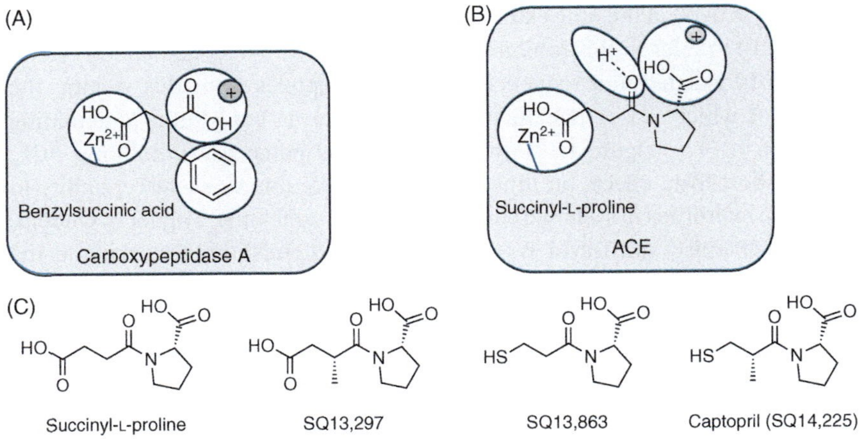

Figure 1.11 (A) Active site model of carboxypeptidase A with benzylsuccinic acid making key contacts with active site subsites, according to the model of Ondetti et al. (1977). (B) Active site model of ACE with succinyl-L-proline making key contacts with active site subsites, also according to the model of Ondetti et al. (1977). (C) Chemical structures of key compounds synthesized and studied by Ondetti and Cushman, ultimately leading to the antihypertensive drug, captopril (Ondetti et al., 1977; Cushman and Ondetti, 1991). In (A) and (B), the small circle enclosing a positive (+) sign is intended to represent an unidentified cation from the protein that serves as a countercharge to the carboxylic acid moiety of the compounds.

tested (Table 1.1). This compound (SQ13,297) displayed a 15-fold improvement in potency relative to succinyl-L-proline. Reasoning that a sulfhydryl moiety would coordinate more strongly to the active site zinc, the terminal carboxylate succinyl-L-proline was substituted by a simple mercapto group, yielding even greater potency (a 2000-fold improvement). Addition of the 2-D-methyl group resulted in SQ14,225, which was marketed as captopril, the first ACE inhibitor to be approved for the treatment of hypertension. The seminal work of Ondetti et al. (1977) was quickly expanded by others, yielding several additional ACE inhibitors, including enalapril (Pattchet et al., 1980) and lisinopril (Figure 1.12) – drugs that are still in use today. Remarkably, it was not until the early 2000s that the crystal structures of human testicular ACE bound by these drugs were solved (Natesh et al., 2003, 2004). The crystal structures confirmed the zinc coordination role of the mercapto group of captopril and the analogous carboxylate groups of enalaprilat (the active form of enalapril) and lisinopril (Figure 1.12). The crystal structures also reveal additional key active site recognition elements of the enzyme that interact with these inhibitors. The power of structural biology in identifying such recognition elements on targets and the use of this information in guiding medicinal chemistry will be a topic of further discussion later in this chapter and in subsequent chapters.

Table 1.1 Effects of potential ACE inhibitors on the biochemical activity of ACE (from rabbit lung) and the biological activity of angiotensin I, angiotensin II, acetylcholine, and bradykinin in excised guinea pig ileum. Data from Ondetti et al. (1977). IC_{50} and AC_{50} are as defined earlier in the text.

	Activity (µg/mL)				
		Excised guinea pig ileum			
Compound	**ACE enzyme activity IC_{50}**	**Angiotensin I IC_{50}**	**Angiotensin II IC_{50}**	**Acetylcholine IC_{50}**	**Bradykinin AC_{50}**
Teprotide (SQ20,881)	1.0	0.068	>32	>32	0.0017
Succinyl-L-proline	135	94	>100	>100	8.0
SQ13,297	12	13	>100	>100	15
SQ13,863	0.04	0.06	>100	>100	0.005
Captopril (SQ14,225)	0.005	0.005	>100	>100	0.0007

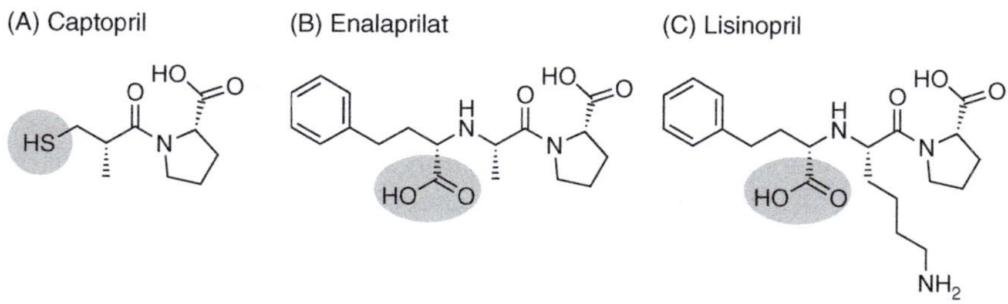

Figure 1.12 The chemical structures of ACE inhibitors in current clinical use as antihypertensives. (A) Captopril, (B) enalaprilat, the active metabolite of the drug enalapril, and (C) lisinopril. The gray circles identify the zinc-coordinating moieties of each inhibitor.

The work of Ondetti and Cushman, leading to the discovery of captopril, and paving the way for additional ACE-targeted therapeutics, is illustrative of a spectrum of early molecular targeted drug discovery campaigns that were pursued during the latter half of the 20th century. Other examples from this period include the targeting of 3-**h**ydroxy-3-**m**ethyl-**g**lutaryl-CoA reductase (HMG-CoA reductase) by statins for cholesterol lowering (see earlier); the targeting of the human immunodeficiency virus (HIV) aspartyl protease by peptidomimetics and of the HIV reverse transcriptase, for treating acquired immunodeficiency syndrome (AIDS); the targeting of steroid 5α-reductase by finasteride for treatment of benign prostate hyperplasia; and many others (see Copeland and Anderson, (2002), and Copeland and Boriack-Sjodin, (2017), for reviews). As described here for captopril, a common theme of these early efforts was the explicit targeting of a specific protein with clear disease association, using a combination of biochemical, biological, and medicinal chemistry insights and expertise. In the 1980s and beyond, these efforts were augmented with atomic-level details of the interactions of small-molecule ligands with their targets, gleaned from X-ray crystallography, nuclear magnetic resonance (NMR) spectroscopy, and other biophysical methods, as we shall describe later in this chapter.

1.8 Genetic Alterations in Disease

The efforts leading to ACE inhibitors, HMG-CoA reductase inhibitor, steroid 5α-reductase inhibitors and other targeted drugs, just discussed, all aimed to attenuate the activity of normal protein functions operating within, and contributing to, specific disease states. With the demonstration of a monogenetic cause of SCA (Section 1.6), and advances in molecular biology and protein sciences, researchers began to search for other examples for which a clear, causal relationship between gene alteration and disease might be established; this remains today an important aspect of target selection, as we shall cover in greater detail in a subsequent chapter. There are a variety of genetic changes that may affect the structure and function of the protein encoded by an affected gene, leading to specific disease. These include mutations – such as that seen in SCA hemoglobin (see earlier), splice variants, deletions and insertions, and chromosomal translocations (see Copeland, (2023), for more details). The example of the chromosomal translocation known as Philadelphia chromosome, and its relationship to chronic myelogenous leukemia (CML), is illustrative of how a genetic alteration can modify protein function, leading to disease, and how knowledge of the genetics and biochemistry of such a disease can lead to a selective molecular targeted treatment for patients.

Chromosomal translocation refers to the situation in which a chromosome is cleaved and a portion of the broken chromosome fuses with a different chromosome. The resulting fused chromosome contains genetic information from both chromosomes, and this in turn encodes for an altered protein, known as a fusion protein. The loss of certain portions of the normal protein and/or the addition of new protein sequence can significantly alter the structure and function of the fusion protein relative to the function of the normal protein. In 1960, scientists based in Philadelphia, at the Fox Chase Cancer Center and the University of Pennsylvania School of Medicine, reported a chromosomal translocation seen among patients with CML (Nowell and Hungerford, 1960). Karyotyping revealed that in many CML patients, one of the two homologs of chromosome 22 is shortened (suggesting a break) and one homolog of chromosome 9 is concomitantly lengthened, due to gene fusion. Subsequent studies demonstrated that the chromosome 9–22 fusion (known as the Philadelphia chromosome,

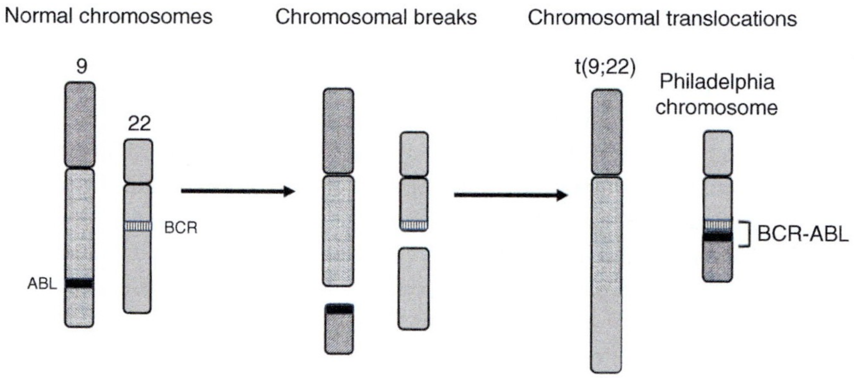

Figure 1.13 Schematic illustration of the chromosomal translocation process leading to the Philadelphia chromosome and the BCR-ABL fusion gene. *Source:* Copeland (2023) / with permission of John Wiley & Sons.

as both authors of the original description of this translocation worked in that city), results in a fusion of the BCR and ABL genes, that are normally located on chromosomes 22 and 9, respectively (Figure 1.13).

Normally, the ABL gene on chromosome 9 encodes for a tyrosine kinase (an enzyme that catalyzes phosphorylation of tyrosine side chains of specific protein substrates) containing an auto-inhibitory domain that keeps the activity of the enzyme in check when not needed by the cell. The fusion protein resulting from the fusion of BCR and ABL genes, however, encodes for a fusion protein, known as Bcr-Abl, in which the auto-inhibitory domain is lost, such that the enzymatic activity of the fusion protein is unchecked and said to be constitutively active. The overactive Bcr-Abl kinase results in the unabated myelocyte proliferation that is a hallmark of CML. Because of this causal relationship between Bcr-Abl enzymatic activity and leukemia, the gene fusion BCR-ABL is considered an oncogene (a gene that has the potential to cause cancer); the BCR-ABL gene fusion is found to occur in about 95% of patients with CML.

During the 1980s, the pharmaceutical company Ciba-Geigy established a program aimed at finding inhibitors of different protein kinases. Screening their library of small-molecule compounds against a panel of kinase enzymes, they succeeded in identifying selective inhibitors for several kinases, including a compound that inhibited the Src and Abl enzymes, CGP57148. During a visit to the Dana-Farber Cancer Center, the lead biochemist for this program, Nicholas Lydon met, Brian Drucker, a faculty member who was studying CML, and agreed to provide a sample of CGP57148 to Drucker (Mukerjee, 2010). Drucker and Lydon tested the compound in cultures of human CML cells and in immunocompromised mice bearing subcutaneous human CML tumors (referred to as a xenograft). They found that cellular proliferation and tumor formation by Bcr-Abl–expressing cells were specifically inhibited by the compound; for example, in colony-forming assays of blood or bone marrow from CML patients, the compound decreased the number of Bcr-Abl colonies by 92–98%, with no effect on normal cell colony formation (Drucker et al., 1996). The compound (now designated ST157) was next tested in CML patients in a phase 1 clinical trial. The drug was found to be well tolerated, and no maximal tolerated dose was identified. Complete hematologic responses were seen in 53 of 54 patients treated daily with doses of 300 mg or more; of these 54 patients, 29 demonstrated cytogenetic responses, including 7 patients who had complete cytogenetic remissions (Drucker et al., 2001). The compound was subsequently approved by the FDA for treatment of CML and is marketed under the brand name Gleevec (Imatinib).

The success of Gleevec stimulated scientists in pharmaceutical companies, biotechnology companies, and academia to broadly adopt the strategy of molecular targeted drug discovery for oncology and many other therapeutic areas.

1.9 Structure-Based Drug Design

In January of 1896, Wilhelm Conrad Roentgen first reported the discovery of a new form of rays of unknown type, which he termed X-rays to emphasize their enigmatic character. His report provocatively demonstrated that these X-rays could be used to reveal images of internal bone structure, including an image of the bones of his wife's hand and the ring that she was wearing at that moment. The discovery of X-rays generated great interest throughout the physics community. Because the wavelength of X-rays is comparable to the distance between atoms in molecules, Max von Laue hypothesized that these rays could be used to determine molecular structures and was among the first scientists to capture on photographic plates, the diffraction patterns from crystals of copper sulfate irradiated by X-rays. The regular, spotted diffraction patterns observed by Laue stimulated William Bragg and his son, Lawrence, to investigate the diffraction patterns of other simple crystalline compounds. In a stroke of insight, Lawrence Bragg realized that the diffraction patterns could be treated as the reflections of X-rays by the plane of atoms in a crystal lattice, allowing their atomic positions to be deduced (Bragg, 1915). For simple salts, the position of spots in the diffraction pattern is sufficient to determine the molecular structure of the compound. The X-rays are mainly diffracted by the electron clouds surrounding atomic nuclei; hence, what the diffraction pattern reveals is a map of electron density from which the molecular configuration can be inferred. However, William Bragg realized that to determine molecular configuration for more complex molecules, one would need to measure not just the position of the spots of the diffraction pattern, but also their intensities, and to process such data through Fourier transformation methods for periodic systems. This led him to invent the diffractometer, an instrument capable of quantifying spot intensity as well as position (Bragg, 1929). This breakthrough, coupled with the abundance of chemical compounds available in pure, crystalline form, led to a compendium of molecular structures, of increasing complexity, to be solved by X-ray crystallography during the 1920s and 1930s and beyond. In fact, this new method formed the basis of Linus Pauling's PhD thesis: *The Determination with X-Rays of the Structure of Crystals* (California Institute of Technology, 1925).

As early as 1840, it was known that crystals of proteins could be obtained in specific cases. Crystals of proteins were anecdotally reported to sporadically occur at abattoirs. F. L. Hunnefeld (1840) was the first scientist to report the occurrence of red crystals of a protein in blood samples from earthworms, which were subsequently determined to be composed of hemoglobin. In 1926, James Sumner reported the first crystals of an enzyme, urease (Sumner, 1926). This work is notable because Sumner went on to show that the crystals were composed of protein, and that when the crystals were dissolved in water, enzymatic activity could be demonstrated. This work provided clear evidence that enzyme activity was due to proteins, a hypothesis that was not broadly accepted at the time and resulted in Sumner being awarded the 1946 Nobel prize in chemistry.

Proteins, however, are large, flexible, highly hydrated molecules, and crystals of proteins require humid conditions to maintain their structural integrity. These factors made the idea of revealing protein structures from X-ray crystallography seem daunting and impractical. This challenge was overcome by Bernel and Crowfoot (1934) who reported clear diffraction patterns from crystals of the enzyme pepsin, by keeping the crystals under humid conditions by sealing them in capillary tubes. The first three-dimensional structure of a protein to be solved by X-ray crystallography was that of sperm whale myoglobin (Kendrew et al., 1958), and this was soon followed by the structure of horse hemoglobin (Perutz et al., 1960).

Over the next three decades, a plethora of other protein structures were solved by X-ray crystallography, including the structures of proteins with natural or synthetic ligands bound. The atomic details afforded by protein–ligand structures yielded great insights into the structures of ligand-binding pockets on proteins and the structural determinants of ligand recognition by these pockets. This same period saw great technical advances in NMR spectroscopy, culminating in the ability to solve solution-based structures of some proteins (Wutrich, 1989). The accumulation of protein structures by X-ray crystallography and NMR naturally led scientists to propose the use of this information prospectively in the design of ligands to optimally fit into specific binding pockets on target proteins. Structure-based drug discovery is often augmented through computational methods, aimed at predicting the docking of ligands of varying structure to a

Cohen (2013), it was during this walk and the subsequent meal that he and Boyer realized that they could use the endonuclease *Eco*R1 (restriction endonuclease 1 of *Escherichia coli*) to cleave plasmid DNA of bacteria at reproducibly specific locations to identify the gene(s) that confer resistance to specific antibiotics. Plasmids are extrachromosomal genetic elements composed of small, circular DNA (in this form, the 3′ and 5′ ends of the DNA are covalently connected through formation of a phosphodiester linkage to generate the circular species) that can pass their constituent genes onto progenies, independent of the main bacterial chromosome. It was already known that the genes conferring antibiotic resistance (known then as resistance factors) existed on bacterial DNA plasmids. During their discussion, Cohen and Boyer also realized that this approach could also allow them to recombine genes from different strains of a bacterium into a single DNA molecule.

In their first paper together, Cohen and Boyer (Cohen et al., 1973) focused on two strains of the bacterium *Escherichia coli* (*E. coli*), one that was resistant to the antibiotic tetracycline (due to the presence in this strain of a plasmid [pSC101] containing the *tet* gene), and another that was resistant to the antibiotic kanamycin (due to the presence of a plasmid [pSC102] containing the *kan* gene). They isolated the plasmids containing the *tet* gene and that containing the *kan* gene and used *Eco*R1 to cleave these two plasmids into fragments. They then combined the collection of fragments from both plasmids and added the enzyme DNA ligase to recombine fragments by reforming phosphodiester bonds between fragments. The collection of this ligated DNA would include ones recombining the *tet* and *kan* genes. They then used cold $CaCl_2$ and heat shock to allow the collection of ligated DNAs to permeate into *E. coli* bacteria. To isolate bacteria that had the *tet/kan* recombination, they then grew the bacteria on culture plates containing both antibiotics, tetracycline and kanamycin. By this selection approach, only bacteria containing the recombined tet/kan gene, or that had been doubly transformed with religated versions of the original two plasmids, would survive. Plasmids from the surviving bacterial colonies were collected, retreated with *Eco*R1, and the fragments separated on agarose gels. From this analysis, they were able to definitively show that indeed some colonies contained the recombinant *tet/kan* gene (Figure 1.15).

Next, Cohen's laboratory showed that not only could they recombine genes from a common species, but they could also recombine genes from two distinct bacterial species: the Gram-positive *Staphylococcus aureus* and the Gram-negative *E. coli* (Chang and Cohen, 1974). In the third paper of this series (Morrow et al., 1974), Cohen and Boyer and coworkers made the quantum leap of demonstrating that DNA from eukaryotic sources (in this case, ribosomal genes from the African clawed frog, *Xenopus laevis*) could be recombined with bacterial DNA and expressed in *E. coli*. This set of three papers thus opened the door for the cloning

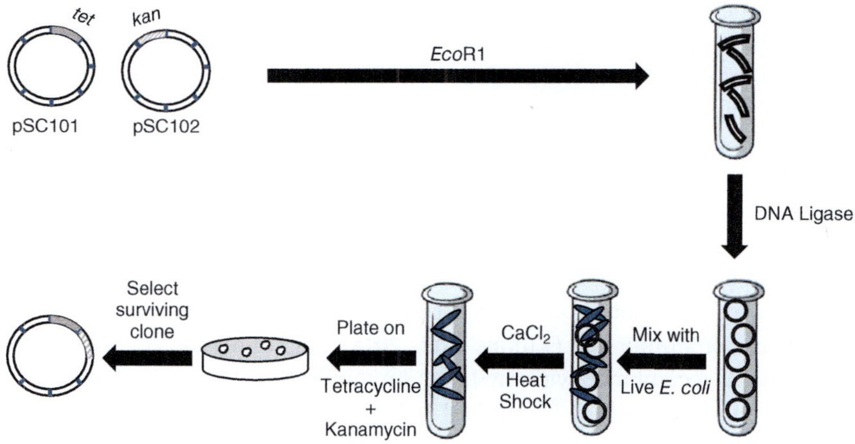

Figure 1.15 Simplified cartoon of the steps used by Cohen et al. (1973) to generate a recombinant plasmid conferring resistance of *E. coli* to the antibiotics tetracycline and kanamycin, due to the presence of the recombined tet and *kan* genes. *Source:* Adapted from Cohen et al. (1973).

of human genes and the expression of the encoded proteins, in bacterial hosts. Indeed, these papers can be thought of as the springboard from which the biotechnology industry emerged. Boyer went a step further to chemically synthesize the gene of a 14-amino-acid human hormone, somatostatin, to fuse this with the gene for *E. coli* β-galactosidase and transform *E. coli* bacteria with the chimeric plasmid (Itakura et al., 1978). His team was able to purify the resulting fusion protein and cleave it with cyanogen bromide to release the purified, human somatostatin. For the first time, a human polypeptide had been produced in a bacterium. In 1976, Boyer, together with business entrepreneur Robert A. Swanson, founded one of the very first biotechnology companies in history, Genentech. Within two years, the company had their first success with the cloning of synthetic genes for human insulin and its expression in *E. coli* (Goeddel et al., 1979). Genentech, and the many other biotechnology companies that followed, went on to produce a variety of peptide and protein products for therapeutic use. In 2009, Genentech was acquired by the Swiss pharmaceutical company, Roche. Roche, founded in 1896, had its start in the synthesis, manufacturing and sale of small organic molecules, such as ascorbic acid (vitamin C). Hence, in some sense, this brings us full circle in our discussion of synthetic molecules as therapeutic agents.

Today, target proteins can be cloned, expressed, and purified by cloning methods in bacteria, yeast, insect cells, and mammalian cells. The decision of what is the best host organism for expression of a target protein depends on the details of physicochemical requirements for producing a fully functional product. For example, does the protein require glycosylation, or other posttranslational modifications, for activity? A useful guide for selecting the best host system for protein expression has recently been provided by Schutz et al. (2023).

The expression of a recombinant target protein in a prokaryotic or eukaryotic host still requires the purification of the target from the myriad endogenous proteins of the host cell. The use of powerful promoter sequences can greatly increase the amount of target protein expressed, relative to that of endogenous proteins, and this can aid purification and overall yield. However, the most significant technical advance for the purification of recombinant proteins has been the use of affinity tags (Kimple et al., 2018). An affinity tag is a peptide or protein that is recombined with the target protein leading to the expression of a fusion protein that can then be purified by binding of the tag to an immobilized complementary ligand. Typically, a protease recognition sequence is also integrated at the interface between the affinity tag and the target protein sequence. This allows for enzymatic cleavage of the affinity tag to release the target protein after affinity purification. Figure 1.16 illustrates a schematic representation of a generic affinity-tagged recombinant protein.

Early examples of affinity tags were large proteins, such as *Staphylococcal* protein A, a 280 amino acid protein that binds with high affinity ($EC_{50} = 10$ nM) to IgG immobilized on Sepharose beads. More recently, smaller proteins or peptides have been used to append an affinity tag to the target protein. Two examples of the latter are polyhistidine tags and Avi tags. A polyhistidine tag is a sequence of multiple histidine residues (most typically six histidine residues) fused with the amino- or carboxyl-terminus of the recombinant target to confer a high-affinity transition metal-binding motif. Most commonly, immobilized nickel is used to affinity purify the recombinant protein. The Avi-tag system consists of a 15-amino-acid

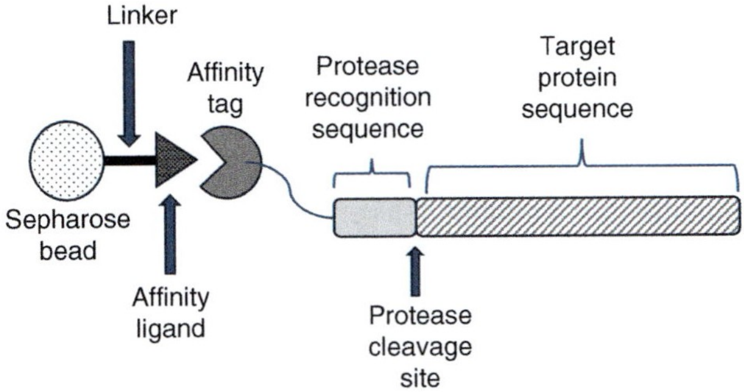

Figure 1.16 Schematic of a generic structure of an affinity tagged target protein, and the corresponding affinity ligand immobilized on Sepharose resin beads. See text for further details.

sequence that is recognized by the enzyme biotin protein ligase, leading to the unique biotinylation of the Avi-tag-target fusion. The fused protein is then isolated using streptavidin (a protein that binds biotin with very high affinity) that is immobilized on Sepharose beads, magnetic beads, or other appropriate resins.

A typical purification process would be to lyse the cells expressing the recombinant, affinity-tagged protein. The cell lysate is then mixed with the resin beads to which the affinity ligand is attached (batch processing) or applied to a chromatography column of the resin beads. The beads are then repeatedly washed with appropriate buffer to elute endogenous proteins, leaving the affinity-tagged target protein bound to the beads. The affinity-tagged protein can then be eluted from the resin by a change of buffers to effect dissociation of the affinity tag from the immobilized affinity ligand; this can be accomplished with changes in pH, changes in ionic strength, addition of soluble affinity ligand or other means. The eluted affinity-tagged protein is then treated with a protease that acts on the engineered protease recognition sequence as a substrate, thereby separating the affinity tag from the target protein sequence. Reapplication of the mixture to the affinity resin allows the affinity tag to be rebound and the isolated target protein to be eluted. Alternatively, the protease may be applied to the resin-bound affinity-tagged protein to effect "on-resin" proteolysis, and the target protein can then be eluted directly. Further chromatographic methods (e.g., size exclusion, ion exchange, etc.) may then be applied to further purify the target protein in a process referred to as "polishing."

1.11 Modern Drug Discovery

As the previous sections of this chapter have demonstrated, there has been a steady evolution of the discovery of medicines from trial-and-error use of natural materials to the isolation of active pharmaceutical ingredients, to an understanding that drug molecules work by binding to and modulating the biological activity of specific bio-macromolecular receptors. The broad acceptance of the receptor theory of drug discovery, the concomitant evolution of our understanding of genetic alterations as progenitors of disease, and the accompanying technical advances in organic chemistry, protein chemistry, genetics, and recombinant DNA methods have conspired to set the scientific and medical communities on the inexorable path toward molecular targeted drug discovery. Screening of compounds for phenotypic effects in cellular or animal models has almost entirely been replaced by drug discovery efforts that begin with compound screens aimed at identifying binders and/or modulators of activity for purified versions of recombinant protein targets. Indeed, over the period of 1999–2013, of 113 first-in-class drugs approved by the FDA, 70% were identified by molecular targeted drug discovery (Eder et al., 2014). We shall discuss these modern methods of compound screening in Chapter 4. We shall also see in that and subsequent chapters that phenotypic assays continue to play critical roles in the validation and optimization of small-molecule target modulators. Yet, one may still ask what, if any, advantages are there in taking a molecular targeted approach to drug discovery. Data that have emerged in the late 20th and 21st centuries suggest that the advantages of molecular targeted drug discovery are manifold. For example, in a comprehensive survey of cancer drug attrition throughout phase 1 and 2 clinical trials, Walker and Newell (2009) showed that targeted small molecules (mainly kinase inhibitors) had much lower attrition rates compared to nontargeted agents. The lower attrition rates seen with targeted therapeutics is likely due to two important factors: (1) target-selective drugs often display fewer adverse effects associated with the collateral modulation of other cellular proteins (what is referred to as off-target toxicity) and (2) targeted drugs often display much greater efficacy. Both factors are thought to emanate from the deeper understanding of the molecular basis of disease that exists today, and which is integral to the process of target validation, as we shall discuss in a subsequent chapter.

The process of hit optimization can also be much simpler and more time- and cost-efficient for targeted versus phenotypic drug screening, especially for intracellular targets (Croston, 2017). This is true for two overarching reasons. First, in screening large compound libraries of diverse chemical structures (diversity screening), hit compounds are typically of modest affinity for targets and very often also display modest cell permeability. Hence, many productive hit compounds would be missed by phenotypic cellular assays. Second, phenotypic readouts can be modulated by compounds that affect different targets in a common biological pathway, often making it difficult for medicinal chemists to realize a cogent and convergent SAR for compound optimization. Finally, the establishment of a causal relationship between drug binding to a specific target, cellular and organismal phenotypic consequence of target activity modulation (pharmacodynamics [PD]), and disease modification is intellectually satisfying and reassuring to physicians, regulatory agencies, and most importantly patients (Copeland and Boriak-Sjodin, 2018). Croston (2017) has reviewed these and other advantages of molecular targeted drug discovery.

Today, the discovery and development of molecular targeted drugs is an exercise in integration of a staggering diversity of scientific disciplines, through a series of well-defined, sequential stages, as illustrated in Figure 1.17. In the chapters and appendices that follow, we shall explore in detail, each of the stages of preclinical drug discovery and development – from Target Discovery through investigational new drug (IND)-enabling studies. It is important for the reader to realize that the activities encompassed in the various stages require a broad diversity of scientific expertise, only a few of which are articulated in Figure 1.17. Contrary to

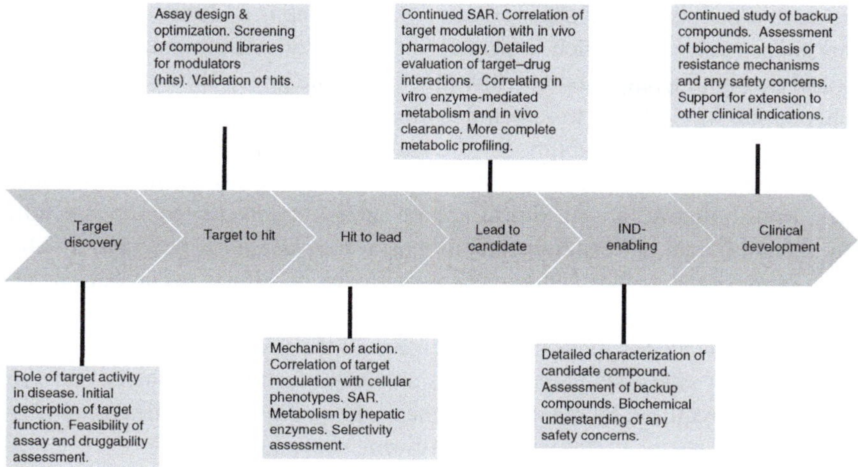

Figure 1.17 A schematic representation of the sequential stages of modern drug discovery, and some of the strategic activities pursued within each stage. *Source:* Copeland (2023) / with permission of John Wiley & Sons.

popular misconceptions, drugs are seldom discovered by individual scientists in a flash of brilliant insight (sometime referred to as "eureka moments"). Rather, drug discovery requires long, hard, often frustrating efforts of matrix project teams, composed of scientists trained in a diversity of disciplines; biochemistry, molecular biology, protein expression and purification, assay development, cell biology, animal pharmacology, drug metabolism, medicinal chemistry, bioinformatics and cheminformatics, structural biology, clinical medicine, and translational medicine are but a small sampling of the myriad disciplines that must be brought together for successful drug discovery.

The first critical stage of drug discovery today is Target Discovery, and this defines the *sine qua non* of molecular targeted drug discovery. During this stage, scientists seek to identify, validate, and select a specific molecular target for a discovery campaign. As we shall see, this is no trivial pursuit. Identification and validation of a molecular target require the establishment of a clear, unambiguous, and causal relationship between a specific biological activity of the target and the disease phenotype, such that modulation of this target activity, by drug intervention, is expected to eradicate, or at least ameliorate, disease. It is also important to understand the essentiality of the target to normal physiology as well as to the disease state, to anticipate whether a selective effect of drug treatment on disease may be assumed. That is, does the biology of the target suggest that there will be a reasonable safety margin, or therapeutic index, for the drug? For targets that pass these and other criteria, the next step is to design a robust and convenient assay with which to screen for compounds or biologics (e.g., antibodies) that bind to and modulate the function of the target. With such an assay in hand, one would begin to screen for initial binders that can be optimized for target affinity, cell permeability (in the case of intracellular targets), resulting biological consequences of target function modulation (PD), stability in whole blood from humans and preclinical animal model organisms (e.g., mouse, rat, dog, monkey), rate of hepatic metabolism, oral bioavailability, pharmacokinetic half-life, and many other factors. These optimization efforts occur during the Hit-to-Lead and Lead-to-Candidate stages, ultimately resulting in a development candidate molecule (or molecules), which would then go into highly prescribed assessments of safety, stability, synthesis and manufacturing methods, and the other criteria required by regulatory agencies, to qualify an investigational new drug (IND) for testing in human subjects during clinical trials.

1.12 Summary

In this chapter, we have explored the evolution of medicine as a molecular science. Starting with early uses of medicinal natural materials, we have seen a march toward purified substances that convey pharmacological effect. As the art of synthetic organic chemistry evolved in parallel, we saw efforts shift from merely isolating active pharmacological ingredients to the systematic study of SARs, as chemists began to "tinker" with the chemical structures of compounds and observe the effects of these excursions in chemical structure on pharmacological effect. The early 20th century yielded an epiphany in our understanding of pharmacology, with the introduction and broad adoption of the receptor theory of drug action. The receptor theory provided a cogent, foundational framework for drug discovery that catalyzed experiments aimed at targeting specific proteins for modulation by drug molecules. Technical advances in organic synthesis, structural biology, recombinant DNA, and protein purification all facilitated more rapid and target-focused drug discovery efforts throughout the rest of the 20th century and beyond. Of course, this chapter is not intended to comprehensively cover every technical advance that has contributed to modern drug discovery. Rather, a curated set of key advances have been articulated, while other, perhaps equally important technical breakthroughs – such as polymerase chain reaction (PCR), next-generation DNA sequencing, RNA silencing methods, and genomics – are briefly covered in subsequent chapter or left to the reader to explore on their own.

Today, molecular targeted drug discovery is well entrenched as an effective and efficient approach to the discovery and development of novel therapeutic agents. We ended this chapter by introducing a cogent, sequential sequence of integrated scientific activities that form the basis of modern drug discovery. We shall now explore each of these stages of drug discovery in turn, in the chapters to follow. First, however, we will introduce the variety of therapeutic modalities that are at play in FDA-approved therapeutic agents today. As we shall come to see, a deep, mechanistic understanding of a target and its role in disease and normal physiology will help inform which of these therapeutic modalities may be most apropos to the specific target and disease. Thus, the marriage of target with desired therapeutic modality becomes a critical first step in the planning and design of drug discovery campaigns.

References

Alberts, A.W. (1988) Discovery, biochemistry and biology of lovastatin. *Am. J. Cardiol.* **62**, J10–J15.

Alberts, A.W., Chen, J., Kuron, G., Hunt, V., Huff, J., Hoffman, C. et al. (1980) Mevinolin: A highly potent competitive inhibitor of hydroxymethylglutaryl-coenzyme a reductase and a cholesterol-lowering agent. *Proc Natl Acad Sci.USA* **77**, 3957–3961.

Bakhle, Y. (1968) Conversion of angiotensin I to angiotensin II by cell-free extracts of dog lung. *Nature* **220**, 919–921.

Baldwin, J.J., Ponticello, G.S., Anderson, P.S., Christy, M.E., Murcko, M.A., Randall, W.C., Schwam, H., Sugrue, M.F., Springer, J.P., Gautheron, P., Grove, J., Mallorga, P., Viader, M.-P., McKeever, B.M. and Navia, M.A. (1989) Thienothiopyran-2-sulfonamides: novel topically active carbonic anhydrase inhibitors for the treatment of glaucoma. *J. Med. Chem.* **32**, 2510–2513.

Bernel, J.D. and Crowfoot, D. (1934) X-ray photographs of crystalline pepsin. *Nature* **133**, 794–795.

Bragg, W.H. (1915) IX. Bakerian Lecture – X-rays and crystal structure. *Philo. Trans. Roy. Soc. London A* **215**, 153–274.

Bragg, W.L. (1929) The determination of parameters in crystal structures by means of Fourier series. *Proc. Roy. Soc. London A* **123**, 537–559.

Byers, L.D. and Wolfenden, R. (1973) Binding of the by-product analog benzylsuccinin by carboxypeptidase A. *Biochemistry* **12**, 2070–2074.

Bynum, W.F. (1970) Chemical structure and pharmacological action: a chapter in the history of 19th century molecular pharmacology. *Bull. Hist. Med.* **44**, 518–538.

Cartwright, A.C. and Armstrong, N.A. (2021) *A History of the Medicines We Take*. Pen & Sword Books, Yorkshire, UK.

Chang, A.C.Y. and Cohen, S.N. (1974) Genome construction between bacterial species *in vitro*: replication and expression of *Staphylococcus* plasmid genes in *Escherichia coli*. *Proc. Nat. Acad. Sci. U.S.A.* **71**, 1030–1034.

Clark, A.J. (1933) *The Mode of Action of Drugs on Cells*. Edward Arnold & Co., London.
Cohen, S.N. (2013) DNA cloning: a personal view after 40 years. *Proc. Nat. Acad. Sci. U.S.A.* **110**, 15521–15529.
Cohen, S.N., Chang, A.C.Y., Boyer, H.W. and Helling, R.B. (1973) Construction of biologically functional bacterial plasmids. *Proc. Nat. Acad. Sci. U.S.A.* **70**, 3240–3244.
Copeland, R.A. (1994) *Methods of Protein Analysis: A Practical Guide to Laboratory Protocols*. Chapman & Hall, New York.
Copeland, R.A. (2023) *Enzymes: A Practical Introduction to Structure, Mechanism, and Data Analysis*. 3rd Edition. Wiley, Hoboken, NJ.
Copeland, R.A. and Anderson, P.S. (2002) Enzymes and enzyme inhibitors. In *Textbook of Drug Design and Discovery* (P. Krogsgaard-Larsen, T. Liljefors and U. Madsen, Eds.). 3rd Edition. Taylor and Francis, New York, pp. 328–363.
Copeland, R.A. and Boriack-Sjodin, P.A. (2017) Enzyme inhibitors: biostructure-based and mechanism-based designs. In *Textbook of Drug Design and Discovery* (P. Krogsgaard-Larsen, T. Liljefors and U. Madsen, Eds.). 5th Edition. CRC Press, Boca Raton, FL, pp. 175–192.
Copeland, R.A. and Boriack-Sjodin, P.A. (2018) The elements of translational chemical biology. *Cell Chem. Biol.* **25**, 128–134.
Croston, G.E. (2017) The utility of target-based discovery. *Expert Opin. Drug Discov.* **12**, 427–429.
Cushman, D.W. and Ondetti, M.A. (1991) History of design of captopril and related inhibitors of angiotensin-converting enzyme. *Hypertension* **17**, 589–592.
David, L., Thakkar, A., Mercado, R. and Engkvist, O. (2020) Molecular representations in AI-driven drug discovery: a review and practical guide. *J. Cheminform.* **12**, 56.
Davis, J.P. and Copeland, R.A. (1996) *Protein Engineering*. In Kirk-Othmer encyclopedia of chemical technology. 4th Edition, Volume **20**. Wiley, New York, pp. 395–427.
Drucker, B.J., Tamura, S., Buchdunger, E., Ohno, S., Segal, G.M., Fanning, S., Zimmerman, J. and Lydon, N.B. (1996) Effects of a selective inhibitor of the Abl tyrosine kinase on the growth of Bcr-Abl positive cells. *Nature Med.* **2**, 561–566.
Drucker, B.J., Talpaz, M., Resta, D.J., Peng, B., Buchdunger, E., Ford, J.M., Lydon, N.B., Kantarjian, H., Capdeville, R., Ohno-Jones, S. and Sawyers, C.L. (2001) Efficacy and safety of a specific inhibitor of the BCR-ABL tyrosine kinase in chronic myeloid leukemia. *N. Engl. J. Med.* **344**, 1031–1037.
Eder, J., Sedrani, R. and Wiesmann, C. (2014) The discovery of first-in-class drugs: origins and evolution. *Nat. Rev. Drug Discov.* **13**, 577–587.
Ehrlich, P. (1913) Address in pathology, on chemiotherapy. *Brit. Med. J.* **2**, 353–259.
Ehrlich, P. ((1959–1960) Papers reprinted in:). In *The Collected Papers of Paul Ehrlich* (F. Himmelweit, M. Marquardt and H.H. Dale, Eds.). Pergamon, London, UK.

Ehrlich, P. and Hata, S. (1910) *die experimented chemotherapie der spirillosen*. Jilius Springer, Berlin.

Farber, S., Diamond, L.K., Mercer, R.D., Sylvester, R.F. and Wolff, J.A. (1948) Temporary remissions in acute leukemia in children produced by folic acid antagonist 4-aminopteroyl-glutamic acid (aminopterin). *New England J.Med.* **238**, 787–793.

Ferreira, S.H., Bartelt, D.C. and Greene, L.J. (1970) Isolation of bradykinin-potentiating peptides from *Bothrops jararaca* venom. *Biochemistry* **9**, 2583–2593.

Futterman, S. (1957) Enzymatic reduction of folic acid and dihydrofolic acid to tetrahydrofolic acid. *J. Biol. Chem.* **228**, 1031–1038.

Goeddel, D.V., Kleid, D.G., Bolivar, F., Heyneker, H.L., Yansura, D.G., Crea, R., Hirose, T., Kraszewski, A., Itakura, K. and Riggs, A.D. (1979) Expression in *Escherichia coli* of chemically synthesized genes for human insulin. *Proc. Nat. Acad. Sci. U.S.A.* **76**, 106–110.

Halperin, E.C. (2004) *Perspect. Biol. Med.* **47**, 1–14.

Henri, V. (1903) *Lois generales de l'action des diastases*. Hermann, Paris.

Herrick, J.B. (1910) Peculiar elongated and sickle-shaped red blood corpuscles in a case of severe anemia. *Arch. Intern. Med.* **6**, 517–521.

Hill, A.V. (1909) The mode of action of nicotine and curari determined by the form of the contraction curve and the method of temperature coefficients. *J. Physiol.* **39**, 374–413.

Huennekens, F.M. (1996) In search of dihydrofolate reductase. *Protein Sci.* **5**, 1201–1208.

Hunnefeld, F.L. (1840) *Der Chemismus in der thierischen Organisation*. Brockhaus, Leipzig, pp. 158–163.

Ingram, V.M. (1957) Gene mutation in human hemoglobin: the chemical difference between normal and sickle-cell hemoglobin. *Nature* **180**, 326–328.

Itakura, K., Hirose, T., Crea, R., Riggs, A.D., Heyneker, H.L., Bolivar, F. and Boyer, H.W. (1978) Expression in *Escherichia coli* of a chemically synthesized gene for the hormone somatostatin. *Science* **198**, 1056–1063.

Kendrew, J.C., Bodo, G., Dintzis, H.M., Parrish, R.G., Wyckoff, H. and Phillips, D.C. (1958) A three-dimensional model of the myoglobin molecule obtained by X-ray analysis. *Nature* **181**, 662–666.

Kimple, M.E., Brill, A.L. and Parker, R.L. (2018) Overview of affinity tags for protein purification. *Curr. Protocols Protein Sci.* **73**, 9.9.1–9.9.23.

J N Langley (1905) On the reaction of cells and of nerve-endings to certain poisons, chiefly as regards the reaction of striated muscle to nicotine and to curari. *J. Physiol.* **33**: 374–413.

Langmuir, I. (1918) The adsorption of gases on plane surfaces of glass, mica and platinum. *J. Am. Chem. Soc.* **40**, 1361–1403.

2

Molecular Targeted Therapeutic Modalities

Key Learning Points

- Currently approved modalities for therapeutic intervention fall into two major categories: biologics and small-molecule drugs.
- Biologics include protein-, nucleic acid–, and cell-based therapeutics. Most current biologic therapeutics, however, consist of protein-based antibodies directed against a specific target protein antigen.
- Small-molecule drugs include several subcategories such as reversible inhibitors/antagonists, covalent inhibitors/antagonists, reversible activators/agonists, protein degraders, and protein correctors.
- Small molecules may or may not compete with the endogenous ligand for the target and may bind at the same location as the endogenous ligand (orthosteric) or at a distinct binding pocket (allosteric).
- Small-molecule drugs can also be appended with other forms, such as prodrugs, antibody–drug conjugates, and peptide–drug conjugates.
- All these modalities have advantages and disadvantages. The best choice of therapeutic modality depends on the biology of the target protein and the disease that one wishes to address.

Before setting out on a drug discovery campaign, researchers must first consider the details of the disease that they wish to address, in as much detail as possible. From this understanding, one can then make rational decisions about what therapeutic modality may be best suited for patients suffering from the disease. In practice, a deep understanding of disease is often an integral part of target selection (Chapter 3). For pedagogic purposes, however, we discuss the topics of therapeutic modality and target selection separately in this and the next chapter.

Molecular Targeted Drug Discovery: A Guide to How Modern Medicines are Created,
First Edition. Robert A. Copeland.

In this chapter, we present common therapeutic modalities that are in current clinical practice today. These fall into two general categories: biologics and small-molecule drugs. Biologic approaches include nucleic acid–based, cell-based, and oncolytic virus–based therapeutic approaches. However, the vast majority of biologics are protein-based antibodies and soluble receptors. Here we shall limit our discussion of biologics to antibodies. Despite the great value that biologics bring to the treatment of specific diseases — especially for immunological diseases and cancer – low-molecular-weight, organic compounds (small-molecule drugs) continue to dominate the landscape of current therapeutics. According to Southey and Brunavs (2023), over 90% of currently marketed drugs are small molecules. This generality holds true even in specific therapeutic areas in which biologics have played a more dominant role. For example, in oncology, 76% of marketed drugs are small molecules (Buvailo, 2018). Among small-molecule drugs, two categories of targets predominate the landscape of marketed drugs: G-protein–coupled receptors (GPCRs) and enzymes. For example, Sun et al. (2017) surveyed FDA-approved anticancer drugs and found that 60% targeted enzymes and another 25% targeted GPCRs; thus, these two target classes accounted for 85% of marketed cancer therapeutics as of the date of publication (2017). There are good physicochemical reasons why these two target classes are such common targets of therapeutics, as we shall discuss further in Chapter 3. While the exact proportions may shift in the future, it is very likely that small-molecule drugs will continue to dominate the therapeutic landscape for the foreseeable future. Hence, the subsequent chapters of this text will focus on the practice of drug discovery and preclinical development for small-molecule drugs. Many of the concepts and practices that we shall discuss for small-molecule drugs are, however, equally germane to the discovery and optimization of biologics. For a review of the development of therapeutic antibodies, the reader is referred to the recent article by Lu et al. (2020).

Biologics and small-molecule drugs each have their advantages and limitations, as summarized in Table 2.1. The choice of which therapeutic modality is most appropriate depends on the nature of the disease in which one wishes to intervene and the nature of the target to be drugged. In some cases, both modalities are viable.

One might think that having a small molecule and a biologic therapeutic against the same target would be redundant. However, in some cases, the combination of both modalities can offer unanticipated clinical synergies. Take, for example, the case of therapeutics targeting the human epidermal growth factor receptor 2 (HER2), a cell surface receptor that is overexpressed in many breast cancer tumors (Slamon et al., 1987). HER2 is a ligand-dependent receptor tyrosine kinase that catalyzes the phosphorylation of intracellular substrate proteins in response to ligand engagement of HER2 at its extracellular ligand-binding domain. Like all members of the epidermal growth factor receptor (EGFR) family, HER2

Table 2.1 Comparison of the advantages and limitations of small-molecule– and protein-based biologic (e.g., antibodies) therapeutics.

Therapeutic modality	Advantages	Limitations
Small molecules	Oral dosing Cell permeable Lower cost	Limited to druggable binding sites Limited period of exclusive marketing rights Potential for off-target effects due to collateral binding to proteins other than the target
Protein biologics	No defined binding pocket required Less generic competition High target specificity	Limited to parenteral administration Higher costs Limited SAR capabilities Limited to extracellular targets Greater potential for immune response

is composed of three major protein domains: an extracellular, ligand-binding domain; a transmembrane domain; and a globular intracellular kinase domain. Small-molecule inhibitors, such as lapatinib (Tykerb), bind to the adenosine triphosphate (ATP)-binding pocket of the kinase domain and thereby block this catalytic activity (Nahta et al., 2007). An alternative therapeutic approach is provided by the HER2 specific antibody, trastuzumab (Herceptin), which binds to the extracellular domain of the receptor (Vogel et al., 2002). Both lapatinib and trastuzumab block the activity of HER2 and thereby block downstream intracellular signaling cascades that rely on HER2 activity. Thus, lapatinib and trastuzumab both block the activity of the same target, but in different biochemical ways. Surprisingly, preclinical and clinical studies suggest that combinations of the two drugs are more effective than one would expect simply from the additive effects of the combined doses of the two. In an open-label, phase III clinical trial, 296 patients with HER2-positive, recurrent breast cancer were randomized to receive lapatinib alone (1,500 mg/d) or a combination of lapatinib (1,000 mg/day) plus trastuzumab (4 mg/kg loading dose followed by 2 mg/kg per week). Patients receiving the combination therapy demonstrated improvements in median progression-free survival (12 weeks compared to 8.1 weeks for lapatinib alone), median overall survival (51.6 compared to 39 weeks), and a doubling (24.7% vs. 12.4%) of clinical benefit rate, defined as response rate plus stable disease for >6 months (Blackwell et al., 2010). Hence, there are situations in which approaching target modulation by both small-molecule and biologic drugs can be beneficial to patients.

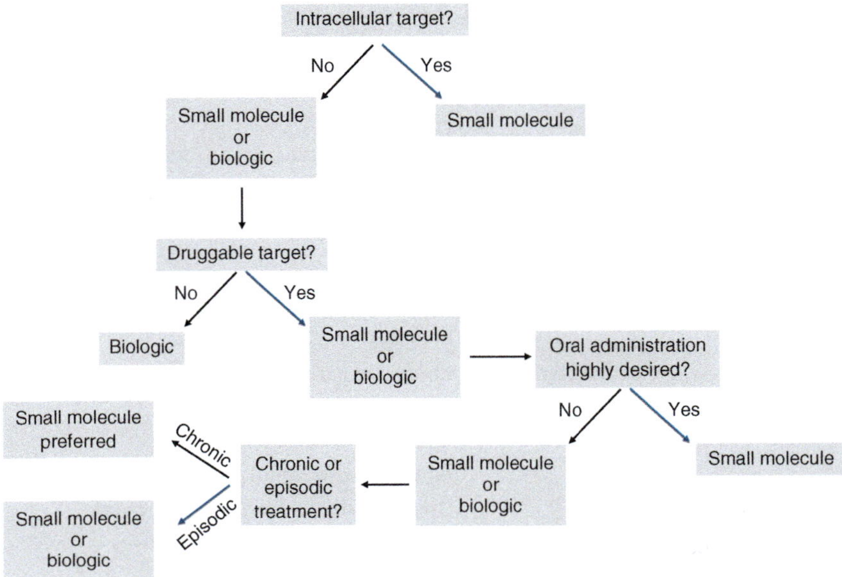

Figure 2.1 Decision tree for prioritizing a biologic or small-molecule therapeutic approach for a molecular target. Note: the term "druggable target" is defined as a protein for which there is evidence of a binding pocket that would likely bind drug-like small molecules. This term will be discussed further in Section 2.3.1 and in Chapter 3.

The choice of whether to pursue a small-molecule or biologic drug, or both, is not always simple. Again, it largely depends on the nature of the disease and the specific target of interest. Many factors can impinge on this decision. A partial decision tree for guiding this decision is provided in Figure 2.1.

2.1 Candidate Drug Profile

The decision of whether to pursue a small-molecule or biologic approach is part of the larger task of developing a candidate drug profile (CDP), based on the disease to be treated and the specific molecular species to be targeted. Regardless of modality, the CDP is a critical mission statement that is established at the early stages of the project and defines the strategic requirements needed to select a compound or biologic for development. Some framing questions that may be asked in developing the CDP are as follows.

1) Is the disease chronic or acute? That is, what may be the required duration of drug treatment for these patients?

2) What frequency of dosing and in what setting will patients tolerate treatment? For example, a patient may be willing to travel to a physician's office or medical center weekly or monthly for intravenous administration to treat serious, life-altering disease. They may be less tolerant of these inconveniences for chronic diseases of less severity.
3) Is the pathobiology of disease driven by intracellular pathways, extracellular pathways, or both? From this understanding, are the targets of intervention likely to occur within the cell, on the cell surface, in the extracellular fluid(s), or some combination thereof?
4) What are current standards of treatment for the disease and any comorbidities with which these patients may likely present? This information may frame what modalities will be tolerable to the addressable patients, considering other treatments that may need to be coadministered.
5) What is the incidence, prevalence, and geographic distribution of the disease? These factors may frame expectations of the acceptable costs, route of administration, and shelf life of the drug product. From this, what is the timing and likelihood of recouping the cost of research and development and making a profit? These are important considerations for any drug discovery entity, but especially so for industrial groups that must satisfy investor needs. It is also important to recognize that the infrastructure for storage, distribution, and patient access to medicine may be very different in different geographic locations. This also may need to be considered in selecting an appropriate therapeutic modality.

These questions, and others, are then used to develop specific requirements for the candidate drug to be developed. For small molecules, project teams may likewise develop profiles for intermediate stages along the path to a development candidate, such as a Validated Hit Profile and a Lead Molecule Profile, to guide decision-making during the target-to-hit and hit-to-lead stages of drug discovery. The criteria for all three of these guiding documents are typically agreed to between the project team members and management, in the early stages of a drug discovery project. We will consider these guiding documents further in subsequent chapters.

2.2 Biologic Therapeutics

Biological agents that are in clinical use today consist of vaccines, cell-based therapies (CAR-T; see Dabas and Danda, 2023 for a review), antibodies, and solubilized receptors; other biological approaches, such as interfering RNAs and clustered regularly interspaced short palindromic repeats (CRISPR) constructs, are under consideration as therapeutic modalities, but these have yet to be proven safe and effective in the clinic. All the currently approved therapeutic biologics are cell- or protein-based and are therefore susceptible to digestion in the gastrointestinal tract. Therefore, biologics are generally administered via parenteral routes; most biologics are administered through intravenous injection (IV), but other parenteral routes, such as intramuscular injection, stereotactic injection, subcutaneous injection, ionophoresis, and intranasal sprays, have also been used successfully in specific cases. Generally, biologics are not cell permeable, thus limiting their utility to cell-surface and extracellular targets. As biomacromolecules, protein-based biologics present an increased potential for immunologic responses than that generally seen with small-molecule drugs. Hence, care must be exercised to minimize this risk. Finally, the cost of production for biologics is considerably higher than that for small-molecule drugs. These limitations notwithstanding, biologics have added great value to the therapeutic armamentarium, especially for treatment of immunologic diseases and for immunomodulation in cancer treatments. Indeed, 78% of total US sales of all antibodies are directed toward the treatment of cancer and autoimmune diseases (Sharma et al., 2023).

Most current therapeutic biologics are antibodies. Antibodies (Abs), also referred to as immunoglobulins (Igs), are glycoproteins produced by plasma B-cells that function to bind and neutralize specific targets referred to as the antigens. The plasma B-cells may produce multiple antibodies recognizing different amino acid sequences (or other structural motifs) on the same target antigen; the specific antigen recognition element that is bound by a particular Ab is referred to as the epitope.

Antibodies are composed of two copies each of two protein subunits – the heavy chain (molecular weight about 50 kDa) and the light chain (molecular weight about 25 kDa) – which are linked together by inter-subunit disulfide bonds to form a heterotetramer, as illustrated in Figure 2.2. Mammals contain five isoforms of the heavy chains, γ, α, μ, ε, and δ, and two isoforms of the light chains, κ and λ. A constant region of the heavy chain determines the antibody isotypes: IgG, IgA, IgM, IgE, and IgD. Of these, the most abundant in humans is the IgG isotype, representing 75% of serum Abs; most therapeutic antibodies in current clinical use are monoclonal IgGs or fragments thereof. The light-chain subunit is composed of two domains: a hypervariable domain, V_L, and a single constant domain, C_L. The heavy chain contains four domains: a hypervariable domain, V_H, and three distinct constant domains C_H1, C_H2, and C_H3. The V_H, V_L, C_L, and C_H1 domains together form a fragment of the antibody molecule known as the **a**ntigen-**b**inding (Fab) **f**ragment, and the paired V_L and V_H domains constitute the **v**ariable (Fv) **f**ragment; the antigen-binding site, known as the paratope, is made up of the amino-termini of these two hypervariable domains. The constant domains C_H2 and C_H3 constitute the **c**rystallizable (Fc) **f**ragment, which is important in determining the half-life and effector function of the Ab. As

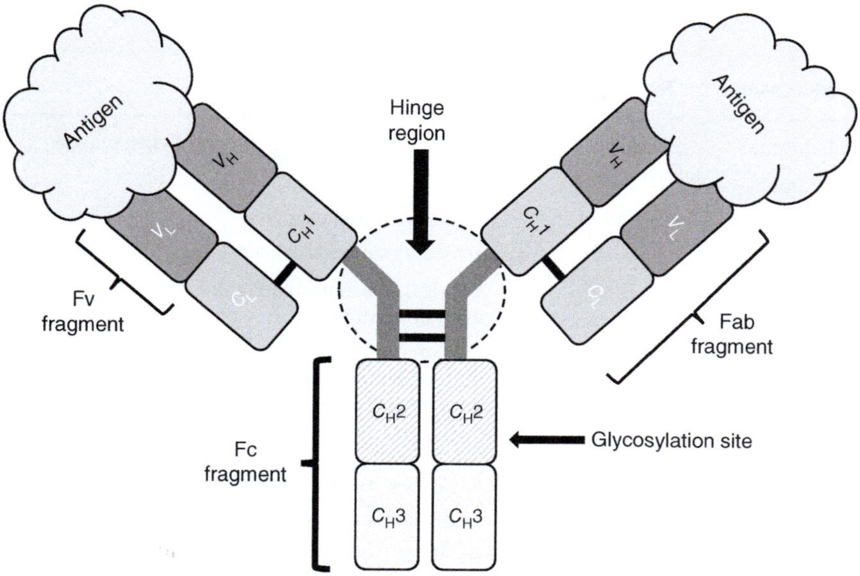

Figure 2.2 The generic structure of an IgG antibody, showing the different domains that make up the heavy chains (denoted by subscript H) and the light chains (denoted by subscript L). As illustrated here, the antigen-binding site (the paratope) is composed of surface areas at the amino-termini of the heavy-chain V_H domain and the light-chain V_L domain. Black lines connecting domains represent inter-domain disulfide bonds.

illustrated in Figure 2.2, the Y-shaped tetrameric IgG thus contains two identical Fab fragments and antigen-binding sites. Interestingly, camelids (camels, llamas, and alpacas) additionally produce IgGs containing only the heavy chains (as a homodimer; molecular weight 90 kDa) and monomeric heavy-chain variable domains (i.e., lacking the constant domains; molecular weight 15 kDa) that are all capable of high-affinity antigen binding. The monomeric heavy-chain variable domain species are referred to as V_HH antibodies, or nanobodies because of their small size. Similarly, a small, variable new antigen receptor (VNAR) protein has been found in sharks that also displays high-affinity antigen binding. These much smaller V_HH and VNAR forms are of great interest as potential therapeutics because they display much greater thermal and pH stability, greater tissue penetration, and higher lot-to-lot consistency as compared to traditional IgGs (Sharma et al., 2023; Hoover et al., 2021; Arbabi-Ghahroudi, 2017). To date, four V_HH antibodies have been approved as therapeutics for the treatment of cancers (LCAR-B38M and envafolimab), rheumatoid arthritis (ozoralizumab), or thrombotic thrombocytopenia purpura (caplacizumab).

The first IgG to be approved by the FDA was muromonab-CD3, as an immunosuppressant for acute transplant rejection. This approval paved the way for the

Table 2.2 Five therapeutic Abs that are responsible for most US sales.

Product name	Target (Antigen)	Therapeutic indication(s)
Adalimimab Infliximab	Tumor necrosis factor (TNF)	Chron's disease; rheumatoid arthritis; plaque psoriasis
Trastuzumab	HER2	HER2(+) breast cancer HER2(+) stomach cancer
Bebacizumab	VEGFA	Cervical, ovarian, colorectal, lung, kidney cancers, and glioblastoma
Rituximab	CD20	Rheumatoid arthritis; non-Hodgkin's lymphoma; chronic lymphocytic leukemia

>100 therapeutic and diagnostic antibodies in use today. As of 2023, most US sales for therapeutic antibodies could be ascribed to five products (referred to as the "Big 5"; Sharma et al., 2023); these are infliximab, adalimumab, trastuzumab, bevacizumab, and rituximab (Table 2.2).

2.2.1 Therapeutic Antibody Discovery

Therapeutic Ab discovery follows a sequential series of stages, much like what was described for small-molecule drug discovery in Chapter 1 (Figure 1.17). For therapeutic Abs, these stages include target discovery, antibody generation, hit-to-lead generation, lead-to-candidate generation, investigational new drug (IND)-enabling studies, and clinical development (Figure 2.3). An excellent review of therapeutic antibody discovery has recently been provided by Hoover et al. (2021).

2.2.1.1 Target Selection and Antibody Generation

The first stage of Ab discovery is target selection and involves identifying and validating a specific disease-associated antigen. The essentiality of a specific antigen can be validated by RNA-silencing methods, such as siRNA or CRISPR, to ensure that knockdown or knockout of the antigen leads to the desired disease-modifying phenotypic outcome. For example, for a cancer therapeutics,

Figure 2.3 Schematic representation of the sequential stages of therapeutic antibody discovery.

knockdown or knockout should produce an antiproliferative phenotype selectively in cancer cells. We will discuss target identification and validation in greater detail in a subsequent chapter.

Once a target antigen has been selected, the recombinant antigen is expressed in an appropriate host cell and purified. The full-length antigen, a peptide from the antigen, or nucleic acids encoding the antigen are then injected into an animal species to elicit an immune response and antibody production. Mice are most used at this stage, but other species, including rats and rabbits, may also be used. Antibodies produced from any of these species must go through a process of humanization, in which amino acid sequences are altered to be more human-like, to thus avoid an antidrug antibody response in dosed patients. Alternatively, transgenic "humanized" animal species can be used for immunization and antibody production. The antigen is co-injected with an adjuvant mixture that enhances the immune response of the organism (see Jones et al. (2005), for details). The full-length antigen is usually the preferred immunogen for this *in vivo* antibody production scheme, as it will present an array of possible epitopes and will present these in their intact three-dimensional conformation.

B-cells from the immunized animal are then harvested from the spleen and/or lymph nodes. These are then mixed and fused with an immortal myeloma cell line to produce B-cell:myeloma fusions referred to as hybridomas. B-cells are limited in their ability to be continuously cultured. By generating hybridomas, one creates a system that can be grown indefinitely to generate antibody. An alternative to hybridoma formation is to use fluorescence-activated cell sorting (FACS) to sort the B-cells for binding to a fluorescently labeled version of the antigen and for other desired phenotypic properties. B-cells thus selected for desired properties are then analyzed to determine the gene sequence for their produced antibody, so that this can then be cloned and recombinantly expressed in mammalian cells, such as Chinese hamster ovary (CHO) or human embryonic kidney 293 (HEK293) cells.

With the modern tools of molecular biology, antibody generation can now also be done *in vitro* using synthetic antibody repertoire technologies, such as bacteriophage (phage), yeast, and mammalian cell display (Rahbarnia et al., 2017; Hoover et al., 2021). In these systems, an antibody sequence is expressed on the surface of a virus or cell and subjected to random mutagenesis to produce an array of antibodies. Because these are surface-located, the phage or cells can be sequentially screened for antigen binding (see next section), using lower and lower concentrations of antigen, to thus enrich for cells or phage expressing high-affinity antibodies.

2.2.1.2 Hit-Finding and Hit-to-Lead

With a pool of antibody-producing cells (or phage) in hand, primary hit-finding methods are used to identify antibodies with the required affinity and selectivity for the target antigen. The minimum affinity and selectivity for the target antigen required to deem a specific antibody a hit will have been predetermined by the project team as part of a Hit Profile (as discussed earlier). The target antigen can be fluorescently labeled and used for FACS-based binning of antibodies by binding affinity. Antigen can alternatively be immobilized on 96-, 384-, or 1536-well microplates, and a variety of assay methods can be used to screen for cell or phage binding, or for semi-purified antibody binding. These methods can be used to bin antibodies by relative binding affinity or in more quantitative formats to determine EC_{50} or K_d values (see Appendix 1). Nontarget antigens can be likewise immobilized and used to determine the selectivity of each antibody for the target antigen. Abs that score as initial hits can be validated in orthogonal binding assays, such as using surface plasmon resonance (SPR) methods to determine the K_d and the kinetic constants k_{on} and k_{off} (Appendix 1). Antibodies that meet the minimum criteria for hit declaration are sequenced and recombinantly expressed in mammalian cells. These hits can then be purified and subjected to additional evaluations to select lead antibodies.

The criteria for lead declaration will vary with the specific needs of the project team. These may include more stringent affinity and selectivity criteria, epitope mapping, and the like.

2.2.1.3 Lead Optimization

Abs that are declared leads are further optimized for multiple additional criteria. With the binding affinity and selectivity of the Ab for antigen already established during the hit-to-lead stage, the Abs are further characterized for the ability to elicit cell killing and/or regulation of cellular function, depending on the desired therapeutic outcome. At this stage, antibodies may also be engineered for improvements in biochemical stability, tissue distribution (in animal models and predicted in humans), systemic half-life (pharmacokinetics), immune effector function, and immunogenicity. Systematic amino acid sequence changes may also be engineered with the goal of deimmunizing or "humanizing": the Ab. Similarly, engineered amino acid changes can increase physical and chemical stability, improve target cell cytotoxicity, and diminish nontarget cell cytotoxicity. Finally, defining and engineering of sites of posttranslational modifications (e.g., glycosylation) to the Ab can also be helpful in enhancing *in vivo* stability/half-life and diminishing untoward characteristics of the Ab. As with hit declaration, the range of specific criteria that must be met to declare a development candidate Ab will vary with the specific therapeutic application and the project team goals. Further details on Ab lead optimization can be found in the review by Hoover et al. (2021).

2.3 Small-Molecule Drugs

Small-molecule drugs continue to dominate the therapeutic landscape across most disease areas. The physicochemical properties of small-molecule drugs must confer on these molecules a number of pharmacological characteristics for utility as therapeutic agents. For example, they must bind with high affinity and specificity to their intended target; for oral administration, they must be absorbed from the intestines into systemic circulation; they must be soluble in aqueous solutions, yet also sufficiently lipophilic to permeate cell membranes to reach intracellular targets; they must be sufficiently stable to metabolic decomposition in the blood, liver, kidneys, and other tissue, to allow sufficient time for the onset of pharmacological effect and for convenient dosing schedules. These and other required properties of small-molecule drugs will be discussed in subsequent chapters, as part of our discussion of hit-to-lead and lead optimization. It follows that the intended targets of small-molecule drugs must contain binding pockets with physicochemical properties that complement those required by the drug for the target to be actionable (druggable). Mechanistically, small-molecule drugs can modulate target function in several different ways that lead to desired pharmacological outcomes. Here, we present some of the more common mechanisms of action of small-molecule drugs.

2.3.1 Target Druggability

For a target to be actionable by a small-molecule drug, the protein must contain a binding pocket that is physiochemically complementary to the stereochemical, electrostatic, hydrophobic, and other structural requirement of drug molecules. In the parlance of drug discovery, such binding pockets are said to be druggable. What then are the physicochemical properties of typical drug-binding pockets on proteins? The salient features of druggable binding pockets have been reviewed by several authors (Liang et al., 1998; Hajduk et al., 2005; Copeland, 2013).

Generally, drug-binding pockets are small cavities or clefts on the surface area of the target. They are typically of small-molecular volume (typically about 1,000 $Å^3$), representing only 1–5% of the overall volume of the protein target (Liang et al., 1998). These pockets characteristically display a high degree of surface roughness, which is defined as the ratio of surface area to volume (Petitt and Bowie, 1999). A high degree of surface roughness confers stereochemical uniqueness to the pocket and facilitates favorable van der Waals interactions between the pocket and the cognate ligand. These sites are generally composed of hydrophobic amino acids to exclude bulk solvent from the pocket, yet they may contain highly ordered water molecules as part of the specific architectural motifs wherein they participate directly or indirectly in ligand interactions. By excluding bulk solvent, these pockets facilitate the formation of stronger intermolecular hydrogen bonds and other electrostatic interactions between the ligand and recognition elements within the pocket. Despite their generally hydrophobic nature, binding pockets may also contain specifically localized charged species, such as amino acid side chains (e.g., lysine, arginine, histidine, glutamic acid, and aspartic acid) and cofactors (e.g., flavins, nicotinamide, pyridoxal phosphate, quinones, hemes, metals, and the like), that form electrostatic interactions (e.g., salt bridges) with charged moieties on ligands. These various factors combine to create considerable surface complexity in drug-binding pockets (Hajduk et al., 2005).

The binding pockets to which small-molecule drugs bind have evolved to bind physiologically relevant small-molecular-weight ligands, such as nucleotides, amino acids, steroids, metabolites, peptides, other proteins, nucleic acids, quinones (and other cofactors), and the like. The interactions of these natural ligands usually modulate the biological activity of the target in either a positive (activating) or negative (inhibiting) way. In some cases, these ligands are chemically transformed by the target, as is the case in the enzymatic conversion of substrate to product molecules. Enzymatic product formation, or conformational changes that attend ligand binding to targets, can set off cascades of downstream biological activity, such as multienzyme metabolic pathways or signal transduction cascades. The extent and duration of such cascades must be carefully regulated to ensure the biological fitness of the cell, tissue, and organism. Regulation often requires rapid responsiveness to induce and then halt the action of the cascade, in response to the changing needs of the cell. The cellular need for facile responsiveness is in large part dictated by the interactions between receptors and their ligands. Hence, one finds that the interactions of receptors with physiological ligands are characterized by four important characteristics: (1) they are reversible; (2) they display moderate binding affinity, typically in the μM to mM range (but there are exceptions in which nM and even pM affinity are observed); (3) they are modulated by changes in the local concentration of ligand; and (4) they are often also modulated by other changes in local environment such as posttranslational modification of the receptor, pH, ionic strength, competing ligand interactions, and the like (Copeland, 2023). We shall revisit the topic of target druggability in our discussion of target selection in Chapter 3.

2.3.2 Reversible Inhibitors/Antagonists

A common mechanism of action (MOA) for small-molecule drugs is to block the pathogenic function of a specific molecular target. Compounds of this MOA are referred to as inhibitors or antagonists, depending on the type of target protein being modulated; generally, compounds that block enzyme function and protein–protein interactions (PPIs) are referred to as inhibitors, and compounds that block the function of GPCRs are referred to as antagonists. However, these terms are merely colloquial and do not reflect any mechanistic distinctions. For convenience, here we shall universally refer to such compounds as inhibitors. Let us use the example of an enzyme target to illustrate some features of inhibitors. Copeland (2013, 2023) uses the schemes shown in Figure 2.4 to illustrate the different modes of inhibitor interaction with a free enzyme molecule or with the binary enzyme–substrate complex (or binary complexes that occur after initial substrate binding along the reaction coordinate from substrate to product molecule).

The free enzyme (E) can combine with substrate to form a binary encounter complex with its substate molecule, symbolized by ES. The binding affinity of E and S to form the ES complex is quantified by the substrate equilibrium dissociation constant, K_S (note, under steady-state conditions, as is typical of enzyme assays used in drug discovery, the term K_S is replaced by the term K_M; see Copeland (2013) and (2023), for further details). The formation of this initial encounter complex sets off a series of conformational changes to the enzyme protein, with the aim of inducing strain in the substate molecule so that it is first transformed into a bound transition-state structure and then, through a potential series of intermediate states, into a bound product structure. These various intermediate forms are not explicitly represented in Figure 2.4 but are collectively captured in the kinetic constant k_{cat} – a composite kinetic constant that represents the reaction rate in going from the initial ES state to the release of product molecule (P) and regeneration of the free enzyme state (E). An inhibitor molecule (I) can combine with the free enzyme state (E) to form the binary complex EI, the affinity of which is quantified by the equilibrium dissociation constant, K_i (sometimes referred to as the inhibition constant). After EI formation, this binary complex can combine with substrate to form a ternary ESI complex. The affinity of the substrate molecule for the EI complex is potentially different than its affinity for the free enzyme (K_S). Hence, the substrate equilibrium dissociation constant K_S (or K_M under steady-state conditions) is modified by a constant α. An alternative kinetic pathway has formation of the ES complex preceding the binding of I to this binary complex, again resulting in the same ternary ESI

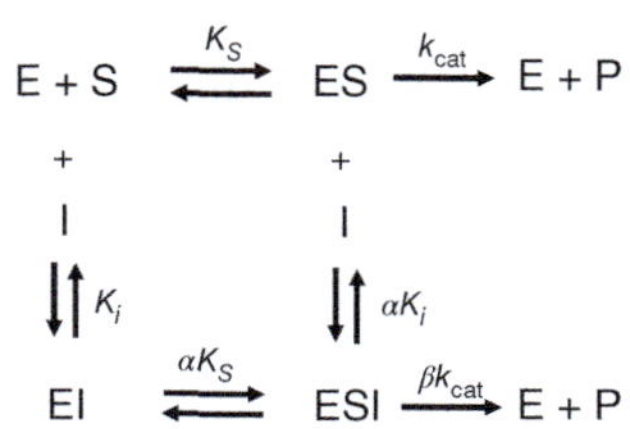

Figure 2.4 Equilibrium scheme for enzyme turnover in the presence and absence of reversible inhibitors. *Source:* Copeland (2023) / with permission of John Wiley & Sons.

complex. In this case, the value of K_S is unperturbed, but that of K_i is modified by the same constant, α, that was just introduced; K_S and K_i are modified by the same constant α in these two kinetic pathways, as they combine to form a thermodynamic cycle from E + S + I to ESI. In some cases, the ESI complex can still go on to form product, albeit at a lower rate in the presence of inhibitor. Hence, the value of k_{cat} is modified by the constant β in the presence of inhibitor.

From the preceding discussion, we see that inhibitory molecules may combine with different forms of an enzyme target and the values of the two constants α and β define different inhibitory modalities. When $\beta = 0$, the inhibitor completely blocks enzyme turnover and is referred to as a dead-end inhibitor. These types of inhibitors are most effective as therapeutic agents. When $\beta < 1$ but >0, the inhibitor only partially blocks enzyme turnover; such inhibitors are referred to as partial inhibitors and are generally not effective as therapeutics (but there are some exceptions to this generality). If β is >1, then the molecule is an activator, rather than an inhibitor; we shall discuss reversible activators separately later.

The value of α also defines different inhibition modalities. When α is infinite, the inhibitor can only bind to the free enzyme state, E, and not to ES or subsequent states. Likewise, in this case, the substrate can bind to E but not to EI. Thus, when $\alpha = \infty$, we have a situation of mutually exclusive binding of S and I to E; inhibitors of this type are said to be competitive with substrate. When $\alpha = 1$, there is no difference in the affinity of substrate for E or EI, and likewise no difference in the affinity of inhibitor for E or ES. Such an inhibitor is said to be noncompetitive with respect to substrate. If $\alpha > 1$ but $< \infty$, the inhibitor can bind to either E or ES, but preferentially binds (i.e., with greater affinity) to E. Such an inhibitor is referred to as a mixed inhibitor and sometime more specifically referred to as a mixed, competitive inhibitor. If $\alpha < 1$, the inhibitor can again bind to either E or ES but, in this case, binds with greater affinity (preferentially) to ES. Such an inhibitor is also referred to as a mixed inhibitor or more specifically as a mixed, uncompetitive inhibitor. Finally, if $\alpha << 1$, the inhibitor binds with much greater affinity to the ES complex, often with unmeasurably little binding to the free E state. Such an inhibitor is referred to as an uncompetitive inhibitor. These different forms of target inhibitors are summarized in Table 2.3.

Copeland (2013, 2023) details experimental approaches to diagnosing inhibitor modality for enzyme inhibitors and the proper analysis of the interactions of enzymes with each modality. For our purposes, the most common way to assess inhibitor potency and modality is by use of the IC_{50} value, a version of the EC_{50}, introduced in Chapter 1, that is specifically used in the case of inhibition of target function. Under appropriate assay conditions (see Copeland, 2013, 2023, and subsequent chapters), the steady-state velocity of an enzyme reaction can be measured as the slope of a linear plot of concentration of product formed ([P]) as a function of reaction time (see later). This velocity in the absence of inhibitor is

Table 2.3 Different modalities of target inhibition based on the values of the constants α and β as defined in Figure 2.4 and the preceding text.

Constant value	Inhibition modality
$\beta < 1$ but >0	Partial inhibition
$\beta = 0$	Dead-end inhibition
$\alpha = \infty$	Competitive inhibition
$\alpha > 1$ but $< \infty$	Mixed, competitive inhibition
$\alpha = 1$	Noncompetitive inhibition
$\alpha < 1$	Mixed, uncompetitive inhibition
$\alpha << 1$	Uncompetitive inhibition

designated v_0. The reaction velocity, under identical conditions, is then measured at various concentrations ($[I]$) of a potential inhibitor. The velocity is reduced by the presence of inhibitor from a value of v_0 to a value of v_i, which depends on the concentration of inhibitor as follows.

$$\frac{v_i}{v_0} = \frac{1}{1 + \frac{[I]}{\mathrm{IC}_{50}}} \tag{2.1}$$

Equation 2.1 describes a descending S-shaped curve of v_i/v_0 with increasing concentration of inhibitor ($[I]$), when plotted on a semilog scale (Figure 2.5). For any dead-end inhibitor, v_i/v_0 should have limits of 1.0 at $[I] = 0$ and zero at $[I] = \infty$. When $[I] = \mathrm{IC}_{50}$, the value of v_i/v_0 is 0.5, as illustrated in Figure 2.5. The IC_{50} is commonly used to compare the relative affinity of compounds for a common target during hit identification (Chapter 4) and to develop structure–activity relationship (SAR) within a pharmacophore series (a series of compounds related by a common core chemical structure that is required for target modulation) during the hit-to-lead and lead-to-candidate phases. However, it is important to understand

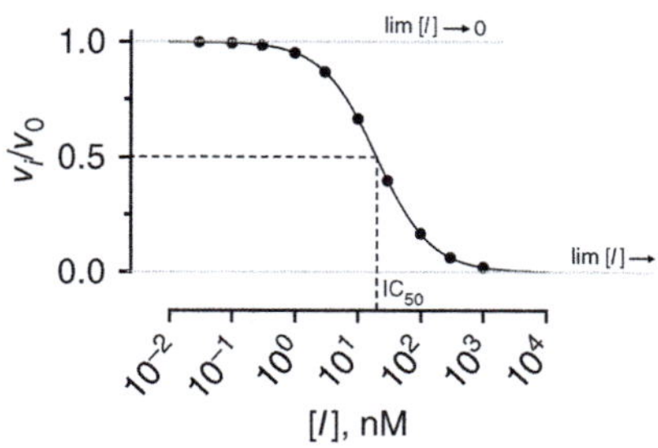

Figure 2.5 Idealized concentration–response plot for enzyme reaction velocity as a function of inhibitor concentration, illustrating the limits of v_i/v_0 as $[I]$ approaches zero and infinity and the determination of IC_{50} value.

that the IC_{50} is a relative measure of target-inhibitor affinity that is highly dependent on several experimental factors, including buffer pH, ionic strength, ion composition, temperature, and the like. Hence, as discussed in Copeland (2013, 2023), one must take great care to ensure that all experimental conditions are held constant in making IC_{50}-based comparisons among compounds within a pharmacophore series, and it is not appropriate to use IC_{50} values in isolation, as a measure of relative compound affinity among compounds from different pharmacophore series that may conform to different inhibition modalities. Copeland (2013, 2023) makes the case that one needs to know the inhibition modality of every pharmacophore series to be compared and then makes comparisons of affinity based on the K_i and/or αK_i value(s). Fortunately, there is a convenient means of determining inhibition modality and equilibrium dissociation constants for inhibitors by measuring the IC_{50} value as a function of substrate concentration relative to the K_M value for that substrate (Cheng and Prusoff, 1973; Copeland, 2013, 2023; Buker et al., 2019).

Cheng and Prusoff (1973) first published equations relating the IC_{50} values of enzymes inhibitors to their K_i and/or αK_i value(s) for different inhibition modalities. For each modality, the equations described the dependence of the IC_{50} value on the concentration of substrate ($[S]$) used in the assay. Buker et al. (2019) rederived these equations in terms of the ratio $[S]/K_M$ for each modality. For our purposes, the equations of Buker et al. (2019) are most convenient and will be presented here. In all cases, the first step is to determine the value of K_M in the absence of inhibitor, by titrating the substrate concentration used in the enzyme assay and plotting the resulting steady-state velocity as a function of substrate concentration, as described later (see Section 2.3.2 and Figure 2.9)

A general equation that is applicable across all inhibition modalities is given by the following.

$$IC_{50} = \frac{1 + \frac{[S]}{K_M}}{\frac{1}{K_i} + \left(\frac{1}{\alpha K_i} \times \frac{[S]}{K_M}\right)} \tag{2.2}$$

Equation 2.2 reduces to simpler forms for noncompetitive, competitive, and uncompetitive modes of inhibition. For noncompetitive inhibition, $\alpha = 1$, and equation 2.2 therefore reduces to the simple equality $IC_{50} = K_i$. For a pure competitive inhibitor, α approaches infinity and thus equation 2.2 reduces to the following.

$$IC_{50} = K_i\left(1 + \frac{[S]}{K_M}\right) \tag{2.3}$$

Table 2.4 Characteristic patterns of effect of $[S]/K_M$ on the IC_{50} value for reversible inhibitors of different modality.

Inhibition modality	Effect on IC_{50}
Competitive	Increases linearly with increasing $[S]/K_M$
Mixed, competitive ($\alpha > 1$)	Increases curvilinearly with increasing $[S]/K_M$
Noncompetitive ($\alpha = 1$)	No change with increasing $[S]/K_M$
Mixed, uncompetitive ($\alpha < 1$)	Decreases curvilinearly with increasing $[S]/K_M$
Uncompetitive	Sharply decreases curvilinearly with increasing $[S]/K_M$

Source: Copeland (2013) / with permission of John Wiley & Sons.

For purely uncompetitive inhibition, K_i approaches infinity but αK_i is finite. Hence, equation 2.2 now reduces to the following.

$$IC_{50} = \alpha K_i \left(1 + \frac{K_M}{[S]}\right) \tag{2.4}$$

Equations 2.1 through 2.4 define clearly distinct patterns for each inhibition modality when the IC_{50} value is plotted as a function of $[S]/K_M$. These diagnostic patterns are summarized in Table 2.4 and Figure 2.6. Once the inhibition modality is determined from the graphic pattern, one can use the appropriate equation (Equations 2.1 through 2.4) to fit the data points and thereby obtain the K_i and/or αK_i value(s) for the compound under study. Construction of the plot of IC_{50} as a function of $[S]/K_M$ requires the simultaneous titration of both $[S]$ and $[I]$ (referred to as a double titration). This is experimentally simple to achieve in the context of multi-well plate assay formats, as described in Buker et al. (2019).

2.3.2.1 Orthosteric and Allosteric Binding Sites

Enzymes and receptors have evolved to bind specific, natural ligands that elicit a physiological activity, as a direct result of binary target–ligand complex formation. The sites of ligand binding have likewise evolved to present a specific, three-dimensional pocket that is structurally complementary to that of the natural ligand. Ligand-binding pockets tend to be of small volume compared to the total volume of the protein, are generally hydrophobic to exclude bulk water and thereby augment hydrogen bonding and electrostatic interactions, and are arranged so that specific amino acid side chains and main chain atoms can form specific recognition elements for structural elements of the cognate ligand (e.g., hydrogen bond donors and acceptors, aromatic rings for π–π, π–cation, and other electronic interactions, and cations and anions for salt bridge formation). The site of natural ligand binding is referred to as the orthosteric pocket (or binding

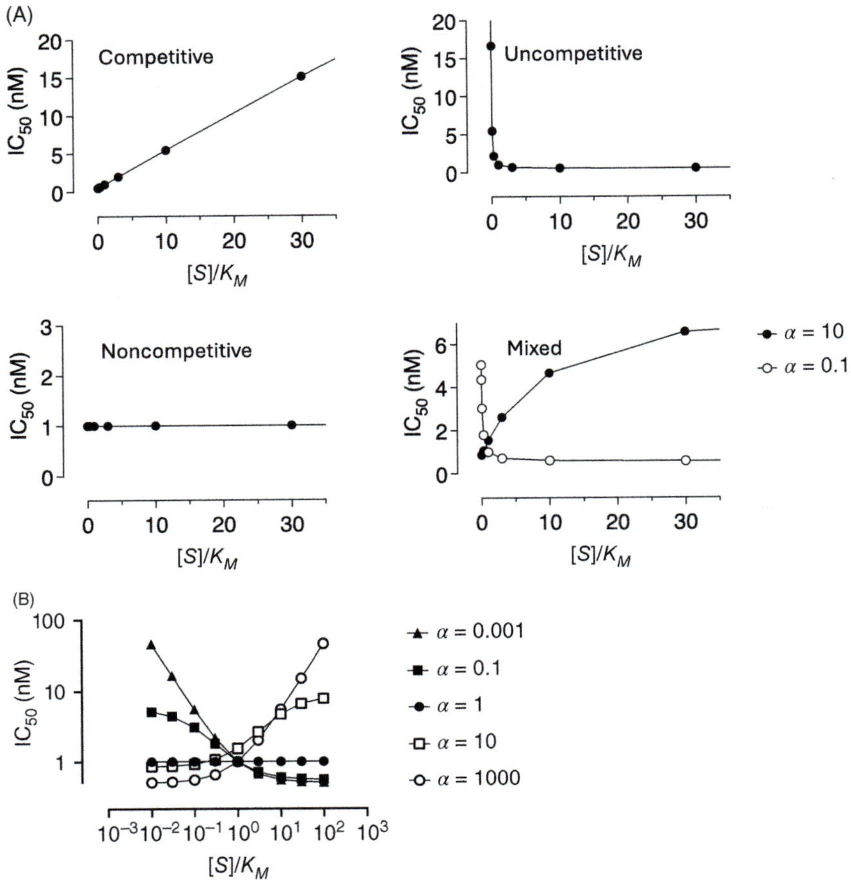

Figure 2.6 Patterns of IC_{50} as a function of $[S]/K_M$ for different inhibition modalities. (A) Patterns for different inhibition modes graphed on linear–linear scales. (B) Pattern for inhibitors with varying values of α graphed on a log–log scale.

site), and natural or synthetic ligands that physically bind within this pocket are referred to as orthosteric ligands. However, the orthosteric pocket is not always the only site on a protein that can bind small molecules. Often, protein activity is conformationally regulated (augmented or diminished) by the binding of small-molecule metabolites (e.g., nucleotide analogues, amino acids, steroid hormones, and the like), cofactors, metal ions, peptides, nucleic acids, or proteins to sites that are distinct from – and often spatially distal to – the orthosteric pocket. These regulatory ligand-binding sites are referred to as allosteric pockets, and the ligands that bind to these sites are known as allosteric ligands. The structural composition of allosteric sites can vary widely among proteins, but, as with orthosteric sites, they have evolved to be structurally complementary to their own cognate ligands. In some cases, the same ligand that binds to the orthosteric site – and thus elicits biochemical activity – also binds allosterically, in what is known as homotropic allostery (Copeland, 2023). More commonly, the structure of the allosteric ligand(s) is distinct from that of the orthosteric ligand, in what is known as heterotropic allostery (Copeland, 2023). Both orthosteric or allosteric binding pockets can be the target sites to which small-molecule drugs bind to impart their pharmacological effect, as we shall see later.

2.3.2.2 Inhibitor Modality is Not an Indication of Inhibitor-Binding Site

A common error on the part of students and some researchers is to equate certain inhibition modalities with specific binding sites on a target. For example, it is tempting, but incorrect, to infer that if an inhibitor is competitive with the natural ligand of a target (e.g., substrate of an enzyme), it must physically bind at the same, orthosteric binding pocket. As reviewed by Buker et al. (2019), while it is often the case that competitive inhibitors do indeed bind within the orthosteric binding pocket, there are also many examples of inhibitors that bind to allosteric sites and induce conformational changes of the protein that disrupt ligand binding at the orthosteric site.

A good example of this is provided by ATP-competitive inhibitors of the kinesin spindle protein (KSP; also referred to as Eg5). Luo et al. (2007) identified a series of biaryl inhibitors of this enzyme that were competitive with the ATP substrate. Two examples of this pharmacophore series are provided by compounds GSK1 and GSK2 (Figure 2.7). The team took two simultaneous approaches to identify the binding pocket for these inhibitors, which many scientists on the project team

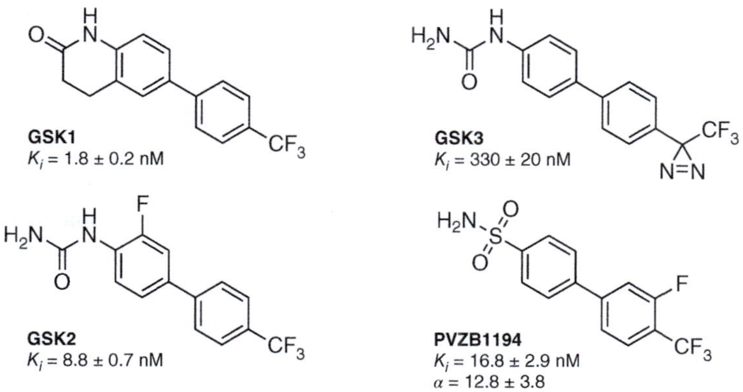

Figure 2.7 Chemical structures of allosteric, ATP-competitive inhibitors of the kinesin spindle protein (KSP or Eg5). Note the photo-cross-linking methyldiazirine moiety incorporated into GSK3 (see text).

assumed was the orthosteric, ATP-binding pocket. The first approach was to grow cancer cells in the presence of sublethal concentrations of compound GSK1 to identify KSP mutations that conferred resistance to the cells. Five mutations were identified by this approach: I19M, A356T, I299F, I332X (where X is any of several amino acids), and V130D. Two of these mutant enzymes (A356T and I299F) were recombinantly expressed and tested to confirm that they indeed resulted in large reductions in GSK1 affinity: the A356T mutation increased the K_i value of GSK1 by over 60-fold, from 1.7 ± 0.4 nM (wild type [WT]) to 109 ± 6 nM (a $\Delta\Delta G_{\text{binding}}$ of 2.45 kcal/mol); the I299F mutation increased the K_i value of GSK1 by almost 40-fold, from 1.7 ± 0.4 nM to 66 ± 5 nM (a $\Delta\Delta G_{\text{binding}}$ of 2.16 kcal/mol). Based on amino acid sequence number, the five mutated residues are clearly spaced widely along the linear sequence of the protein. However, when these residues are mapped onto the X-ray crystal structure of the KSP–adenosine diphosphate (ADP) complex, they all cluster at an interface between alpha helices $\alpha 4$ and $\alpha 6$. The second approach used by Luo et al. (2007) was to replace the phenyltrifluoromethyl group of GSK1 and GSK2 with a phenyltrifluoromethyldiazirine. This imparted a photo-cross-linkable (Pham et al., 2013) diazirine moiety into the compound which, when irradiated with ultraviolet/visible light, generates a reactive species that forms a covalent bond with the protein (GSK3, K_i under ambient light conditions = 330 ± 20 nM). This compound, as well as all the compounds in the pharmacophore series, was competitive with the ATP substrate, but uncompetitive with the second substrate of the enzyme, microtubules (MTs). Therefore, GSK3 was equilibrated with the binary KSP:MT complex before exposing the mixture to ultraviolet light to initiate photo-cross-linking. The KSP was then separated from the MT by gel electrophoresis, subjected to proteolytic digestion, and the resulting peptides were analyzed by mass spectral methods. The data allowed Luo et al. (2007) to unambiguously identify Leu295 as the site of photo-cross-linking. This amino acid occurs in the middle of alpha helix $\alpha 4$, and its side chain protrudes into the $\alpha 4$–$\alpha 6$ interface identified by the mutagenesis studies summarized earlier. Molecular dynamic modeling suggests that a pocket forms at the $\alpha 4$–$\alpha 6$ interface into which GSK1 can be fit, with the trifluoromethyl group buried in a hydrophobic area flanked by residues Phe102, Leu266, Leu295, Ile332, and Tyr352. At the time of the studies by Luo et al. (2007), the inability to crystallize a KSP:MT binary complex precluded attempts to verify the binding site of GSK1 and its analogues through structural biology. However, in 2015, Yokoyama et al. (2015) found that a much simpler biaryl compound, PVZB1194 (Figure 2.7), was a mixed, competitive inhibitor of KSP (K_i = 16.8 ± 2.9 nM; $\alpha = 12.8 \pm 3.8$), and thus able to bind to the enzyme in the absence of MT. They went on to solve the crystal structure of the KSP:PVZB1194 complex and largely confirm the binding site inferred by Luo et al. (2007). Luo et al. (2007) speculated that helix $\alpha 6$ acts as a rigid lever connecting the inhibitor-binding pocket to the orthosteric ATP-binding site. The concerted ratcheting of helix $\alpha 4$ and shift in $\alpha 6$ are critical conformational changes that attend the transition from the KSP:ADP complex to the ternary KSP:ATP:MT complex. Hence, inhibitor binding at the $\alpha 4$–$\alpha 6$ interface was predicted to perturb the structure of the ATP-binding site and thereby abrogate ATP binding. The crystallographic data of Yokoyama et al. (2015) also corroborated this hypothesized mechanism of ATP competitive inhibition by compounds binding to a distal allosteric site.

Similarly, mixed, noncompetitive, and even uncompetitive inhibitors may bind to either orthosteric or allosteric sites to effect inhibition (Buker et al., 2019; Copeland, 2023). For example, Chan-Penebre et al. (2015) reported small-molecule inhibitors of the protein arginine methyltransferase, PRMT5, that were uncompetitive with the methyl-donating substrate *S*-adenosylmethionine (SAM). One of these compounds, EPZ015666 (Figure 2.8A), was co-crystallized with the enzyme also bound by SAM, by *S*-adenosylhomocysteine (SAH, the product of enzyme turnover) or by the SAM-mimetic inhibitor, sinefungin. The crystal structure of the ternary complexes with SAM and sinefungin demonstrated that, in addition to recognition elements on the enzyme itself, the aryl ring of the tetrahydroisoquinoline moiety of the inhibitor (highlighted in Figure 2.8A) engages in a critical π–cation interaction with the cationic methyl group of SAM and sinefungin; no such interaction is possible with the product of enzyme turnover, SAH, which lacks this methyl group. Consistent with these

(A)

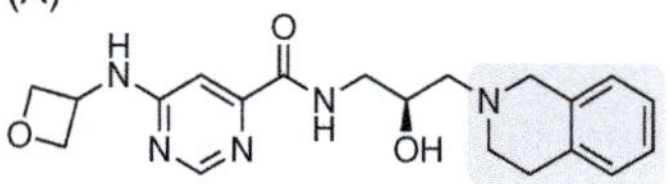

EPZ015666

(B)

Enzyme form	K_d (nM)
SAM-bound	<1
Sinefungin-bound	<1
SAH-bound	171

(C)

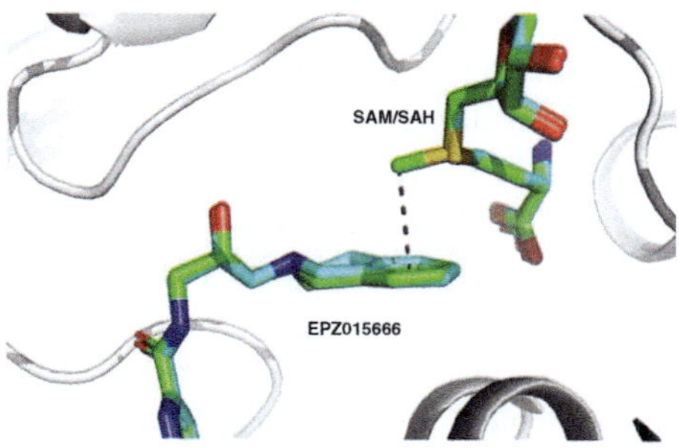

Figure 2.8 (A) Chemical structure of EPZ015666, with the tetrahydroisoquinoline moiety highlighted in gray. (B) Equilibrium dissociation constants for EPZ01566 binding to different ligand-bound forms of PRMT5, as measured by SPR. (C) Superposition of the co-crystal structures of PRMT5 with SAM and EPZ015666 bound and that with SAH and EPZ015666 bound, illustrating the π–cation interaction (dashed line) between the cationic methyl group of SAM and the tetrahydroisoquinoline of EPZ015666. Note also the identical positioning of the substrate SAM and the product SAH in the enzyme active site. *Source:* Buker et al. (2019) / with permission from SAGE PUBLICATIONS.

crystallographic data, the affinity of the compound, as measured by SPR was reduced by ca. 200-fold for the SAH-bound enzyme – lacking the charged methyl group – relative to its affinity for the PRMT5:SAM complex (Figure 2.8B and C).

2.3.3 Reversible Agonists and Activators

The term agonist is defined as a substance that combines with a specific receptor and, as a direct consequence of formation of the agonist–receptor binary complex, initiates a physiological response. During our review of the contributions of A. V. Hill in Chapter 1, we saw a specific example of a receptor agonist in nicotine, which binds to the nicotinic acetylcholine receptor in muscle to initiate muscle contractions. In modern biochemistry, the term agonist is generally reserved for ligands that bind to membrane-bound receptors, such as GPCRs. Important, contemporary examples of this are drugs that act as agonists for the GPCR glucagon-like peptide-1 (GLP-1) receptor (GLP-1R). By agonizing GLP-1R, these drugs regulate blood glucose levels and lipid metabolism, enhance insulin secretion, inhibit glucagon release, delay gastric emptying, and suppress appetite. Hence, these drugs have proven to be valuable treatments for a spectrum of diseases including diabetes, obesity, cardiovascular disease, nonalcoholic fatty liver disease, and others (Zheng et al., 2024). However, the term agonist can be used equally well to describe ligands of other target classes that stimulate a physiological response. For example, the substrate(s) of an enzyme that bind to form an initial ES complex and from there set in motion the catalytic steps that ultimately lead to product formation (vide supra) could be considered as agonists of enzyme catalysis.

The binding of an agonist to its cognate receptor can be quantified by the equilibrium dissociation constant for the resulting binary complex, K_d, or, as discussed in Chapter 1, by the EC_{50} value. This, however, is an incomplete description of the action of a receptor agonist because it only reports the affinity of the initial receptor-ligand (RL) complex and not on the magnitude of the corresponding physiological response. What is needed is a single metric that captures the contributions of both ligand affinity and effectiveness in eliciting a physiological response. This is provided by the ratio of E_{max} over EC_{50} (see Chapter 1 and Figure 1.5), which defines the relative efficiency of the agonist.

$$\text{Relative Agonist Efficiency} = \frac{E_{\max}}{\text{EC}_{50}} \tag{2.5}$$

We use the term *relative* agonist efficiency here because the EC_{50} is an experimentally determined parameter that may be greatly influenced by experimental conditions and not a fundamental thermodynamic constant of the system under study. Nevertheless, if experimental conditions and total receptor concentration are held constant, Equation 2.5 provides a convenient way to experimentally quantify and compare the effectiveness of different compounds as agonists of a target receptor. Whenever it is possible to measure the K_d value for a receptor–agonist complex, the agonist efficiency is best quantified as the ratio of $E_{\max}$ over K_d.

$$\text{Agonist Efficiency} = \frac{E_{\max}}{K_{\text{d}}} \tag{2.6}$$

Let us again illustrate this with the example of an enzyme target, for which the agonist is a natural or synthetic substrate molecule that is converted by the enzyme into a product. Enzyme reactions are typically studied under steady-state conditions (see Copeland (2013, 2023), and later chapters of this text), where the generation of product molecules is a linear function of reaction time (Figure 2.9A). The rate, or velocity, of the enzymatic reaction is simply the slope of the linear [Product] as a function of time curve and is symbolized by v. If we measure velocity at varying concentrations of substrate ([S]), we will see that the velocity increases with increasing substrate, as seen in Figure 2.9A. If we replot velocity (v) as a function of substrate concentration ([S]), we find that the relationship is well modeled by a hyperbolic function (referred to in the literature as the Michaelis–Menten equation; see Copeland, 2013, 2023), related to Equation 1.5 (Figure 2.9B):

$$v = \frac{V_{\max}}{1 + \frac{K_M}{[S]}} \tag{2.7}$$

Here, we are using the steady-state parameter, K_M (the Michaelis constant) in place of K_S, because the experiments were performed under steady-state, rather

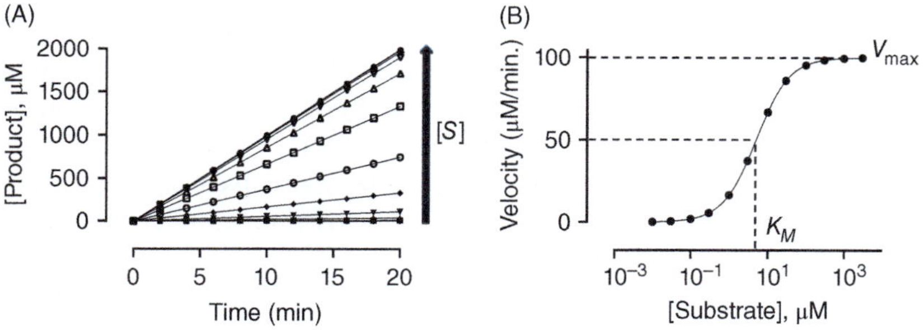

Figure 2.9 (A) Reaction progress curves for an enzyme reaction measured under steady-state conditions at varying concentrations of substrate ([S]). The thick, ascending arrow is meant to indicate the increasing concentrations of substrate used. (B) Semilog replot of the reaction velocity (measured as the slopes of the linear plots in A) as a function of substrate concentration and fit to the Michaelis–Menten equation. The value of K_M is determined from the midpoint of the fitted data and the $V_{\max}$ is determined from the high ligand concentration plateau, as indicated.

than equilibrium, conditions. The value V_{max} is a special case of the more generic term, E_{max}, that we have described earlier. In the specific case of steady-state enzyme reactions, V_{max} can be shown to be equivalent to the concentration of total enzyme ($[E]$) multiplied by the composite kinetic constant, k_{cat}.

$$V_{max} = k_{cat} \times [E] \tag{2.8}$$

Thus, from the semilog scale replot of v as a function of $[S]$ (Figure 2.9B), we can determine the steady-state Michaelis constant, K_M, as the concentration of substrate that yields a steady-state velocity equal to ½ V_{max} (the midpoint or EC_{50} value) and the maximum velocity at infinite substrate concentration, V_{max} (the high ligand plateau value or E_{max} value). From the V_{max} value and the known concentration of enzyme used, we can then compute the k_{cat} value from equation 2.8. Hence, from this set of experiments, we can determine the two fundamental kinetic parameters associated with a specific substrate (agonist) for our target enzyme. We can now quantify the efficiency of enzyme catalysis for this specific enzyme–substrate pair as the ratio of k_{cat} over K_M, as a specific case of the more general form shown in equation 2.6.

$$\text{Catalytic Efficiency} = \frac{k_{cat}}{K_M} \tag{2.9}$$

In some pharmacological applications, one wishes to discover synthetic compounds that act as alternative agonists to the naturally occurring agonists of specific, target receptors. Such synthetic agonists will almost always engage the target by occupying the same binding pocket (also referred to as binding site) on the receptor as the natural agonist. As stated above, this binding pocket is referred to as the orthosteric site with respect to the natural agonist, and ligands that bind to this site are said to be orthosteric binders. Therefore, in most cases, a molecule of receptor can bind either the natural agonist or a synthetic agonist, but not both simultaneously. Hence, the two agonists are mutually exclusive to one another, or in other words, competitive with each other (see earlier).

As we shall see further in Chapter 4, hit-finding efforts typically begin with assays of compound binding to, or impact on the activity of, recombinant targets in a cell-free assay system. In the case of enzymes and other soluble protein targets, this may be in the form of a purified, recombinant protein in an appropriate buffer. For membrane-bound targets, such as GPCRs, the recombinant target is typically present within membrane fragments, membrane vesicles, or detergent micelles. If such cell-free systems are used to measure binding of compounds to the receptor, then the E_{max} term in equation 2.1 is defined by the total concentration of receptor in the sample, $[R]_{total}$ (see Appendix 1 and Copeland, 2023). However, at some point during hit-finding, hit-to-lead, and lead-to-candidate activities (Chapters 4–6), one will wish to ensure and quantify that a potential agonist not only binds to the target, but also elicits the downstream signaling expected for an agonist. Hence, cell-based assays will be employed to measure a convenient analyte of downstream signaling (a measure of cellular pharmacodynamics). In some cases, the cell system used for these signaling assays may be disease-associated cells that express endogenous, physiological levels of the target. In other cases, cells may be engineered to express the target at higher than physiological levels. An important caveat in such assay systems is that both the E_{max} and EC_{50} terms of equation 2.1 depend on total target concentration and will therefore vary from cell line to cell line. Hence, caution must be exercised in making comparisons among data generated in different cell lines within the same laboratory, between data generated in the laboratory and literature values, and the like. The reason for this variation in E_{max} and EC_{50} is well explained by the operational model of pharmacological agonism, developed by Black and Leff (1983) (see also Klotz, 1997 and Kenakin, 2022). The interested reader is also referred to a set of excellent tutorial videos on this topic, presented by Dr. Sam Hoare of Pharmechanics (pharmechanics.com). We shall return to this important point in our discussions of pharmacodynamic assays in subsequent chapters.

An activator is a substance that augments the physiological response that is elicited by a receptor–agonist complex. In some cases, the target receptor does not elicit any physiological response in the absence of a naturally occurring activator; in other cases, there is a basal level of physiological response that is augmented in the presence of an activator. Target activation can occur by several mechanisms in cells, including various enzyme-catalyzed posttranslational modifications of the target (e.g., phosphorylation, acetylation, ubiquitinoylation, and proteolytic processing; see Copeland, (2023), for a more complete description of these forms of target activity regulation). What is most germane to our discussion is target activation by the reversible binding of small-molecule activators, such as natural metabolites and peptides, and synthetic compounds.

Small-molecule, reversible activators can function by several different mechanisms. The compound can bind to an extended orthosteric site, so long as this does not interfere with agonist binding (although this is a rare mechanism of activation). More commonly, activators bind to other sites on the target protein, referred to as allosteric sites, that when bound by a ligand, either communicate conformationally with the orthosteric site or otherwise elicit conformational changes that augment response. Hence, activators may bind orthosterically, but in most cases bind allosterically to their target. Allosteric activator–binding sites may occur within the same protein domain as that of the orthosteric site, within other domains of the same protein subunit, or even on other protein subunits of a target composed of a multi-subunit complex (Copeland, 2023).

The binding of an activator to a target can affect response by lowering the K_d value for the receptor–agonist binary complex, increasing the value of E_{max}, or some combination of both. Hence, quantitative assessment of activator efficiency must account for any changes in either of these parameters, much as we saw for agonist efficiency. When the basal activity is nonzero, this must be accounted for as well. Typically, analysis of activator effect is initially performed at a fixed concentration of agonist, often at or near the K_d value for that agonist. At this fixed concentration of agonist, the activity of target can be measured in the absence (basal) and presence of varying concentrations of activator. Let us say that we are again studying an enzyme target for which the experimental measure of activity is the velocity of product formation. If we fix the substrate concentration (say at $[S] = K_M$), the effect of activator concentration ($[A]$) on velocity can be described by the following equation (Copeland, 2023).

$$v = \frac{v_A - v_0}{1 + \frac{K_A}{[A]}} + v_0 \tag{2.10}$$

where v_A is the maximum velocity at $[\mathrm{A}] = \infty$, v_0 is the basal velocity in the absence of activator A, and K_A is the K_d value for activator binding to the enzyme target. The efficiency of activation is then given as (Copeland, 2023)

$$\text{Activator Efficiency} = \frac{v_A - v_0}{K_A} \tag{2.11}$$

Some researchers also define the fold-activation as the ratio of v_{A} over v_0 (Copeland, 2023) as another measure of activator effectiveness.

In pharmaceutical applications, one is typically looking to either augment or inhibit the activity of a target by identifying synthetic small-molecule activators or inhibitors that bind to a natural activator–binding pocket. In most cases, the natural activators that bind to this target pocket are themselves, small-molecule metabolites, cofactors, peptides, and the like, which reversibly bind to the target to form a binary target–activator complex, in what is referred to as intermolecular activation (Copeland, 2023). In some cases, however, the natural ligand for the activator-binding pocket is a posttranslational modification that resides within the target protein itself, in what is referred to as intramolecular activation or inhibition. For example, the kinase cAbl 1b is posttranslationally modified by having a myristoyl group appended to the amino-terminal glycine of the protein. This myristoyl group can bind within a hydrophobic pocket elsewhere on the folded protein and thereby inactivate (inhibit) the enzymatic activity of cAbl 1b in what is referred to as an autoinhibitory regulation mechanism. Dissociation of the intramolecular myristoyl group from its binding pocket results in a conformational change to the protein that activates its enzymatic activity. Hence, controlling the equilibrium between the inactive and active forms of the enzyme provides a mechanism of regulatory control. Drug discovery researchers have identified synthetic small molecules that bind to the myristoyl-binding pocket of this enzyme, resulting in either activation (Yang et al., 2011) or inhibition (Wylie et al., 2017) of this kinase.

2.3.4 Reversible Inhibitors of Protein–Protein Interactions

The function of some target proteins is greatly influenced by the binding of other proteins to the target. In some cases, these other proteins are absolutely required for function; in other cases, the binding of other proteins augments or diminishes target function directly; and in still other cases, auxiliary proteins act to localize the function of the target protein to specific sites on other macromolecules or subcellular locations. Table 2.5 provides some examples of target proteins that depend on additional auxiliary proteins to modify their function. The concept of auxiliary proteins causing inhibition or activation of target protein function is easily understood. Let us consider a specific example of the, perhaps less familiar, situation of auxiliary protein binding causing ectopic localization of a target protein function.

The enzyme DOT1L is a protein methyltransferase that specifically catalyzes the methylation of lysine 79 (K79) on histone H3 (H3), one of the histones that forms the protein core of the protein–DNA complex known as chromatin. Methylation

Table 2.5 Examples of proteins that are functionally modulated by the binding of auxiliary proteins.

Functional protein subunit	Auxiliary protein	Modulation of function
Cyclin dependent kinases (CDKs)	Cyclins	Activation
EZH2	AEBP2, EED, RbAp48, and SUZ12	Activation
cGMP phosphodiesterase	GTP-bound α-transducin	Activation
SCF family E3 ligases	Hrt1	Activation
Aurora A kinase	TPX2	Activation
Aurora B kinase	INCENP	Activation
Barnase	Barstar	Inhibition
Trypsin	BPTI	Inhibition
MMP9	TIMP-1	Inhibition
Caspace-3	XIAP	Inhibition
DOT1L	MLL–AF4 fusion	Ectopic localization

Source: Copeland (2023) / with permission of John Wiley & Sons.

of H3K79 by DOT1L can lead to transcriptional activation of genes in proximity to the methylated H3K79 site on chromatin. The enzyme mixed lineage leukemia (MLL) is another protein methyltransferase that normally methylates lysine 4 of histone H3 (H3K4). In the acute leukemia known as mixed-lineage leukemia, the gene encoding the MLL enzyme undergoes a chromosomal translocation, leading to the encoding of a fusion protein between a portion of the MLL protein and any of several other proteins; the most abundant fusion seen in mixed lineage leukemia is with the protein AF4, creating the MLL–AF4 fusion protein. The portion of the MLL protein that remains in these fusion proteins retains the ability to localize the MLL fusion to genes that are normally transcriptionally regulated by MLL-catalyzed H3K4 methylation. However, a common feature of the MLL fusion proteins is the loss of the catalytic domain of MLL, so that these fusion proteins lack the ability to carry out the normal methylation function of MLL. The various fusion partners, such as AF4, share another common feature: formation of a protein–protein complex with DOT1L. Hence, genes that are normally regulated by H3K4 methylation are instead activated by H3K79 methylation, catalyzed by the ectopically localized DOT1L resulting from the MLL–AF4:DOT1L complex at these gene locations. Among the genes aberrantly activated by this mechanism is the HOX cluster of genes; this aberrant activation of HOX genes is directly leukemogenic (Copeland et al., 2009; Brent and Armstrong, 2011). Hence, the pathogenic alteration responsible for mixed lineage leukemia does not change the function of DOT1L, but rather changes the gene locations on chromatin where DOT1L function is carried out. For this reason, DOT1L has become a key molecular target for small-molecule therapeutics aimed at treating mixed lineage leukemia patients (Daigle et al., 2011).

If a pathogenic target function is dependent on the formation of a complex with other proteins, it stands to reason that modulating complex formation or stability may be a reasonable approach to therapeutic intervention. However, it has generally been considered difficult to modulate PPIs with small-molecule drugs, because of the distinct physicochemical nature of PPIs compared to that of small ligand-binding interactions with target receptors.

When a small-molecule ligand binds to a target, it typically occupies a surface area of about 424 ± 208 Å^2. As discussed earlier, small-molecule ligands bind in specific pockets within the target protein that are characteristically hydrophobic, exclude bulk solvent, and display a high degree of surface roughness; characteristics that ensure relatively high fidelity for binding specific ligands (but see Copeland, (2023), for an account of the role of active site promiscuity in the evolution of neofunctionalization in enzymes). In stark contrast, proteins interface other proteins over much larger surface areas averaging about $1{,}600 \pm 400$ Å^2. These surfaces are generally flatter (i.e., displaying minimal surface roughness) and more solvent exposed, hence composed largely of hydrophilic amino acids.

Thus, the free energy of binding for PPIs is generally dispersed over a large, flat surface area and driven more by electrostatic interactions and expansive networks of hydrogen bonds, rather than by the hydrophobic forces that drive much of the energy of small-molecule ligand–target complex formation (LoConte et al., 1999; Arkin and Wells, 2004; Copeland, 2023). The interfacial surface used for PPIs also tend to display higher promiscuity, binding multiple other proteins using the same binding surface (Copeland, 2023).

These generalities notwithstanding, PPIs can, in favorable cases, be modulated by small-molecule ligands. In a seminal series of studies, James A. Wells and coworkers have demonstrated that PPIs can exhibit "hot spots" – smaller, more hydrophobic patches on a protein–protein interfacial surface(s) that contribute disproportionately large amounts of the overall free energy of binding ($\Delta G_{\text{binding}}$) to PPIs (see Wells and McClendon (2007), for a review). In the first of these studies, Cunningham and Wells (1993) explored complex formation between human growth hormone (hGH) and site 1 of the extracellular binding domain of its cognate receptor, human growth hormone binding protein (hGHbp). These two proteins form a high-affinity complex ($K_d = 300$ pM; $\Delta G_{\text{binding}} = -12.3$ kcal/mol) that initiates an hGH-dependent signal transduction pathway that results in transcriptional activation of multiple genes. From the X-ray crystal structures of free hGH and that of the binary hGH:hGHbp complex, they identified 31 amino acid side chains, larger than alanine, that become buried due to complex formation with the receptor, representing a total buried surface area on each partner protein of ca. 1,300 Å^2. They refer to this buried surface as the "structural epitope" of binding. They then mutated 30 of these 31 amino acids individually to alanine and measured the resulting change in association rate constant (k_{on}), dissociation rate constant (k_{off}), and binding affinity ($K_d = k_{\text{off}}/k_{\text{on}}$) to immobilized hGHbp by SPR. One of the 31 mutant proteins, K41A, could not be expressed at sufficient levels for purification and SPR measurements, but the effect of this mutant on hGHbp-binding affinity had previously been determined from radioimmunoassay data (Lowman and Wells, 1993). From the K_d values of each mutant (K_{mut}) compared to that of the WT hormone (K_{WT}), they calculated the change in binding free energy associated with each mutation from the following formula.

$$\Delta\Delta G_{\text{binding}} = RT \ln\left(\frac{K_{\text{mut}}}{K_{\text{WT}}}\right) \tag{2.12}$$

The results of these experiments are summarized in the histogram of $\Delta\Delta G_{\text{binding}}$ for each mutation in Figure 2.10A and show that many of the mutations did not cause a meaningful change in binding free energy, and in a few cases, the mutation to alanine enhanced binding free energy (i.e., resulting in negative values of $\Delta\Delta G_{\text{binding}}$). Cunningham and Wells (1993) defined a significant diminution of binding free energy to be > +1 kcal/mol (dashed horizontal line in Figure 2.10A),

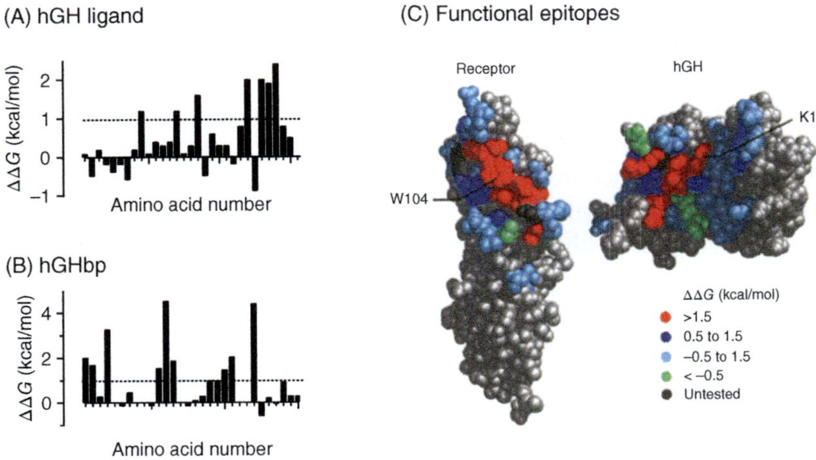

Figure 2.10 Mapping of the functional epitopes on hGH ligand and its cognate receptor. (A) Map of $\Delta\Delta G_{binding}$ associated with mutation of amino acids within the structural epitope of hGH ligand individually to alanine. The dashed horizontal line illustrates the threshold for significant change in binding free energy (> +1 kcal/mol) as defined by Cunningham and Wells. *Source:* Adapted from Cunningham and Wells (1993). (B) As in A but for the structural epitope on hGHbp. *Source:* Adapted from Clackson and Wells (1995). (C) Structural map of the functional epitopes on the hGH ligand and hGHbp. *Source:* Reproduced from Clackson and Wells (1995), with permission of American Association for the Advancement of Science.

corresponding to about a 5-fold increase in K_d value for the mutant relative to the WT hormone. Of the 31 amino acids defining the structural epitope of hGH, eight showed a significant change in binding free energy, as summarized in Table 2.6; these eight amino acids (22% of the 31 components of the structural epitope) contribute 85% of the overall binding free energy of the hGH:hGHbp complex; an additional six residues (P48, E56, Q68, D171, I179, and R183) account for most of the remaining binding free energy. Thus, Cunningham and Wells (1993) defined the eight amino acids listed in Table 2.6 as the "functional epitope" of binding for hGH. It is interesting to note that very little change in the association rate constant (k_{on}) were observed across all these mutants. The changes in binding free energy were almost exclusively driven by changes in the dissociation rate constant (k_{off}), consistent with the drug–target residence model (Copeland et al., 2006).

Next, Clackson and Wells (1995) performed the corresponding set of experiments on the cognate receptor, hGHbp. Here they identified a structural epitope composed of 33 amino acid side chains. They made individual alanine mutants for 30 of the 33 residues, choosing not to mutate G168, as there is no aliphatic side chain on glycine, nor C108 and C122 that together form a structurally

Table 2.6 Critical amino acid residues of the functional epitopes of hGH and hGHbp and the free energy of binding changes associated with their mutation to alanine.

hGH functional epitope amino acids		hGHbp functional epitope amino acids	
Residue	**$\Delta\Delta G_{binding}$ (kcal/mol)**	**Residue**	**$\Delta\Delta G_{binding}$ (kcal/mol)[a)]**
K41	1.1	R43	2.0
L45	1.2	E44	1.7
P61	1.2	I103	1.6
R64	1.6	W104	4.5
K172	2.0	I105	1.9
T175	2.0	P106	3.3
F176	1.9	D126	1.0
R178	2.4	E127	1.0
		D164	1.5
		I165	2.1
		W169	4.5

a) Values estimated from graphs published by Clackson and Wells (1995).
Source: Cunningham and Wells (1993) and Clackson and Wells (1995).

significant intramolecular disulfide bond. The results of this set of experiments are also summarized as a histogram in Figure 2.10B. Like the results of the hGH mutational study, they found that only a portion of the structural epitope component amino acids made up the much smaller functional epitope – in this case, 11 of 33 amino acids (33%; Table 2.6). The constituent amino acids of both the hGH and hGHbp functional epitopes are mixtures of hydrophobic and polar amino acids, forming a hydrophobic core–binding surface that is surrounded by critical electrostatic interactions among the polar amino acid side chains. Both the size and nature of the functional epitopes (Figure 2.10C) – with respect to the hydrophobic core, well-defined structural water molecules, and electrostatic interactions – are reminiscent of druggable binding sites on other targets (vide supra), thus portending the potential druggability of such hot spots on PPI-binding surfaces.

The concept of druggable hot spots as critical features of PPIs, developed by Wells and others, has stimulated significant interest within the biotechnology and pharmaceutical industry, as well as stimulating further studies of such hot spots within academia. This has led to several drugs, based on inhibition of PPIs, that have entered clinical studies, as exemplified in Table 2.7 (see also Scott et al., 2016).

Table 2.7 Some examples of clinical-stage small-molecule inhibitors of protein–protein interactions.

Protein–Protein interaction target	Drug	Indication	Status
HDM2/p53	Idasanutlin	Acute myeloid leukemia	Phase III
BCL-2/BAX	Venetoclax (ABT-199)	Chronic lymphocytic leukemia; Acute myeloid leukemia	Approved
BCL-2, BCL-X_L, BCL-W/BAX	Navitoclax (ABT-263)	Primary myelofibrosis and several cancers	Phase I/II
XIAP/caspase-9	LCL-161	Multiple myeloma	Phase II
Gp120/CCR5	Maraviroc	HIV infection	Approved
LFA-1/ICAM-1	Lifitegrast	Dry eye	Phase IV
β-catenin/CBP	PRI-724	Liver cirrhosis	Phase I/III
BRD4/Histone	Apabetalone (RVX-208)	Cardiovascular disease	Phase III
Menin/KMT2	Revumenib (SNDX-5613)	Acute myeloid leukemia	Phase I/II
EED/Histone	ORIC-944	Metastatic prostate cancer	Phase I
PD1/PD-L1	CA-170	Prostatic neoplasia	Phase II

The first PPI inhibitor–based drug to be approved by the FDA was venetoclax, a leukemia treatment that acts by binding to the protein BCL-2 and thereby blocks the formation of a PPI between BCL-2 and another protein, BAX. BCL-2 is a member of a class of proteins referred to as pro-survival proteins, which also includes the related proteins BCL-X_L, BCL-W, A1/BFL-1, and MCL-1. These proteins act to block programmed cell death (apoptosis) by binding to and thereby restraining the pro-apoptotic activity of the proteins BAX and BAK. More specifically, the pro-apoptotic proteins contain a helical region known as the BCL-2 Homology 3 (BH3) motif, which binds within the BH3-binding groove of the pro-survival proteins to form the apoptosis-inhibiting PPIs. It was hypothesized that small-molecule mimetics of the BH3 helical motif would bind to the BH3-binding groove of pro-survival proteins, such as BCL-2, and thereby block complex formation with the pro-apoptotic proteins, such as BAX. Thus, the unfettered BAX or BAK would induce apoptosis in cancer cells. The first small-molecule BH3 mimetic to be tested clinically was navitoclax (ABT-263), a compound that bound to the P2 and P4 subsites within the BH3-binding grooves of multiple pro-survival proteins, including BCL-2, BCL-X_L and BCL-W (Oltersdorf et al., 2005; Tse et al., 2008). However, navitoclax resulted in dose-limiting thrombocytopenia (a reduction in platelet count) which

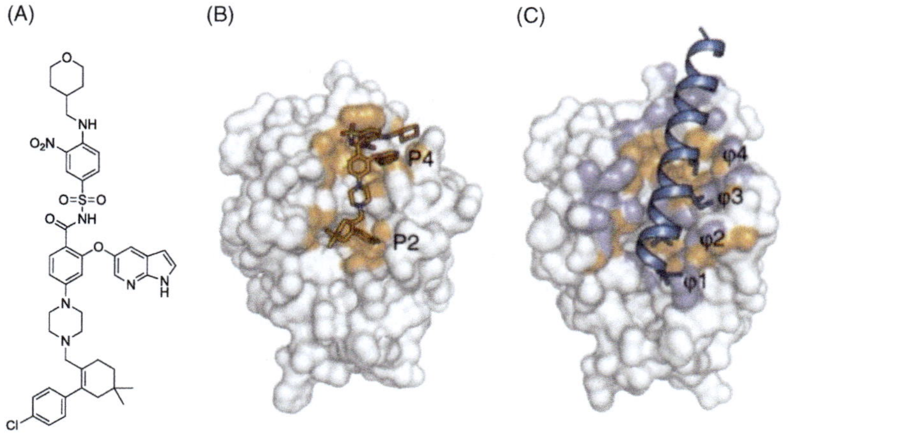

Figure 2.11 (A) Chemical structure of venetoclax (ABT-199). (B) Crystal structure of Bcl-2 with venetoclax bound in the BH3-binding groove. *Source:* Birkinshaw et al. (2019) / Springer Nature / CC BY 4.0. (C) Crystal structure of Bcl-2 with the helical peptide representing the BH3 motif of Bax bound within the BH3-binding groove. *Source:* Birkinshaw et al. (2019) / Springer Nature / CC BY 4.0.

precluded further clinical progression. The thrombocytopenia was shown to be mechanism-based and specifically due to inhibition of BCL-X_L (Mason et al., 2007). Hence, next-generation compounds sought to selectively bind to BCL-2 to avoid thrombocytopenia. These efforts led to venetoclax (ABT-199), a selective BCL-2 inhibitor that induced apoptosis in tumor cells while sparing platelets (Souers et al., 2013). Figure 2.11 illustrates the chemical structure of venetoclax (A) and the crystal structures of BCL-2 with the drug (B), or a helical peptide representing the BH3 motif of BAX (C), bound within the BH3-binding groove of BCL-2 (Birkinshaw et al., 2019).

2.3.5 Small-Molecule Modulators of Protein Homeostasis

Nuclear-encoded cellular proteins are synthesized on ribosomes within the endoplasmic reticulum (ER). Once the polypeptide chain is synthesized, it undergoes initial protein folding and quality control checks within the ER. Further folding and posttranslational modification then takes place in the Golgi apparatus, before the protein is trafficked to its final cellular location by a combination of several mechanisms. Proper protein folding and trafficking are essential to ensure correct and complete protein function. In several diseases, mutations within a protein result in improper or incomplete folding within the ER. When this occurs, the misfolded protein is shuttled to the cytosol for degradation through the ubiquitin–proteosome pathway, as a mechanism for ridding the cell of dysfunctional proteins.

Ubiquitin is a small (76 amino acids) protein that can be enzymatically appended to the ε-nitrogen of lysine side chains on other proteins. The cellular protein ubiquitination pathway consists of a three-enzyme cascade that results in the formation of poly-ubiquitin chains being added to proteins at specific lysine side chains. The initial enzyme of the pathway is the E1 ubiquitin-activating enzyme. In an ATP hydrolysis-dependent manner, E1 forms a covalent complex with the small protein ubiquitin (Ub), via thioester bond formation. The Ub-charged E1 then transfers the Ub to the next enzyme in the cascade, E2. The Ub–E2 is then used as a co-substrate, together with the protein targeted for ubiquitination, for the E3 ubiquitin ligase enzyme that catalyzes formation of an initial isopeptide bond between a lysine side chain on the target protein and the C-terminus of the Ub derived from the E2–Ub complex. The E3 enzyme can then ubiquitinate the K48 residue of this first Ub (on the target protein), to form a dimeric target–Ub–Ub (Ub_2) species, and this process can be continued to form a poly-Ub chain composed of n Ub molecules on the target protein (Ub_n); poly-Ub chains of $n \geq 4$ mark the target protein for proteosome-mediated degradation (see below). Human cells contain only two E1 enzymes, about 40 E2 enzymes and many E3 ligases, such that target protein specificity is conferred by a combination of E2 and E3 ligase identities (see Burslem and Crews (2017), (2020) and Copeland, (2023) for further details). Poly-Ub chains of $n \geq 4$ serve as the primary recognition signal for target protein binding to the 26S proteosome, a large complex of threonine protease subunits that is responsible for the unfolding, deubiquitination, and proteolytic digestion of the protein. In this manner, the proteosome acts as a quality control mechanism for ridding cells of damaged, dysfunctional, and no-longer-needed proteins.

Protein synthesis by the ribosome/ER/Golgi pathway, the trafficking of proteins to their proper cellular location, and the opposing action of the ubiquitin–proteosome pathway must be carefully controlled to establish an appropriate level of cellular protein homeostasis. If a protein homeostatic setpoint is dysregulated, by mutations that cause misfolding, by amplified protein expression, or by aberrant proteosome activity, a pathological state of the cell can ensue. Modern drug discovery efforts have resulted in several small-molecule drug approaches to combat these pathologies. The proteosome inhibitor bortezomib (Paramore and Frantz, 2003; Kouroukis et al., 2014), for example, is used to treat patients with multiple myeloma (an example of a reversible enzyme inhibitor, as discussed in Section 2.3.2). More recently, small-molecule correctors and potentiators of misfolded proteins have been developed to treat diseases, such as cystic fibrosis (CF), and small molecules that selectively induce proteosome-mediated degradation of a target protein have entered the pharmacological armamentarium, as we shall describe next.

2.3.5.1 Small-Molecule Correctors and Potentiators of Misfolded Proteins

A series of studies during the 1950s demonstrated that enzymes and other proteins must fold into a specific three-dimensional tertiary structure to elicit their full biological function. During this same period, it was shown that a folded protein could be unfolded (denatured) by changes in temperature or pH, or by the introduction of high concentrations of chaotropic compounds, such as urea. Most remarkably, if these perturbations are removed, for example by dialysis to remove the urea from the protein solution, purified proteins will often spontaneously refold and regain activity. Hence, protein function requires proper folding, and all the information required for protein folding must be contained within the amino acid sequence of the polypeptide (see Copeland, 1994, for a review). It has also become clear that small molecules that bind to specific pockets on proteins can often stabilize the folded protein structure to thermal or chemical denaturation (see Copeland, 2023, for a review). Thus, it appears that the folded and unfolded states of a protein are in equilibrium with one another (this is referred to as the two-state model of protein folding; it is clearly a gross simplification, as it has become clear that other intermediate states of folding can occur, but this simple model nevertheless suffices for explaining much of protein chemistry), and the position of this equilibrium (favoring the folded or unfolded state) can be affected by environmental conditions and by the binding of small-molecule ligands. These observations suggest that it might be possible to rescue the activity of misfolded proteins in cells by the application of small-molecule drugs. Recently this possibility has been realized (Burslem and Crews, 2017).

A small molecule can affect the activity of a misfolded protein in two distinct ways. Compound binding could shift the equilibrium to favor the folded state of the protein and thereby facilitate trafficking of the now-folded protein from the ER to its proper cellular location. A compound that function by this mechanism is referred to as a protein *corrector*; some authors also refer to such compounds as *pharmacological chaperones*. Alternatively, a compound could bind to a misfolded protein and augment its inherent activity. Compounds that work by this mechanism are referred to as protein *potentiators*. As we shall see, both correctors and potentiators of misfolded protein function have been discovered and found to work synergistically with one another. The first examples of protein correctors and potentiators to become approved drugs were targeted toward misfolding mutations of the cystic fibrosis transmembrane conductance regulator (CFTR).

CFTR is a chloride ion channel that exists within the apical membrane of bronchial epithelial cells, where it plays a critical role in the maintenance of salt and fluid homeostasis (Mijinders et al., 2017). The channel is a member

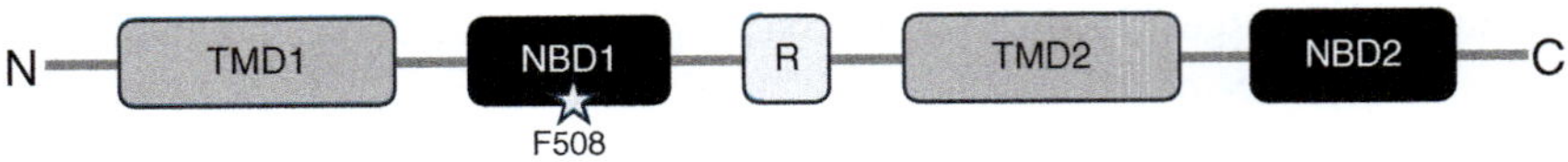

Figure 2.12 Schematic diagram of the domain structure of the CFTR ABC transporter protein. The star within domain NBD1 illustrates the approximate location of phenylalanine 508 (F508) that is deleted in about 80–85% of patients with cystic fibrosis.

of the **A**TP-**b**inding **c**assette (ABC) transporter superfamily of proteins and is composed of two transmembrane domains (TMD1 and TMD2) each containing six membrane-traversing α-helices, two cytoplasmic nucleotide-binding domains (NBD1 and NBD2), and an intervening regulatory domain (R), as illustrated in Figure 2.12.

Patients that present with CF invariably have biallelic mutations in the CFTR gene that cause misfolding of the protein, retention in the ER, and/or diminished ion channel function; individuals that carry only a single mutant allele are asymptomatic carriers of CF. Analysis of approximately 1,000 CFTR mutations demonstrates a strong correlation between the degree of mutational impairment of CFTR function and the severity of lung disease in patients; these data also suggest that restoration of mutant function to 5–30% of WT CFTR would suffice to improve lung function in patients (Van Goor et al., 2006). The most common mutation observed in CF patients is deletion of phenylalanine 508 (F508del), a mutation found in 80–85% of all CF patients. About 50% of CF patients are homozygous for the F508del mutation, while most of the rest (30–40%) have one copy of the F508del mutant and one copy of another CFTR mutation, such as G552D that likewise causes misfolding and trafficking defects (Van Goor et al., 2006).

Deletion of phenylalanine 508, within the NBD1 domain, causes misfolding of the domain and domain-assembly defects. F508 interacts with the 4th transmembrane α-helix of TMD2, and the loss of this side chain creates a cavity, thereby disrupting the NBD1–TMD2 interface that is critical for folding and ion transport (Mijinders et al., 2017). The F508del form of CFTR causes most of the protein to be retained in the ER and eventually degraded. A small fraction of the F508del CFTR escapes the ER and is transported to the apical membrane but has a significantly reduced half-life at the cell surface membrane compared to the WT CFTR.

A team at Vertex Pharmaceuticals performed high-throughput screening (HTS; see Chapter 4) of diverse chemical libraries to identify correctors and potentiators of the CFTR WT, G552D, and F508del mutants. The first screening effort (Van Goor et al., 2006) tested 122,000 compounds in a cell-based assay of NIH/3 T3 cells expressing F508del–CFTR, that measured compound-induced changes in anion flux through the cell membrane to identify CFTR potentiators (Van Goor et al., 2006). This screen identified 10 chemical scaffolds as hits that were assessed further. The first hit to be published was VRT-532 (Figure 2.13), which demonstrated

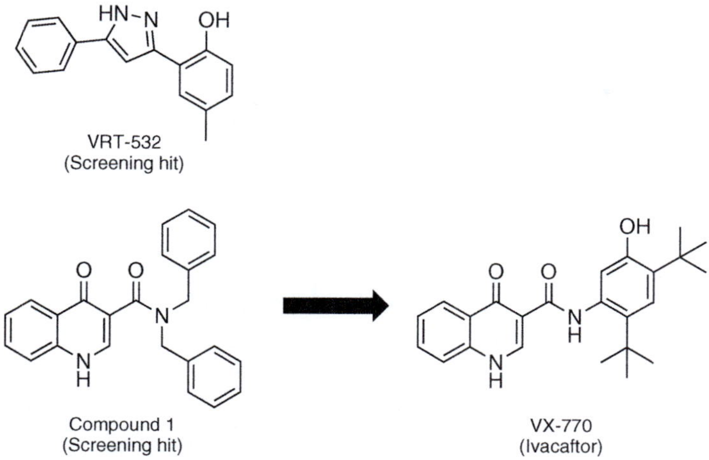

Figure 2.13 CFTR potentiators identified from high-throughput screening and the clinical compound VX-770 (Ivacaftor) that resulted from hit optimization.

an EC_{50} of 2.4 ± 0.2 μM in the cell assay and an E_{max} of 106 ± 3% (Van Goor et al., 2006). However, a second screening hit, referred to as "compound 1" (Figure 2.13, EC_{50} = 1.5 μM; Hadida et al., 2014), was subsequently optimized to yield VX-770 (Ivacaftor, Figure 2.13; Van Goor et al., 2009; Jih and Hwang, 2013; Hadida et al., 2014), the first CFTR potentiator to be approved as a treatment for CF. The same team also screened a library of 164,000 compounds in a similar assay to identify CFTR correctors that would enhance trafficking of F508del–CFTR to the cell surface membrane. This screening effort yielded 13 distinct chemical scaffolds as hits, including the quinazolinone series exemplified by screening hit VRT-422 (Figure 2.14), which displayed an EC_{50} of 3.7 ± 0.9 μM and an E_{max} of 47 ± 4% (Van Goor et al., 2006). Subsequent SAR efforts ultimately led to three CFTR corrector molecules being approved for the treatment of CF: VX-445 (Elexacaftor), VX-661 (Tezacaftor), and VX-809 (Lumacaftor), the structures of which are shown in Figure 2.14. Biochemical and cryo-electron microscopy studies have shown these correctors interact with the TMD1 domain of CFTR (Fiedorczuk and Chen, 2022). Lumacaftor (VX-809) and tezacaftor (VX-661) are structurally related compounds, sharing a common 1,3-benzodioxol-cyclopropane carboxamide (BCC) headgroup. These compounds are categorized as type-1 correctors and do not show additive effects when coadministered (see later). Cyro-electron microscopy has shown that VX-809 binds to a narrow pocket within the transmembrane region of TMD1, formed by components of TMD1 transmembrane helices 1, 2, 3, and 6. The compound interacts with the pocket largely through van der Waals interactions with hydrophobic amino acid side chains but also forms a salt bridge with a lysine residue within the pocket (K68). The BCC headgroup forms a classic "lock and

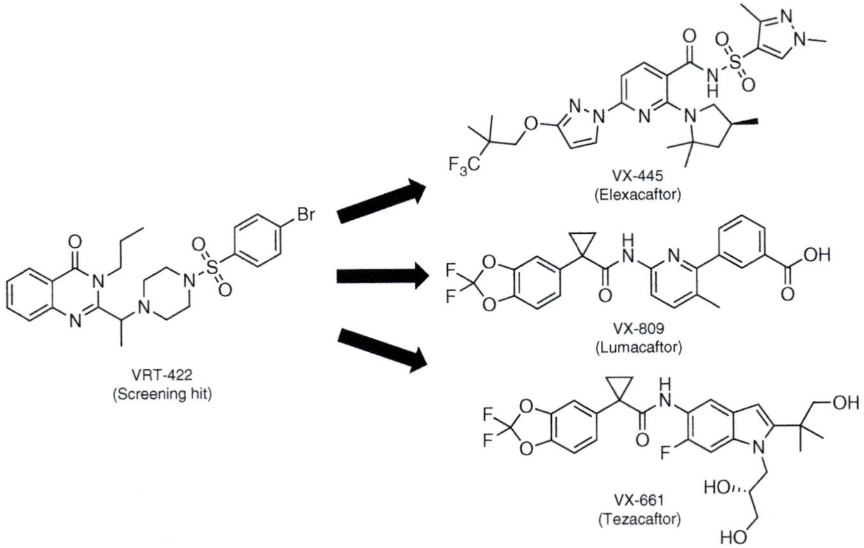

Figure 2.14 CFTR correctors identified from high-throughput screening and optimized to yield three clinical candidates.

key" interaction with these binding pocket components, whereas the polar half of the molecule extends outside the pocket, bringing together the cytoplasmic ends of helices 1 and 6 through interactions with residues 70–74 (of helix 1) and L365 and I368 (of helix 6).

Combinations of type-1 correctors, structurally distinct correctors, and potentiators have shown additive effects that synergistically improve CFTR function and CF patient health. For example, Trikafta®, a combination of elexacaftor (VX-445, a distinct category of corrector), tezacaftor (VX-661, a type-1 corrector), and ivacaftor (VX-770, a potentiator), has been approved for the treatment of CF patients aged 2 years and older with F508del mutation in CFTR (Bacalhau et al., 2023). These drugs together have produced a sea change in the treatment of CF patients and have greatly enhanced and extended the lives of these patients. The success of these CFTR correctors and potentiators have stimulated great interest in applying the same approaches to other protein classes, as reviewed recently by Burslem and Crews (2017) and Radomsky et al. (2024).

2.3.5.2 Inducers of Protein–Protein Interactions

Molecular Glues The term molecular glue (MG) refers to a small molecule that induces the formation of a novel PPI or greatly stabilizes a naturally occurring PPI, by creating a ternary complex between the two proteins and the MG. In the 1990s, it was discovered that two clinical immunosuppressive drugs, cyclosporin

A and FK506, functioned as molecular glues, inducing a PPI between the enzyme calcineurin and cyclophilin or FK506-binding protein, respectively. The formation of these PPIs inhibits the enzymatic activity of calcineurin, thereby blocking T-cell activation; this is the mechanism of immunosuppression by these drugs (Liu et al., 1991). The discovery of this MOA for cyclosporin A and FK506 marks the first pharmacological examples of MGs. Today, most MG drug discovery efforts are focused on inducing ubiquitin/proteosome-mediated degradation of a pathogenic target protein by forming a PPI between the target protein and one of the enzymes responsible for protein ubiquitination, collectively known as the E3 ligases. The first example of this strategy of MG-induced target protein degradation was retrospectively discovered for the immunomodulatory drug (ImD) thalidomide.

Thalidomide was first marketed in the late 1950s in Europe as a treatment for anxiety, sleep disruption, and morning sickness during pregnancy. This tragically led to many birth defects because of the then-unrealized teratogenicity of thalidomide (a teratogen is a substance that causes nonheritable birth defects) and, as a result, the compound was withdrawn from the market. In 1998, however, thalidomide was approved by the FDA as a treatment for multiple myeloma in adult patients. Several thalidomide analogues also have subsequently been approved, or are in clinical trials, for treatment of myeloma (Figure 2.15), despite their MOA being unknown at the time. This group of molecules is collectively referred to as ImDs for reasons that will become clear shortly. In a series of papers published between 2010 and 2015, the MOA of ImDs was clearly elucidated (see Burslem and Crews (2017), for a review). Thalidomide, and its analogues, act to form a protein–protein complex between the protein cereblon and any of several proteins important in immune response, such as members of the Ikaros family of zinc finger protein transcription factors (e.g., IKZF1 and IKZF2). Cereblon is the substrate-binding subunit of the multiprotein E3 ligase complex known as CRL4 (together with three other subunits: DDB1, Cullin-4, and RBX1). Hence,

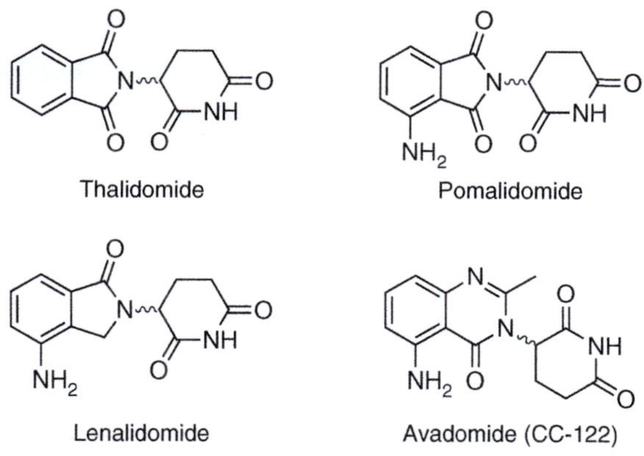

Figure 2.15 Chemical structures of clinical-stage immunomodulatory drugs (ImDs) that act as molecular glues. Thalidomide, pomalidomide, and lenalidomide are FDA-approved drugs for multiple myeloma, whereas avadomide is currently in clinical trials.

formation of the ImD-mediated complex between cereblon and IKZF1/3 results in CRL4-mediated poly-ubiquitination, and subsequent proteosome-mediated degradation, of the IKZFs. The loss of the IKZFs diminishes transcription of a program of genes, leading to apoptotic killing of myeloma cells.

Studies in the absence of ImDs failed to detect any innate interactions between cereblon and the IKZF proteins, suggesting that the ImDs served to create novel PPIs between cereblon and the IKZFs (and other proteins); hence, these novel substrate proteins for catalytic ubiquitination by the CRL4 complex are referred to as neo-substrates. More generally, compounds that act to form any novel PPIs are referred to as MGs. However, Cao et al. (2022) have argued that the neo-substrate concept is flawed and that what MGs do is greatly augment (or reinforce) the stability of preexisting PPIs, even if these interactions are too weak to be experimentally observed in the absence of the molecular glue. This seems to make thermodynamic sense (as the equivalent of an uncompetitive mode of enzyme activation; see above) and is supported by recent molecular dynamic simulations of the impact of lenalidomide on the strength of hydrogen bonds between cereblon and the neo-substrate casein kinase 1A (Minarro-Lleonar et al., 2023). Whether MGs create neomorphic PPIs or greatly enhance existing PPIs, structural studies of protein–protein–MG ternary complexes make clear that in general the binding pockets for the MG are interfacial – composed of recognition elements from both partner proteins of the PPI – and that the MGs not only create proximal interactions between the two proteins, but also create more long-range binding surface interactions, as illustrated schematically in Figure 2.16.

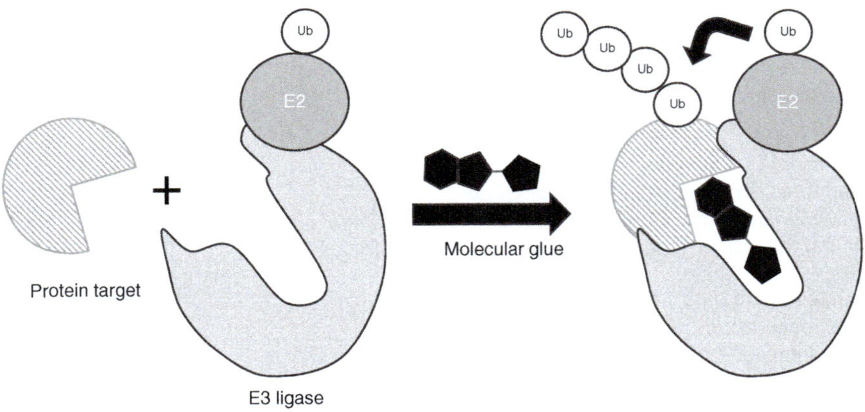

Figure 2.16 Cartoon depiction of the mechanism of molecular glue (MG)-mediated formation of a PPI between a target protein of interest and an E3 ligase. Formation of the ternary protein–MG–E3 ligase complex leads to E3 ligase-catalyzed poly-ubiquitination of the target protein and its subsequent degradation by the proteosome.

The concept of MGs has stimulated great interest within the chemical biology, biotechnology, and pharmaceutical communities, with the goal of targeting specific, pathogenic proteins for degradation by formation of a target protein–E3 ligase–MG ternary complex. Indeed, several examples of successful MG discovery campaigns have now advanced to clinical studies, as recently reviewed by Kong and Jones (2023) and by Sasso et al. (2023). An excellent recent review of the field of MGs is provided by Konstatinidou and Arkin (2024).

Proteolysis Targeting Chimeras (PROTACs) An alternative to MGs for inducing novel PPIs between target proteins and an E3 ligase – with the goal of ubiquitin/proteosome-mediated degradation of the target protein – is by creation of a heterobifunctional compound. The strategy here is to covalently combine a ligand for the target protein, a ligand for the E3 ligase, and a linker between the two ligands, to create a single molecule that would bind simultaneously to the target protein and the E3 ligase, thereby inducing spatial proximity between the two proteins. Molecules that adhere to this general strategy are referred to in the literature as heterobifunctionals, heterobivalents, and proteolysis targeting chimeras (PROTACs).

The first demonstration that one could chemically recruit an unorthodox substrate protein to an E3 ligase with a heterobifunctional species was the recruitment of the enzyme methionine aminopeptidase-2 (MetAP-2) to an SCF family E3 ligase, and the resultant ubiquitination of MetAp-2 by Sakanoto et al. (2001). In this case, the ligand binding to the E3 ligase was a di-phosphorylated 10 amino acid peptide (D-R-H-D-pS-G-L-D-pS-M; where pS indicates a serine residue that is phosphorylated) derived from the known E3 ligase substrate $I_{\kappa}B_{\alpha}$. The ligand for MetAP-2 was the known potent inhibitor of this enzyme, ovalicin, and the two ligands were covalently linked by amide formation at each end of a heptane chain (the resulting heterobifunctional compound was designated PROTAC1). Sakamoto et al. demonstrated *in vitro* that the compound bound to MetAP-2 and could recruit the isolated E3 ligase complex to form a ternary MetAP-2:PROTAC1:E3 ligase complex. They further showed that PROTAC1 could induce ubiquitination of MetAP-2 *in vitro*, with addition of ATP, recombinant E1, E2 (Cdc34), and ubiquitin to the mixture of MetAP-2, E3 ligase complex, and PROTAC1, and that the presence of PROTAC1 was required for MetAP-2 ubiquitination. Finally, Sakamoto et al. demonstrated that PROTAC1 could induce degradation of MetAP-2 when mixed with a cellular extract from *Xenopus* eggs.

The seminal studies by Sakanoto et al. (2001) were followed by the development of the first PROTAC composed entirely of small, organic molecules and the first demonstration of intracellular target protein degradation by a PROTAC (Schneekloth et al., 2008). In this study, a known inhibitor of p53 binding to the E3 ligase MDM2, Nutlin-3 (binding IC_{50} = 90 nM; Vassilev et al., 2004),

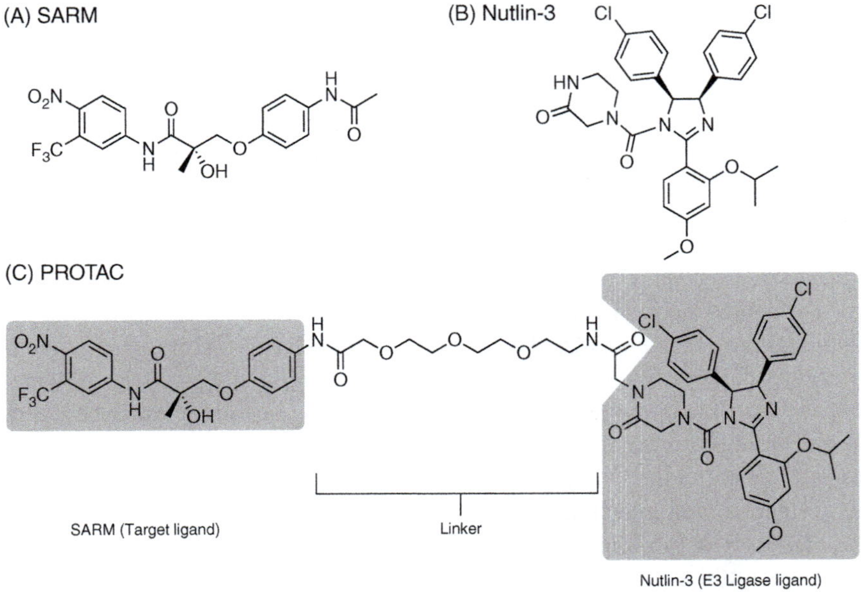

Figure 2.17 (A) Chemical structure of the SARM first reported by Marhefka et al. (2004). *Source:* Adapted from Marhefka et al. (2004). (B) Chemical structure of Nutlin-3, an inhibitor of MDM2 binding to p53, first reported by Vassilev et al. (2004). *Source:* Adapted from Vassilev et al. (2004). (C) The PROTAC produced by Schneekloth et al. (2008) that links the SARM and Nutlin-3 ligands and induces intracellular degradation of the androgen receptor. *Source:* Adapted from Schneekloth et al. (2008).

and a known selective androgen receptor modulator (SARM; $K_i = 4.9 \pm 0.2$ nM; Marhefka et al., 2004) were covalently conjugated via a short polyethylene glycol linker (Figure 2.17). The resulting PROTAC was demonstrated to induce intracellular degradation of the androgen receptor when applied to HeLa cells in culture. This work established PROTACs as a bona fide mechanism of selective target protein degradation within a cellular context and opened a floodgate of additional studies, ultimately leading to several PROTAC-based drugs entering clinical trials, as recently discussed in Copeland (2023).

***In Vitro* Assessment of Molecular Glues and PROTACs** The predicate to target protein degradation induced by MGs or PROTACs is clearly formation of a ternary complex between the target protein, the E3 ligase, and the molecular inducer of PPI. Hence, one might assume that cell-free assays that measure ternary complex formation with recombinant proteins would be a simple and accurate way to measure the affinity and efficiency of complex formation by small-molecule inducers. This is indeed the case, and assays based on, for example, time-resolved fluorescence resonance energy transfer, SPR, and the like have been used to screen diversity libraries for MGs (PROTACs, in contrast, are generally designed from existing knowledge of ligands for each protein of the intended PPI) and for SAR studies aimed at optimizing both MGs and PROTACs (Holdgate et al., 2024). However, the consensus experience of the community has shown that while ternary complex formation is a prerequisite for targeted protein degradation, it is not of itself sufficient. The ternary complex that is induced by the small molecule must not just bring the two proteins into proximity with one another but must orient their engagement in a manner that is conformationally competent for ubiquitination of the target protein. In other words, the target protein must bind within the ternary complex in a manner that makes it a productive substrate for the E3 ligase. Nonproductive binding poses of the protein target, within the ternary complex, do not merely fail to be productively ubiquitinated but can also act as inhibitors of the reaction (a form of enzyme inhibition known as substrate inhibition; Copeland, 2023). Hence, biochemical/biophysical assays of ternary complex formation can be productively used as a first step in MG and PROTAC discovery campaigns, but hits from such assays must be validated further in assays of compound-mediated target protein ubiquitination and compound-mediated target protein degradation. Early in discovery efforts, when compound affinity and cell permeability may be modest, one may choose to use a cell-free assay of target protein ubiquitination, as was originally done by Sakanoto et al. (2001). Ultimately, however, as compound optimization continues, one will need to demonstrate compound-mediated target protein ubiquitination (in the presence of a proteosome inhibitor to block degradation) and degradation intracellularly. The need for orthogonal biochemical/biophysical assays to validate hits from an initial assay format and the need to establish convergent SAR trends between cell-free and intracellular assays of target engagement are not unique to MGs and PROTACs but are a common need of all small-molecule drug discovery efforts, as we shall discuss further in subsequent chapters.

An additional complication in the analysis of PROTACs is that they often do not follow the conventional, saturating Hill–Langmuir concentration–response behavior that we introduced in Chapter 1. Instead, formation of ternary complex as a function of PROTAC concentration displays a biphasic pattern in which the concentration of ternary complex first increases with increasing [PROTAC], but rather than reaching a saturating plateau, it displays a maximum after which the concentration of ternary complex declines (Figure 2.18). This pattern is referred to as the hook effect and is thought to result from formation of individual binary complexes of the PROTAC with the target protein and with the E3 ligase, at the expense of ternary complex formation, as schematically indicated in Figure 2.18. The hook effect can be observed in cell-free and intracellular assays of ternary

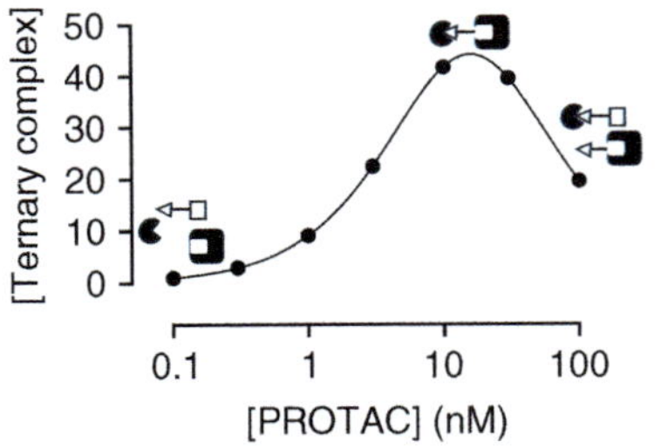

Figure 2.18 Concentration–response plot of ternary complex formation as a function of PROTAC concentration, illustrating the "hook effect" as discussed in the text. At low PROTAC concentration, the three components of the ternary complex (target protein, represented by the "Pacman" circle; E3 ligase, represented by the rectangle; and heterobifunctional PROTAC compound, represented by the shape with a triangle at one end and a square at the other end) exist mainly in their respective free (unbound) forms. As [PROTAC] is increased, the concentration of the ternary complex increases, reaching a maximum between 10 and 50 nM. At higher concentrations of PROTAC, the ternary complex concentration is diminished as it is out-competed for compound by formation of binary target–PROTAC and E3 ligase–PROTAC complexes.

complex formation, target protein ubiquitination, and in some cases target protein degradation. This is a confounding factor in the analysis of compound concentration–response relationships and SAR, which requires the use of more specialized equations for data analysis (Douglass et al., 2014).

Alternative Forms of PPI and Other Macromolecular Complex Induction Until now, our focus on MGs and PROTACS has been on formation of a ternary complex involving a target protein and an E3 ligase as a means of promoting proteosome-mediated degradation of the target protein. However, as we discussed in Section 2.3.5.2.1, early examples of MGs did not form complexes with E3 ligases but rather induced other forms of PPIs with distinct, attendant biological consequences. Recently, the field of induced protein proximity (i.e., MGs and PROTACs) has come full circle in this regard, as researchers have begun to explore MGs and heterobifunctional small molecules to induce complexes between target proteins and enzymes that perform a variety of other posttranslational protein modifications. This emerging area of research has recently been reviewed by Heitel (2023). For example, heterobifunctional molecules have been made with the aim of inducing a complex between a ubiquitinated target protein (destined for proteasomal degradation) and one of the deubiquitinase enzymes (referred to as DUBs), to deubiquitinate and thereby stabilize, rather than degrade, the target protein; molecules that work by this approach have been referred to as DUBTACs. In a like manner, heterobifunctional small molecules are being developed to create complexes between target proteins and kinases, phosphorylases, methyltransferases, acetyltransferases, glycosylases, and a range of other enzymes that catalyze specific posttranslational modifications (Heitel, 2023). Additionally, researchers are experimenting with designing PROTACs in which the protein target ligand is replaced with other species that may form high-affinity, selective interactions with proteins of interest, such as peptides, antibodies, aptamers, and nucleic acids. Finally, current research is also exploring the targeting of RNAs, rather than proteins, for degradation using heterobifunctional molecules composed of an RNA-binding moiety, a linker, and a ligand that binds to an RNase enzyme. It will be interesting to see how these various alternative approaches evolve and which of these may eventually advance to clinical evaluation.

2.3.6 Irreversible Covalent Inactivators

One of the benefits of reversible modes of target modulation is that the degree and duration of modulation can be controlled by titration of the modulator and changes in dosing schedule. In this manner, sufficient modulation of a target can be accomplished to achieve the desired pharmacological impact, while any adverse effects can be mitigated. In some cases, however, a high degree and duration of target inhibition may be required to effect the desired pharmacological outcome. In such cases, MGs and PROTACs that induce target degradation may be advantageous to achieve robust loss of target activity that is only recovered by new protein synthesis in the cell. An alternative approach in these cases is provided by covalent inactivation of the target, leading to irreversible ablation of target activity; as with MGs and PROTACs, irreversible, covalent inactivation of a target is only overcome by new protein synthesis by the cell.

There is a long history of the use of covalent inactivators as therapeutics. Indeed, willow bark salicin and its modern derivative, aspirin (see Chapter 1), derive their analgesic and antipyretic effects from covalent inactivation of the enzyme prostaglandin synthase (also known as cyclooxygenase or COX). Likewise, the β-lactam antibiotics, exemplified by penicillin, also act by covalent inactivation of bacterial transpeptidases that are required for proper cell wall biosynthesis (see Chapter 1 and Copeland, 2013). Generally, drugs that work as covalent inactivators consist of a small organic molecule appended with an electrophile moiety (often referred to as a "warhead") that undergoes nucleophilic attack by an appropriate amino acid nucleophile (e.g., cysteine) to form a covalent bond between the warhead and the nucleophilic amino acid side chain. The side chain sulfhydryl of cysteine is the most common point of attack for covalent inactivator drugs, but other amino acid side chains, such as lysine, histidine, and others, can also be covalently modified by specific electrophilic moieties (Copeland, 2013).

Singh et al. (2011), Copeland (2013), and Raouf (2024) have reviewed the use of covalent inactivators as drugs. Copeland (2013) provides examples of multiple

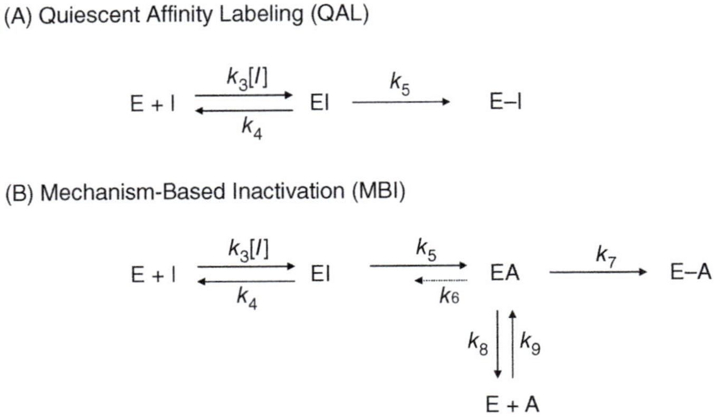

Figure 2.19 Two pharmacologically tractable mechanisms of irreversible, covalent inactivation of target proteins. (A) the QAL mechanism and (B) the MBI mechanism. In this figure, E represents the target enzyme, I is the compound, EI is the initial, reversible binary complex between E and I, and E–I represents a covalent bond linking E and I, which leads to irreversible inactivation of the target protein. In the MBI mechanism, EI formation leads to enzymatic transformation of I to A, which is the reactive species. The noncovalent EA complex can dissociate or go on to form a covalent E–A species. Note that the kinetic rate constants start with k_3, because k_1 and k_2 are reserved for the reaction of the enzyme with its natural substrate. *Source:* Copeland (2013) / with permission of John Wiley & Sons.

mechanisms of covalent reactions with protein targets. Of these, two mechanisms are most amenable to pharmaceutical applications: the quiescent affinity labeling (QAL) mechanism and the mechanism-based inactivation (MBI) mechanism, both of which are illustrated in Figure 2.19 (in this figure, E is used to denote an enzyme target, but the mechanisms are more generally applicable to any target protein).

All covalent inactivation mechanisms involve a chemical reaction that results in formation of an irreversible covalent bond between the compound electrophile and the protein nucleophile. However, indiscriminate reaction of a compound electrophile with any protein nucleophile would not provide a safe and effective mechanism for therapeutic use. What distinguishes the mechanisms shown in Figure 2.19 is that they share a common initiating step: reversible binding of the compound to a specific pocket on the target, to form a noncovalent binary complex. When this reversible binding step is a requisite predicate to covalent bond formation, it offers an opportunity for medicinal chemistry to confer significant target selectivity to covalent inactivators, a property that is critical for drug development.

After formation of the reversible, initial encounter complex between the compound and target, the QAL and MBI mechanisms diverge. In the QAL mechanism, after reversible binding the inherent electrophilicity of the compound warhead, in the confines of the binding pocket and the resultant proximity to the nucleophile ensure covalent bond formation. As we shall see below, this offers an opportunity to trade off the warhead electrophilicity for reversible binding affinity to achieve significant target selectivity. In other words, the chemical step of covalent bond formation can be slow (less efficient) if the binding affinity is high, especially when binding affinity is driven by a slow rate of reversible compound dissociation (a long drug–target residence time; see subsequent chapters and Copeland et al., 2006) to ensure sufficient time for the covalent chemistry to ensue.

In the MBI mechanism, the compound (I in Figure 2.19B) is not itself an inactivator of the enzyme but instead is recognized by the enzyme as a potential substrate. After binding to form a reversible binary EI complex, enzyme catalysis transforms the initial compound I into a bound product (A), which is noncovalently bound to the enzyme active site but now contains an electrophilic species that is susceptible to nucleophilic attack. Compound A may dissociate from the enzyme, as shown in Figure 2.19. Alternatively, if compound A remains bound to the enzyme for sufficient time, it can undergo nucleophilic attack to form a covalent bond and thereby irreversibly inactivate the enzyme. Note that whereas the QAL mechanism can occur with any target protein at any binding pocket containing an appropriately placed nucleophile, the MBI mechanism can only occur within the active site of enzyme targets, as this mechanism inherently relies on the intrinsic catalytic activity of the enzyme. As summarized in Table 2.8, both the QAL and MBI mechanisms have resulted in clinical-stage therapeutics.

2.3.6.1 *In Vitro* Analysis of Irreversible Covalent Inactivators

In our discussions of various reversible mechanisms, we have evaluated the potency of target modulation by compounds in terms of the equilibrium dissociation constant, K_d, or experimental measures related to the K_d, such as EC_{50} and IC_{50}. Such measures are inappropriate in the case of irreversible target inhibition because here the reactions do not approach equilibrium but rather continue until all the target is covalently inhibited (in the case that the target concentration is limiting) or all the modulatory compound is exhausted (in the less common case where compound concentration is limiting). Hence, IC_{50}s and like measures should never be used to evaluate the potency of irreversible, covalent inactivators in cell-free assays. Instead, we must evaluate inactivators by other means, relying on the kinetics of the irreversible reaction.

For most applications to drug discovery, we are dealing with assays in which the target concentration is much lower than that of the modulatory compound under study. Let us consider an enzyme target that is assayed by measuring the course of product formation as we described in Figure 2.9. If we fix the enzyme and substrate concentrations, we expect to see a linear increase in product

Table 2.8 Examples of clinical-stage therapeutics based on the QAL or MBI mechanism of irreversible, covalent inactivation of a target protein.

Mechanism	Drug or drug class	Target protein	Clinical indication
QAL	Aspirin	COX1 and COX2	Anti-inflammatory Antithrombosis
	Penicillins	Bacterial peptidoglycan transpeptidases	Antibiotic
	Cephalosporins	Bacterial peptidoglycan transpeptidases	Antibiotic
	Penems	Bacterial peptidoglycan transpeptidases	Antibiotic
	CI-1033	ErbB1 and ErbB2	Cancer
	EKB-569	ErbB1 and ErbB2	Cancer
	AG7088	Rhinovirus 3C protease	Common colds
MBI	Allopurinol	Xanthine oxidase	Gout
	Clavulanic acid, sulbactam	β-lactamase	Antibiotic resistance
	Eflornithine	Ornithine decarboxylase	Protozoan infections
	Finasteride, dutasteride	Steroid 5α-reductase	Benign prostate hyperplasia
	5-Fluorouracil	Thymidylate synthase	Cancer
	Formestane, exemestane	Aromatase	Cancer
	Selegiline	Monoamine oxidase B	Parkinson's disease
	Tranylcypromine	Monoamine oxidase	Depression
	Trifluridine	Thymidylate synthase	Herpes infection
	Vigabatrin	Gamma-aminobutyric acid (GABA) transaminase	Epilepsy

Source: Copeland (2013) / with permission of John Wiley & Sons.

concentration with reaction time in the absence of an inactivator, much like we saw in Figure 2.9. In the presence of a known concentration of an irreversible, covalent inactivator, the time course of product formation will no longer be linear. Instead, as illustrated in Figure 2.20, a plot of product as a function of reaction time will now be curvilinear, as the covalent reaction will typically occur slowly on the time scale of enzyme turnover.

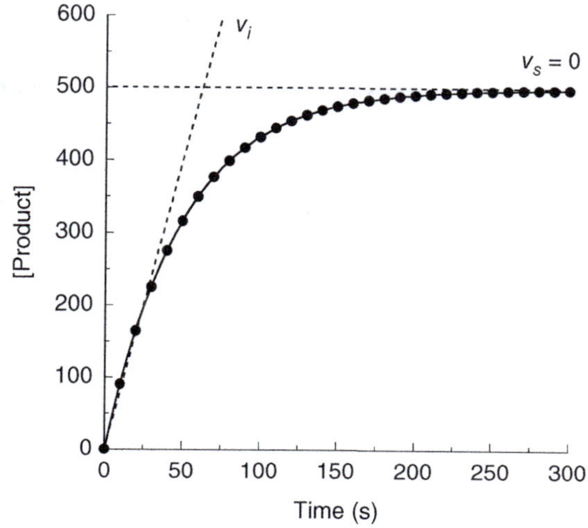

Figure 2.20 Typical time course of product formation as a function of reaction time for an enzyme in the presence of a fixed concentration of an irreversible, covalent inhibitor. See text for further details. *Source:* Copeland (2013) / with permission of John Wiley & Sons.

At early time points in the reaction, product formation will continue to be linear with time as the enzyme remains largely unmodified by the inactivator; the slope of the product concentration as a function of time data will provide an estimate of the initial velocity, v_i, which in most cases will be similar in magnitude to, but not necessarily identical to that of v_0, the reaction velocity in the absence of inactivator. As time goes on, however, the covalent reaction ensues, and the plot begins to curve until all molecules of enzyme have been covalently inactivated. At this point, no further product formation can occur. Hence, the final velocity, v_s, is equal to zero. The plot illustrated in Figure 2.20 represents a first-order, irreversible reaction, which can be described by an observed rate constant, k_{obs}. The value of k_{obs} can be determined by fitting data, such as that in Figure 2.20, to the following equation.

$$[P] = \frac{v_i}{k_{obs}}\left[1 - e^{-k_{obs}t}\right] \tag{2.13}$$

where $[P]$ is product concentration, and t is reaction time.

If one were to measure k_{obs} at various concentrations of inactivator, $[I]$, one would find that k_{obs} behaves as a saturating, hyperbolic function of inactivator concentration, according to equation 2.14.

$$k_{obs} = \frac{k_{inact} \times [I]}{[I] + K_I} \tag{2.14}$$

As we have seen for other cases of saturating, hyperbolic functions, a semilog scale plot of k_{obs} as a function of inactivator concentration ([Inactivator]), yields an S-shaped Hill–Langmuir isotherm (Figure 2.21) for which the high-concentration plateau provides an estimate of k_{inact}, the highest rate of covalent inactivation achieved at infinite inactivator concentration, and the half-maximal value of k_{obs} provides an estimate of K_I. The ratio of k_{inact} over K_I yields a second-order rate constant with units of reciprocal time, reciprocal molarity (typically, s, M_{-1}, or min., M_{-1}) that is the best measure of irreversible inactivator effectiveness.

The value of k_{inact} reports on the speed of covalent inactivator and is related to the electrophilicity of the compound warhead. The value of K_I can be easily mistaken for an equilibrium dissociation constant for the initial, reversible encounter complex EI (i.e., mistakenly equating K_I with K_i). However, this is not mathematically correct. The value of K_I is determined not just by the ratio of k_4/k_3 in the schemes shown in Figure 2.19 but is also influenced by the kinetic constants for reactions after EI formation, depending on the mechanism of inactivation. Nevertheless, the value of K_I is commonly used by medicinal chemists as a metric by which to assess changes in reversible binding affinity that result from changes in compound structure. Thus, the analysis just described offers medicinal chemists three important metrics with which to follow SAR during compound-optimization campaigns: k_{inact}, K_I, and k_{inact}/K_I. To optimize the overall efficiency of inactivation (as measured by k_{inact}/K_I), one can in principle, increase the value of k_{inact} (i.e., increase reactivity of the warhead), decrease the value of K_I (i.e., increase reversible affinity), or both. However, as we have already stated, in practice increasing warhead reactivity (k_{inact}) generally results in a greater probability of nonselective covalent inactivation of proteins other than the target protein, and this can be a source of toxicity due to collateral (off-target) protein inactivation (discussed further in subsequent chapters). Generally, one finds that compound optimization is often best accomplished by a lower value of k_{inact} (a less reactive warhead) and driving overall efficiency by minimizing K_I.

There has been renewed interest in the use of irreversible, covalent inactivators as therapeutics over the past decade. While there are many advantages to target inactivation, there are also several disadvantages. Key among the disadvantages of this modality is the inadvertent collateral inactivation of nontarget proteins, leading to various toxicities including idiosyncratic immune responses. Additional information on the advantages and disadvantages of irreversible, covalent inactivators can be found in Copeland (2013), Singh et al. (2011), Raouf (2024), and Schaefer and Cheng (2023).

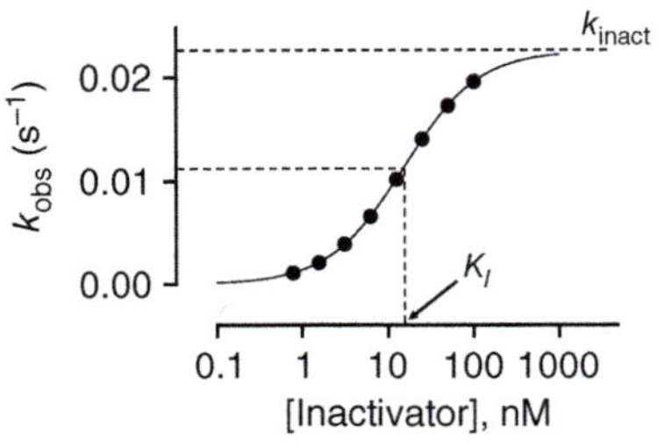

Figure 2.21 Semilog scale plot of k_{obs} as a function of inactivator concentration for an irreversible, covalent inactivator. Data fit to equation 2.14.

2.3.6.2 KRAS^{G12C} Inactivators: A Case Study

A recent success story of irreversible, covalent inactivators as drugs comes from efforts to inhibit the KRAS^{G12C} mutant protein for cancer indications. KRAS is one of the RAS proteins (along with HRAS and NRAS) that are small, membrane-bound GTPase enzymes. The RAS proteins cycle between a GTP-bound active state and a GDP-bound inactive state to regulate several signal transduction pathways critical to cell proliferation. In their active, GTP-bound conformations, RAS proteins bind to and thereby activate the proteins RAF, Pi3K, and RALGDS, to initiate distinct signal transduction cascades that results in transcription of different gene programs, all stimulating cell proliferation. In normal cells, a balance between active, GTP-bound and inactive, GDP-bound RAS is regulated by guanine nucleotide exchange factors (GEFs), such as son of sevenless (SOS) and GTPase-activating proteins (GAPs). GEFs accelerate the otherwise-slow rate of GDP dissociation from RAS, allowing more facile exchange of GTP for GDP. In opposition to the action of the GEFs, the GAPs accelerate the intrinsically slow rate of GTP hydrolysis of the RAS proteins, thereby returning RAS to its inactive, GDP-bound conformation. Somatic, activating mutations of the RAS proteins bypass this regulatory system, leading to unbridled signal transduction, transcriptional activation, and attendant cell proliferation in human cancers. Indeed, mutations in KRAS are the most common activating mutations in human cancers, found in approximately 17% of cancer patients (Kwan et al., 2022). Of these oncogenic KRAS mutations, 81% involve changes of glycine 12 (G12) to a different amino acid. The most predominant of the G12 mutations is G12C, which represents about 12% of all KRAS-mutant cancers and is particularly prominent in lung, colorectal, and pancreatic adenocarcinomas (Kwan et al., 2022).

Clearly the KRAS^{G12C} mutation is an attractive target in terms of disease association. However, enthusiasm for this protein as a drug discovery target was dampened by several factors. RAS proteins display a very high affinity for GTP and GDP (K_Ms about 20 pM; John et al., 1990), and the cellular concentration of GTP is estimated to be about 500 μM. Thus, $[S]/K_M$ in the cell would be about 2.5×10^7 -fold and, considering the Cheng–Prusoff relationship (equation 2.3), this suggests that developing a pharmacologically tractable, nucleotide-competitive inhibitor would be virtually impossible. The crystal structures of RAS proteins (Milburn et al., 1990) do not reveal any allosteric pockets that would be amenable to small-molecule inhibitor binding. Hence, the overall view of the drug discovery community was that KRAS mutant oncoproteins were not tractable targets for small-molecule therapeutics (deemed undruggable, a term that we will discuss further in Chapter 3).

Despite these daunting challenges, the groups of James Wells and Kevan Shokat (both at University of California, San Francisco) decided to take a novel approach to identifying covalent inactivators of KRAS^{G12C}, taking advantage of the unique cysteine nucleophile imparted to the protein by the missense mutation (Ostrem

6H05
(Screening hit)

Compound 6
(Optimized disulfide)

Compound 12
(Optimized acrylamide)

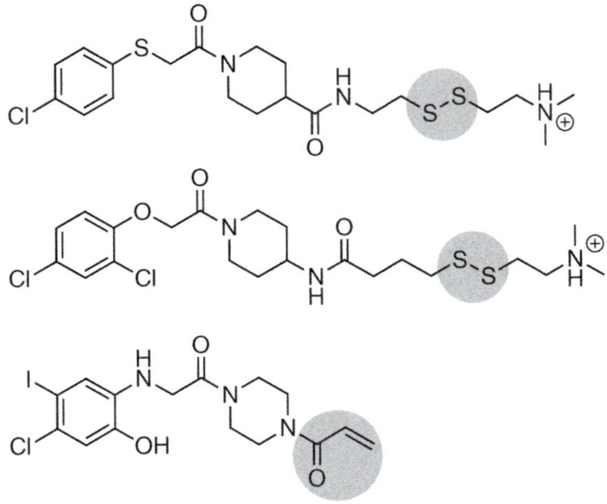

Figure 2.22 Chemical structures of irreversible, covalent inactivators of $KRAS^{G12C}$ as reported in Ostrem et al. (2013). The electrophilic moiety responsible for covalent bond formation with cysteine 12 of the protein (i.e., the warhead) for each compound is highlighted by a gray circle. *Source:* Adapted from Omstrem et al. (2013).

et al., 2013). Using a library of 480 disulfide-fragments, referred to as tethers, at a concentration of 100 μM, they screened 1 μM GDP-bound $KRAS^{G12C}$ to identify compounds that formed a new disulfide bond with cysteine 12. Fragment 6H05 (Figure 2.22) was a top hit from this screen, demonstrating 94 ± 1% modification of the protein. Replacement of the sulfhydryl of 6H05 by a methylene group ablated binding, demonstrating the requirement for disulfide formation. Modification of the linker group resulted in compound **6** (Figure 2.22), which demonstrated improved binding. A co-crystal structure was then obtained of compound **6** bound to a recombinant form of GDP-bound $KRAS^{G12C}$ in which the three other cysteines of KRAS were replaced (C51S, C80L, C118S) to ensure uniform binding of the compound to cysteine 12.

The 1.29 Å structure demonstrated clearly that compound **6** did form a disulfide bond with cysteine 12 of the protein, but surprisingly also showed the formation of a neomorphic pocket adjacent to cysteine 12, into which portions of the compound bound (Figure 2.23). This pocket is not observed in other RAS structures and seems to be induced by compound interactions. The pocket is referred to as the switch-II pocket (S-IIP) and is located between the central β-sheet and α2 (switch-II) and α3 helices of RAS. Compound binding appears to induce significant reordering of switch-II to form the S-IIP, and clear hydrogen bonding and hydrophobic interactions between the compound and amino acids of the pocket are evident from the crystal structure (Ostrem et al., 2013). The conformational changes that attend covalent inactivator

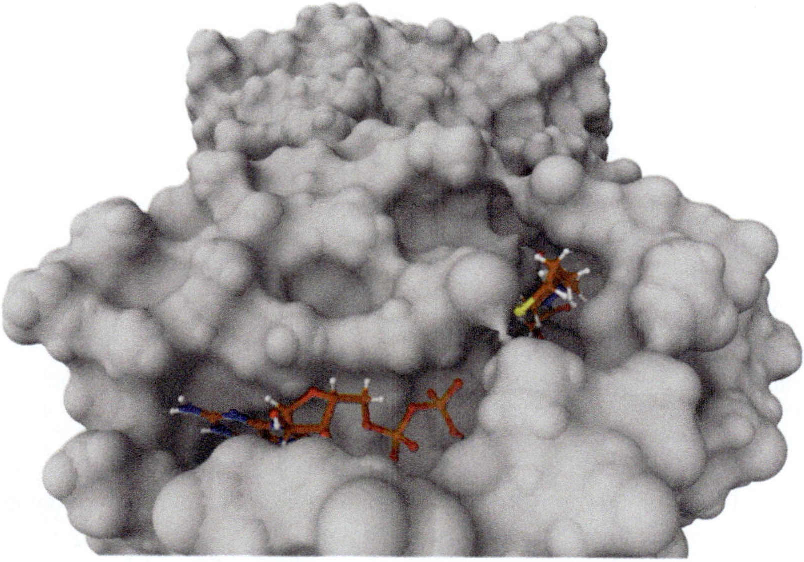

Figure 2.23 Crystal structure of $KRAS^{G12C}$ with compound **6** bound in the neomorphic switch-II pocket, as reported by Ostrem et al. (2013). PDB ID: 4LUC. *Source:* Adapted from Omstrem et al. (2013).

binding to the S-IIP have two distinct functional consequences: they shift the nucleotide equilibrium to favor the GDP-bound, inactive conformation of the proteins and they disrupt binding of effector proteins, such as SOS, to $KRAS^{G12C}$ (Ostrem et al., 2013).

Ostrem et al. next substituted the reactive disulfide of the compounds with carbon-based electrophiles acrylamides and vinyl sulfonamides as the warheads for irreversible, covalent modification of cysteine 12. This resulted in the optimized acrylamide-containing compound **12** (Figure 2.22). Compound **12** selectively modified $KRAS^{G12C}$ while demonstrating no modification of bovine serum albumin nor WT KRAS. This compound also showed selective antiproliferative activity in lung cancer cell lines harboring the $KRAS^{G12C}$ mutant (H1792, H358, H23, and Calu-1 cells), but little effect on lung cancer cells with WT KRAS (H1437, H1299, and A549 cells). For example, compound **12** displayed an EC_{50} of 0.32 μM for cell killing in H1792 cells, but <20% inhibition of cell proliferation of KRAS WT cells at the highest concentration tested (10 μM).

The seminal studies by Ostrem et al. (2013) demonstrated that $KRAS^{G12C}$ could be drugged by irreversible inactivators selective for modification of cysteine 12. This work was followed by many compounds entering the clinic that work by similar mechanisms of inactivation; for example, as of 2024, more than 100

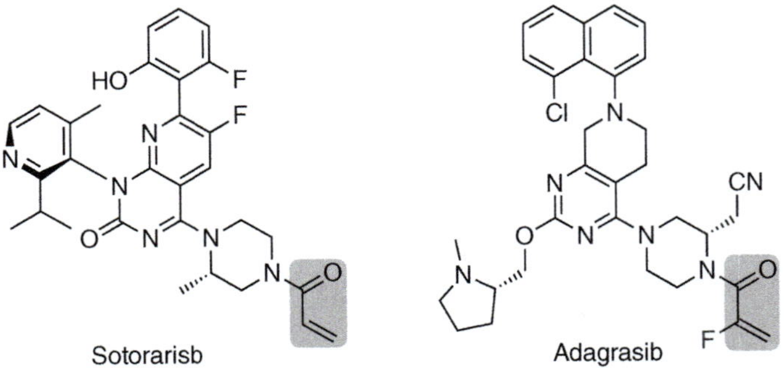

Figure 2.24 Chemical structures of the two FDA-approved irreversible, covalent inactivators of $KRAS^{G12C}$, sotorasib, and adagrasib. The electrophilic warheads responsible for covalent bond formation with cysteine 12 for each compound are highlighted in gray rectangles.

clinical trials of over a dozen $KRAS^{G12C}$ inhibitors – as monotherapy or in combination – were listed on clinicaltrials.gov. To date (i.e., 2025), two compounds that work by this mechanism have been approved by the FDA: Adagrasib and Sotorasib (Figure 2.24).

2.4 Antibody–Drug Conjugates (ADCs)

A relatively new therapeutic modality has been developed for the treatment of cancer patients that combines the cell specificity of monoclonal antibodies (mAbs) with the powerful cell killing effects of traditional, small-molecule cancer chemotherapeutics. The main issue with traditional chemotherapeutics has been that they indiscriminately cause damage to normal cells, as well as cancer cells, and so they lack the therapeutic index (the ratio of toxic over efficacious doses or systemic exposures) needed to treat cancer patients without excessive side effects. This lack of cancer cell specificity can be overcome by covalently linking a small-molecule chemotherapeutic to a mAb that recognizes cell-surface antigens that are specifically or preferentially expressed on cancer cells. Linked mAb–chemotherapy composite molecules of this type are referred to as antibody–drug conjugates (ADCs; Liu et al., 2024).

The principle behind ADCs is to use the cancer cell antigen specificity of the mAb as a homing mechanism to deliver the cytotoxic payload (the chemotherapy) uniquely or preferentially to cancer cells, thereby sparing normal cells from the effects of the chemotherapy. ADCs thus require three component parts to effect cell-selective cytotoxicity: a human-derived mAb, a linker component, and a cytotoxic drug (Li et al., 2021; Drago et al., 2021). The mAb portion of the ADC binds to a cell-surface antigen that is preferentially expressed on cancer cells. The binding event induces endocytosis, bringing the ADC into the cell interior and trafficking it first to early endosomes and then to lysosomes. The acidity and concentration of proteolytic enzymes within these subcellular organelles results in the mAb degradation, thereby releasing the free chemotherapeutic to the cell cytoplasm where it can elicit its cytotoxic effects.

ADCs are still early in their clinical validation, but 13 ADCs have already been approved by the FDA for the treatment of various cancer indications (Table 2.9), and more than 100 ADCs are currently in clinical development. To date, the ADC

Table 2.9 FDA-approved ADCs.

ADC Brand name	Antigen targeted	Antibody	Antibody isotype	Payload[a)]	Tumor type[b)]
Mylotarg	CD33	Gemtuzumab	$IgG4_K$	Calicheamicin	AML
Adcetris	CD30	Brentuximab	$IgG1_K$	MMAE	HL, ALCL
Kadcyla	HER2	Trastuzumab	$IgG1_K$	DM1	Breast cancer
Besponsa	CD22	Inotuzumab	$IgG4_K$	Calicheamicin	ALL
Lumoxiti	CD22	Moxetumomab	$IgG4_K$	PE-38	HCL
Polivy	CD79b	Polatuzumab	$IgG1_K$	MMAE	DLBCL
Padcev	Nectin-4	Enfortumab	$IgG1_K$	MMAE	UC
Enhertu	HER2	Trastuzumab	$IgG1_K$	DXD	Breast cancer
Trodelvy	TROP2	Sacituzumab	$IgG1_K$	SN-38	Breast cancer, uterine cancer
Aidixi	HER2	Disitamab	$IgG1_K$	MMAE	GC, UC
Zynlonta	CD19	Loncastuximab	$IgG1_K$	PBD	DLBCL
Tivdak	Tissue Factor	Tisotumab	$IgG1_K$	MMAE	Cervical cancer
Elahere	Folate Receptor α	Mirvetuximab	$IgG1_K$	DMA	Ovarian cancer

a) Abbreviations: MMAE, monomethyl auristatin E; DM1, emtansine; PE-38, a 38 kDa fragment of *Pseudomonas* exotoxin A; DXD, an exatecan derivative; SN-38, an active metabolite of irinotecan; PBD, pyrrolobenzodiazepine dimer toxin; DMA, a microtubule inhibitor.

b) Abbreviations: AML, acute myeloid leukemia; HL, Hodgkin's lymphoma; ALCL, anaplastic large cell lymphoma; ALL, acute lymphoblastic leukemia; HCL, hairy cell leukemia; DLBCL, diffuse large B-cell lymphoma; UC, urothelial carcinoma; GC, gastric cancer.

Source: Adapted from Liu et al. (2024).

strategy has relied exclusively on mAb–cytotoxic conjugates. Some companies are, however, exploring the utility of linking mAbs to more targeted therapeutics for cancer and other disease indications; to date, these efforts have all been limited to preclinical studies. An excellent review of the current clinical landscape of ADCs can be found in Liu et al. (2024).

2.5 Summary

In this chapter, we have described the major modalities of therapeutic intervention that have achieved regulatory approval as treatments for human diseases. Several other targeted modalities, such as RNA-targeted small molecules, CRISPR, and antisense oligonucleotide drugs, were not covered here, but are nevertheless interesting additions to the pharmacological armamentarium. We have focused our attention in this chapter on mAb and protein-targeted small-molecule drugs, as these make up the vast majority of approved therapeutics.

We have seen in this chapter that small-molecule drugs continue to dominate the clinical landscape across most disease indications, and that these drugs can work through a variety of biochemical mechanisms of target modulation: reversible inhibitors/antagonists, reversible agonists/activators, reversible promoters or inhibitors of PPIs, correctors and potentiators of misfolded proteins, and irreversible, covalent inactivation of targets. We have further seen that small-molecule drugs can be covalently linked together to form heterobifunctional molecules that induce proximity between protein pairs. Finally, we have also seen that small-molecule drugs can be covalently linked to mAbs to form ADCs that have proven useful as anticancer therapeutics.

In the chapters and appendices that follow, we will restrict our attention to the systematic execution of drug discovery and preclinical development for small-molecule drugs. We focus on small-molecule drugs not only because of their continued dominance in clinical medicine, but also because the systematic approaches used throughout the various stages of small-molecule drug discovery are exemplary for the discovery of other drug modalities as well.

Prior to initiating a molecular targeted drug discovery campaign, however, one must select and validate a specific molecular target for which modulation of its biological function is likely to elicit a pharmacologically desirable, phenotypic outcome. Hence, we will next present a detailed description of target selection and validation for drug discovery.

References

Arbabi-Ghahroudi, M. (2017) Camelid single-domain antibodies: historical perspective and future outlook. *Front. Immunol.* **8**, 1589.

Arkin, M.R. and Wells, J.A. (2004) Small -molecule inhibitors of protein–protein interactions: progressing towards the dream. *Nature Rev. Drug Discov.* **3**, 301–317.

Bacalhau, M., Camargo, M., Magalhaes-Ghiotto, G.A.V., Drumond, S., Castelletti, C.H.M. and Lopes-Pacheco, M. (2023) Elexacaftor-Tezacaftor-Ivacaftor: a life-changing triple combination of CFTR modulator drugs for cystic fibrosis. *Pharmaceuticals (Basal)* **16**, 410.

Birkinshaw, R.W., Gong, J.-N., Luo, C.S., Lio, D., White, C.A., Anderson, M.A., Blombery, P., Lessene, G., Majewski, I.J., Thijssen, R. and Roberts, A.W. (2019) Structures of BCL-2 in complex with venetoclax reveal the molecular basis of resistance mutations. *Nature Commun.* **10**, 2385.

Black, J.W. and Leff, P. (1983) Operational models of pharmacological agonism. *Brit. J. Pharmacol.* **220**, 141–162.

Blackwell, K.L., Burstein, H.J., Storniolo, A.M., Rugo, H., Sledge, G., Koehler, M., Ellis, C., Casey, M., Vukelja, S., Bischoff, J. and Baselga, J. (2010) Randomized study of lapatinib alone or in combination with trastuzumab in women with ErbB2-positive, trastuzumab-refractory metastatic breast cancer. *J. Clin. Oncol.* **28**, 1124–1130.

Brent, K.M. and Armstrong, S.A. (2011) A role for DOT1L in *MLL*-rearranged leukemias. *Epigenomics* **3**, 667–670.

Buker, S.M., Boriack-Sjodin, P.A. and Copeland, R.A. (2019) Enzyme-inhibitor interactions and a simple, rapid method for determining inhibition modality. *SLAS Discov.* **24**, 515–522.

Burslem, G.M. and Crews, C.M. (2017) Small-molecule modulation of protein homeostasis. *Chem. Rev.* **117**, 11269–11301.

Burslem, G.M. and Crews, C.M. (2020) Proteolysis-targeting chimeras as therapeutics and tools for biological discovery. *Cell* **181**, 102–114.

Buvailo, A. (2018) Will biologics surpass small molecules in the pharmaceutical race? *Biopharm. Insights.* **3**, 2023. https://www.biopharmatrend.com/business-intelligence/will-small-molecules-sustain-pharmaceutical-race-with-biologics/.

Cao, S., Kang, S., Mao, H., Yao, J., Gu, L. and Zheng, N. (2022) Defining molecular glues with a dual-nanobody cannabidiol sensor. *Nature Commun.* **13**, 1–14.

Chan-Penebre, E., Kulpast, K.G., Majer, C.R., Boriack-Sjodin, P.A. et al. (2015) A selective inhibitor of PRMT5 with *in vivo* and *in vitro* potency in MCL models. *Nature Chem. Biol.* **11**, 432–437.

Cheng, Y. and Prusoff, W.H. (1973) Relationship between the inhibition constant (K_i) and the concentration of inhibitor which causes 50 percent inhibition (I_{50}) of an enzymatic reaction. *Biochem. Pharmacol.* **22**, 3099–3108.

Clackson, T. and Wells, J.A. (1995) A hot spot of binding energy in a hormone-receptor interface. *Science* **267**, 383–386.

Copeland, R.A. (1994) *Methods for Protein Analysis: A Practical Guide to Laboratory Protocols.* Chapman and Hall, New York.

Copeland, R.A. (2013) *Evaluation of Enzyme Inhibitors in Drug Discovery: A Guide for Medicinal Chemists and Pharmacologists*. 2nd Edition. Wiley.

Copeland, R.A. (2023) *Enzymes: A Practical Introduction to Structure, Mechanism and Data Analysis*. 3rd Edition. Wiley.

Copeland, R.A., Pompliano, D.L. and Meek, T.D. (2006) Drug-target residence time and its implications for lead optimization. *Nature Rev. Drug Discov.* **5**, 730–739.

Copeland, R.A., Solomon, M.E. and Richon, V.M. (2009) Protein methyltransferases as a target class for drug discovery. *Nature Rev. Drug Discov.* **8**, 724–732.

Cunningham, B.C. and Wells, J.A. (1993) Comparison of a structural and a functional epitope. *J. Mol. Biol.* **234**, 554–563.

Dabas, P. and Danda, A. (2023) Revolutionizing cancer treatment: a comprehensive review of CAR-T cell therapy. *Med. Oncol.* **40**, 275.

Daigle, S.R., Olhava, E.J., Therkelsen, C.A., Sneeringer, C.J., Song, J., Johnston, L.D., Scott, M.P., Smith, J.J., Xiao, Y. and Jin, L. (2011) Selective killing of mixed lineage leukemia cells by a potent small-molecule DOT1L inhibitor. *Cancer Cell* **20**, 53–65.

Douglass, E.F., Miller, C.J., Sparer, G., Shapiro, H. and Spiegel, D.A. (2014) A comprehensive mathematical model for three-body binding equilibria. *J. Am. Chem. Soc.* **135**, 6092–6099.

Drago, J.Z., Modi, S. and Chandarlapaty, S. (2021) Unlocking the potential of antibody-drug conjugates for cancer therapy. *Nature Rev. Clin. Oncol.* **18**, 327–344.

Fiedorczuk, K. and Chen, J. (2022) Mechanism of CFTR correction by type 1 folding correctors. *Cell* **185**, 158–168.

Hadida, S., Van Goor, F., Zhou, J., Arumugam, V. et al. (2014) Discovery of N-(2,4-di-tert-butyl-5-hydroxyphenyl)-4-oxo-1,4-dihydroquinoline-3-carboxamide (VX-770, invacaftor), a potent and orally bioavailable CFTR potentiator. *J. Med. Chem.* **57**, 9776–9795.

Hajduk, P.J., Huth, J.R. and Fesik, S.W. (2005) Druggability indices for protein targets derived from NMR-based screening data. *J. Med. Chem.* **48**, 2518–2525.

Heital, P. (2023) Emerging TACnology: heterobifunctional small molecule inducers of targeted posttranslational protein modifications. *Molecules* **28**, 690.

Holdgate, G.A., Bardelle, C., Berry, S.K., Lanne, A. and Cuomo, M.E. (2024) Screening for molecular glues: challenges and opportunities. *SLAS Discov.* **29** (100), 136.

Hoover, J.M., Prinslow, E.G., Teigler, J.E., Truppo, M.D. and LaPorte, S.L. (2021) Therapeutic antibody discovery. In *Remington: The Science and Practice of Pharmacy* (A. Adejare, Ed.). 23rd Edition. Academic Press, New York, pp. 417–436.

Jih, K.-Y. and Hwang, T.-C. (2013) Vx-770 potentiates CFTR function by promoting decoupling between the gating cycle and ATP hydrolysis cycle. *Proc. Natl. Acad. Sci. USA* **110**, 4404–4409.

John, J., Sohmen, R., Feuerstein, J., Linke, R., Wittinghofer, A. and Goody, R.S. (1990) Kinetics of interaction of nucleotides with nucleotide-free H-ras p21. *Biochemistry* **29**, 6058–6065.

Jones, L.T.S., Peek, L.J., Power, J., Markham, A., Yazzie, B. and Middaugh, C.R. (2005) Effects of adsorption to aluminum salt adjuvants on the structure and stability of model protein antigens. *J. Biol. Chem.* **280**, 13406–13414.

Kenakin, T.P. (2022) *A Pharmacology Primer*. 6th Edition. Academic Press, New York.

Klotz, I.M. (1997) *Ligand-receptor energetics: a guide for the perplexed*. Wiley, Hoboken, NJ.

Kong, N.R. and Jones, L.H. (2023) Clinical translation of targeted protein degraders. *Clin. Pharmacol. Ther.* **114**, 558–568.

Konstantinidou, M. and Arkin, M.R. (2024) Molecular glues for protein–protein interactions: progressing toward a new dream. *Cell Chem. Biol.* **31**, 1064–1088.

Kouroukis, T.C., Baldassarre, F.G., Haynes, A.E., Imrie, K., Reece, D.E. and Cheung, M.C. (2014) Bortezomib in multiple myeloma: systematic review and clinical considerations. *Curr. Oncol.* **21**, e573–e603.

Kwan, A.K., Piazza, G.A., Keeton, A.B. and Leite, C.A. (2022) The path to the clinic: a comprehensive review on direct $KRAS^{G12C}$ inhibitors. *J. Exp.Clin.Cancer Res.* **41**, 27.

Li, W.Q., Guo, H.F., Li, Y., Zhang, Y.F. and Cui, J.W. (2021) The promising role of antibody-drug conjugate in cancer therapy: combining targeting ability with cytotoxicity effectively. *Cancer Med.* **10**, 4677–4696.

Liang, J., Edelsbrunner, H. and Woodward, C. (1998) Anatomy of protein pockets and cavities: measurement of binding site geometry and implications for ligand design. *Protein Sci.* **7**, 1884–1897.

Liu, J., Farmer, J.D., Lane, W.S., Friedman, J., Weissman, I. and Schreiber, S.L. (1991) Calcineurin is a common target of cyclophilin-cyclosporin A and FKBP-FK506 complexes. *Cell* **66**, 807–815.

Liu, K., Li, M., Li, Y., Li, Y., Chen, Z., Tang, Y., Yang, M., Deng, G. and Liu, H. (2024) A review of the clinical efficacy of FDA-approved antibody-drug conjugates in human cancers. *Mol. Cancer* **23**, 62.

LoConte, L., Chothia, C. and Janin, J. (1999) The atomic structure of protein–protein recognition sites. *J. Mol. Biol.* **285**, 2177–2198.

Lowman, H.B. and Wells, J.A. (1993) Affinity maturation of human growth hormone by monovalent phage display. *J. Mol. Biol.* **234**, 564–578.

Lu, R.-M., Hwang, Y.-C., Liu, I.-J., Lee, C.-C., Tsai, H.-Z., Li, H.-J. and Wu, H.-C. (2020) Development of therapeutic antibodies for the treatment of diseases. *J. Biomed. Sci.* **27**, 1.

Luo, L., Parrish, C.A., Nevins, N., McNulty, D.E., Chaudhari, A.M., Carson, J.D., Sudakin, V., Shaw, A.N., Lehr, R., Zhao, H. and Sweitzer, S. (2007) ATP-competitive inhibitors of the mitotic kinesin KSP that function via an allosteric mechanism. *Nature Chem. Biol.* **3**, 722–726.

Marhefka, C.A., Gao, G., Chung, K., Kim, J., He, Y., Yin, D., Bohl, C., Dalton, J.T. and Miller, D.D. (2004) Design, synthesis, and biological characterization of metabolically stable selective androgen receptor modulators. *J. Med. Chem.* **47**, 993–998.

Mason, K.D., Carpinelli, M.R., Fletcher, J.I., Collinge, J.E., Hilton, A.A., Ellis, S., Kelly, P.N., Ekert, P.G., Metcalf, D., Roberts, A.W. and Huang, D.C. (2007) Programmed anuclear cell death delimits platelet life span. *Cell* **128**, 1173–1186.

Mijinders, M., Kleizen, B. and Braakman, I. (2017) Correcting CFTR folding defects by small-molecule correctors to cure cystic fibrosis. *Curr. Opin. Pharmacol.* **34**, 83–90.

Milburn, M.V., Tong, L., deVos, A.M., Brunger, A., Yamaizumi, Z., Nishimura, S. and Kim, S.H. (1990) Molecular switch for signal transduction: structural differences between active and inactive forms of protooncogenic ras proteins. *Science* **247**, 939–945.

Minarro-Llenar, M., Bertan-Mostazo, A., Duro, J., Barril, X. and Juarez-Jimenez, J. (2023) Lenalidomide stabilizes protein–protein complexes by turning labile intermolecular H-bonds into robust interactions. *J. Med. Chem.* **66**, 6037–6046.

Nahta, R., Yuan, L.X., Du, Y. and Esteva, F.J. (2007) Lapatinib induces apoptosis in trastuzumab-resistant breast cancer cells: effects on insulin-like growth factor-1 signaling. *Mol. Cancer Ther.* **6**, 667–674.

Oltersdorf, T., Elmore, S.W., Shoemaker, A.R., Armstrong, R.C., Augeri, D.J., Belli, B.A., Bruncko, M., Deckwerth, T.L., Dinges, J., Hajduk, P.J. and Joseph, M.K. (2005) An inhibitor of Bcl-2 family proteins induces regression of solid tumors. *Nature* **435**, 677–681.

Ostrem, J.M., Peters, U., Sos, M.L., Wells, J.A. and Shokat, K.M. (2013) K-Ras(G12C) inhibitors allosterically control GTP affinity and effector interactions. *Nature* **503**, 548–551.

Paramore, A. and Frantz, S. (2003) Bortezomib. *Nature Rev. Drug Discov* **2**, 611–612.

Petitt, F.K. and Bowie, J.U. (1999) Protein surface roughness and small molecular binding sites. *J. Mol. Biol.* **285**, 1377–1382.

Pham, N.D., Parker, R.B. and Kohler, J.J. (2013) Photocrosslinking approaches to interactome mapping. *Curr. Opin. Chem. Biol.* **17**, 90–101.

Radomsky, T., Anderson, R.C., Millar, R.P. and Newton, C.L. (2024) Restoring function to inactivating G protein-coupled receptor variants in the hypothalamic–pituitary-gonadal axis. *J. Neuroendocrinol.* **36**, e13418.

Rahbarnia, L., Farajnia, S., Babaei, H., Majidi, J., Veisi, K., Ahmadzadeh, V. and Akbari, B. (2017) Evolution of phage display technology: from discovery to application. *J. Drug Target* **25**, 216–224.

Raouf, Y.S. (2024) Covalent inhibitors: to infinity and beyond. *J. Med. Chem.* **67**, 10513.

Sakanoto, K.M., Kim, K.B., Kumagai, A., Mercurio, F., Crews, C.M. and Deshaies, R.J. (2001) Protacs: chimeric molecules that target proteins to the Skp1-Cullin-F Box complex for ubiquitination and degradation. *Proc. Natl. Acad. Sci. USA* **98**, 8554–8559.

Sasso, J.M., Tenchov, R., Wang, D., Johnson, L.S., Wang, X. and Zhou, Q.A. (2023) Molecular glues: the adhesive connecting targeted protein degradation to the clinic. *Biochemistry* **62**, 601–623.

Schaefer, D. and Cheng, X. (2023) Recent advances in covalent drug discovery. *Pharmaceuticals* **16**, 663.

Schneekloth, A.R., Pucheault, M., Tae, H.S. and Crews, C.M. (2008) Targeted intracellular protein degradation induced by a small molecule: en route to chemical proteomics. *Bioorg. Med. Chem. Lett.* **18**, 5904–5908.

Scott, D.E., Bayly, A.R., Abell, C. and Skidmore, J. (2016) Small molecules, big targets: drug discovery faces the protein–protein interaction challenge. *Nature Rev. Drug Discov.* **15**, 533–550.

Sharma, P., Joshi, R.V., Pritchard, R., Xu, K. and Eicher, M.A. (2023) Therapeutic antibodies in medicine. *Molecules* **28**, 6438.

Singh, J., Petters, R.C., Baillie, T.A. and Whitty, A. (2011) The resurgence of covalent drugs. *Nature Rev. Drug Discov.* **10**, 307–317.

Slamon, D.J., Clark, G.M., Wong, S.G., Levin, W.J., Ullrich, A. and McGuire, W.L. (1987) Human breast cancer: correlation of relapse and survival with amplification of the HER-2/*neu* oncogene. *Science* **235**, 177–182.

Souers, A.J., Leverson, J.D., Boghaert, E.R., Ackler, S., Catron, N.D., Chen, J., Dayton, B.D., Ding, H., Enschede, S.H., Fairbrother, W.J. and Huang, D.C. (2013) ABT-199, a potent and selective BCL-2 inhibitor, achieves antitumor activity while sparing platelets. *Nature Med.* **19**, 202–208.

Southey, M.W.Y. and Brunavs, M. (2023) Introduction to small molecule drug discovery and preclinical development. *Frontiers Drug Discov.* **3**, 1314077.

Sun, J., Wei, Q., Zhou, Y., Wang, J., Liu, Q. and Xu, H. (2017). *BMC Syst. Biol.* **11**, 87.

Tse, C., Shoemaker, A.R., Adickes, J., Anderson, M.G., Chen, J., Jin, S., Johnson, E.F., Marsh, K.C., Mitten, M.J., Nimmer, P. and Roberts, L. (2008) ABT-263: a potent and orally bioavailable Bcl-2 family inhibitor. *Cancer Res.* **68**, 3421–3428.

Van Goor, F., Straley, K.S., Cao, D., Gonzalez, J., Hadida, S., Hazlewood, A., Joubran, J., Knapp, T., Makings, L.R., Miller, M. and Neuberger, T. (2006) Rescue of DF08-CFTR trafficking and gating in human cystic fibrosis airway primary cultures by small molecules. *Am. J. Physiol. Lung Cell Mol. Physiol.* **290**, L1117–L1130.

Van Goor, F., Hadida, S., Grootenhuis, P.D.J., Burton, B., Cao, D., Neuberger, T., Turnbull, A., Singh, A., Joubran, J., Hazlewood, A. and Zhou, J. (2009) Rescue of CF airway epithelial cell function *in vitro* by a CFTR potentiator, VX-770. *Proc. Natl. Acad. Sci. USA* **106**, 18825–18830.

Vassilev, L.T., Vu, B.T., Graves, B., Carvajal, D., Podlaski, F., Filipovic, Z., Kong, N., Kammlott, U., Lukacs, C., Klein, C. and Fotouhi, N. (2004) *In vivo* activation of the p53 pathway by small-molecule antagonists of MDM2. *Science* **303**, 844–848.

Vogel, C.L., Cobleigh, M.A., Tripathy, D., Gutheil, J.C., Harris, L.N., Fehrenbacher, L., Slamon, D.J., Murphy, M., Novotny, W.F., Burchmore, M. and Shak, S. (2002)

Efficacy and safety of trastuzumab as a single agent in first-line treatment of HER2-overexpressing metastatic breast cancer. *J. Clin. Oncol.* **20**, 719–726.

Wells, J.A. and McClendon, C.L. (2007) Reaching for high-hanging fruit in drug discovery at protein–protein interfaces. *Nature* **450**, 1001–1009.

Wylie, A.A., Schoepfer, J., Jahnke, W., Cowan-Jacob, S.W., Loo, A., Furet, P., Marzinzik, A.L., Pelle, X., Donovan, J., Zhu, W. and Buonamici, S. (2017) The allosteric inhibitor ABL001 enables dual targeting of BCR-ABL1. *Nature* **543**, 733–737.

Yang, J., Campbasso, N., Biju, M.P., Fisher, K., Pan, X.-Q., Cottom, J., Galbraith, S., Ho, T., Zhang, H., Hong, X. and Ward, P. (2011) Discovery and characterization of a cell-permeable small-molecule c-Abl kinase activator that binds to the myristoyl binding site. *Chem. Biol.* **18**, 177–186.

Yokoyama, H., Sawada, J., Katoh, S., Matsuno, K., Ogo, N., Ishikawa, Y., Hashimoto, H., Fujii, S. and Asai, A. (2015) Structural basis of new allosteric inhibition in kinesin spindle protein Eg5. *ACS Chem. Biol.* **10**, 1128–1136.

Zheng, Z., Zong, Y., Ma, Y., Tian, Y., Pang, Y., Zhang, C. and Gao, J. (2024) Glucagon-like peptide-1 receptor: mechanisms and advances in therapy. *Signal Transduct.Target Ther.* **9**, 234. DOI: 10.1038/s41392-024-01931-z.

3

Target Selection

Key Learning Points

- Target selection is a critical first step in any molecular targeted drug discovery campaign. To be considered an appropriate target for therapeutic intervention, the target must display a set of key features as presented below.
- First, it must be established that the function of the target plays a critical role in the disease in which one wishes to intervene, such that modulation of target function by a drug will elicit a disease-modifying phenotypic outcome.
- Next, it must be established that the function of the target is selectively essential for the disease state and thus dispensable in normal cells/tissues. This differential essentiality is required to ensure a reasonable *therapeutic index*.
- Additionally, the target must be actionable. Generally, for small-molecule drugs, this requires that the target contain binding pocket (s) to which drug molecules may bind and that these pockets play a key role in target function, so that drug binding is likely to modulate target activity. Targets that meet this requirement are often said to be *druggable*.
- Finally, there must be clear evidence of an unmet medical need among patients that may be fulfilled by a drug that modulates the specific target.

The entire concept of molecular targeted drug discovery centers on the identification and validation of specific targets for drug intervention. The term target validation, however, is somewhat controversial, meaning different things to different researchers. Most drug discovery scientists would agree that true target validation is not established until clinical benefit is achieved with a drug designed to modulate the activity of a specific target. Some would state that complete target validation must also include a demonstration that the clinical benefit of a drug is

Molecular Targeted Drug Discovery: A Guide to How Modern Medicines are Created,
First Edition. Robert A. Copeland.

directly due to modulation of the intended target. For our purposes, however, what is most important is instead to ask, "what is the minimum level of target validation required to initiate a drug discovery program" and thus expend the considerable resources that attend drug discovery.

In this chapter, we will explore the features of a molecular target (again, mainly proteins) that qualify it as worthy of pursuit by a drug discovery team. These features can be grouped into four key categories:

- Disease relevance
- Differential essentiality in disease and normal states
- Actionable by drug therapies
- Unmet medical need

Validation of a target requires that all these categorical features be fulfilled. For pedagogic purposes, we shall discuss each of these categories in turn, although in practice there are no clear lines of demarcation between them. Rather, they represent a continuum of target characterization and validation. Below we exemplify the approaches typically taken in each category for reversible, small-molecule drugs, as these are illustrative of the considerations generally used for all forms of molecular targeted therapeutics. The reader should note, however, that for certain molecular targeted drug modalities (see Chapter 2), some of these requirements may be relaxed. For example, in the case of an antibody–drug conjugate (ADC), it may be less critical that the target's function is selectively essential in disease versus normal cells, because here the antibody portion of the ADC acts as a homing device to ensure that chemotherapy is only effectively delivered to disease cells that express the cognate antigen on their surfaces. Likewise, in discussing target actionability, we will see that the drug-binding pocket must play a critical role in target function, so that binding of drug leads to modulation of that function. In the case of heterobifunctional Proteolysis-targeting chimera (PROTACs) and glues (Chapter 2), the need for functional association of the binding pocket is less critical; in these cases, ligand binding – without functional consequence – may suffice to bring the target and E3 ligase into proximity for target protein ubiquitination and subsequent proteasome-mediated degradation. Nevertheless, by focusing on the approaches used for reversible, small-molecule drugs, we can illustrate more completely the relevant concepts that must be integrated for proper target validation.

3.1 Disease Relevance

The first categorical feature that must be satisfied for target selection is disease relevance. What is meant by this term is that one can demonstrate a clear, causal relationship between a specific function of a protein (or other biomacromolecular target) and a disease phenotype to be modulated to effect a desired therapeutic outcome. For example, localized inflammation at the site of an injury is typified by pain, fever, or localized increased temperature, and swelling of tissue. It is known that these phenotypic symptoms are all triggered by augmented cellular synthesis of prostaglandins in the tissue surrounding the site of injury. The enzyme prostaglandin synthase (also referred to as cyclooxygenase or COX) plays a critical role in cellular synthesis of prostaglandins by catalyzing the initial step of converting arachidonic acid to prostaglandin H2 (PGH2), a key precursor of multiple inflammatory prostaglandins. Thus, it would seem logical that one could diminish the discomforts associated with inflammation by inhibiting the enzymatic function of prostaglandin synthase. Indeed, this is the molecular mechanism by which aspirin and other nonsteroidal anti-inflammatory drugs (NSAIDs, such as ibuprofen, naproxen, and the like) elicit their pharmacological effects.

In the example of NSAIDs, the desired pharmacological outcome is relief of specific symptoms of disease, and this is sufficient to provide great benefit to patients suffering from inflammatory disease. Likewise, symptom relief may be the desired pharmacological outcome in other ailments as well, for example, specific neurological, cardiovascular, and psychiatric diseases. Consider, as another example, the use of antithrombotic drugs to prevent thrombosis in patients with various cardiovascular disorders. Patients with cardiovascular dysfunction are at greater risk of thrombosis. However, treating these patients with strong anticoagulant therapies puts them at risk of excessive bleeding, as blood clotting is essential for wound healing. The blood coagulation cascade involves a complex series of sequential proteolytic enzyme activation steps, through two distinct pathways known as the contact pathway and the tissue factor pathway (Copeland 2023). Researchers studying these pathways found that the two pathways converge at a specific enzyme activation step: the conversion of the inactive protein factor X to the enzymatically active species factor Xa (Wong et al. 2011; Copeland 2023). Separate studies also identified factor Xa as the molecular target associated with the anticlotting activity of low-molecular-weight heparins. Based on these observations, researchers hypothesized that selective inhibition of factor Xa would provide a means of controlling thrombosis and, by titrating the degree of factor Xa inhibition, spare patients from excessive bleeding. Today, several selective factor Xa small-molecule inhibitors are used to treat cardiovascular disease patients and others at elevated risk of thrombosis, such as apixaban, edoxaban, and rivaroxaban (Cabral and Ansell 2015). These drugs do not modify the underlying cardiovascular disease per se, but rather function to prevent the thrombosis often associated with the underlying disease.

As the examples above illustrate, for many diseases, one can identify and validate targets for drug intervention by meticulous study of pathway dependencies of disease and normal human cells. In other cases, the genetics of familial forms of a disease can provide important clues to identify specific proteins or pathways to target in more sporadic forms of the same (or similar) disease. For example, mutations in the genes for BRCA1 and BRCA2 in familial forms of breast and ovarian cancers, as well as in the epidermal growth factor receptor (EGFR) in familial lung cancer, have revealed the importance of the protein products of these genes in sporadic forms of breast, ovarian, and lung cancers, respectively. Similarly, mutations in the amyloid precursor protein (APP) found in early-onset familial Alzheimer's disease have led to the identification of two proteolytic enzymes involved in APP processing as potential targets for sporadic Alzheimer's disease.

In the case of infectious diseases and cancer, the desired pharmacological outcome is selective killing of the infectious agent or the cancer, while sparing normal human cells. For these applications, one needs to identify either a target that is unique to the disease cells or that differs from the same or similar target(s) in normal human cells, in such a way that the drug molecule can selectively modulate the target form associated with the disease cells with minimal impact on the target form associated with normal human cells. Often, the difference(s) between a disease-associated target and its normal human congener originate in genetic differences at the DNA (gene) level (Claussnitzer et al. 2020). In other cases, the difference is not in the genetics of the target itself, but instead results from genetic changes affecting other cellular proteins that create a unique dependence of the disease cell on the function of the target (a situation referred to as synthetic lethality and discussed further below). Let us look at some examples of such genetic differences between disease and normal states.

3.1.1 Genetic Alterations in Disease

3.1.1.1 Differences in Orthologue Sequences

In infectious diseases, a critical protein of the infectious organism may have human orthologue(s) or human proteins of related biochemical mechanism. Often, however, amino acid differences between the target in the infectious organism and the human protein(s) can provide a basis for selective small-molecule binding and modulation of target activity. Examples of this include several antiviral drugs, such as inhibitors of the NS3/4A serine protease of hepatitis C virus, which selectively inhibit the viral enzyme over human serine proteases, and inhibitors of the human immunodeficiency virus (HIV) aspartyl protease that likewise selectively inhibit this enzyme over human aspartyl proteases. Similarly, ivermectin, a cure for river blindness and other diseases caused by round worms, selectively activates ligand-gated channels in the round worms with minimal effect on related human channel proteins. Many additional examples of selective therapeutics for infectious diseases, based on structural differences between the infectious organism and human orthologues of the target protein, can be found in the scientific literature (see summaries in Copeland, 2013, 2023). An illustrative example of targeting a unique structural difference between an infectious organism and human protein for drug discovery is provided by the target identification of the dihydroorotate dehydrogenase (DHODase) enzyme of the Gram-negative bacterium *Helicobacter pylori*.

H. pylori is known to cause gastritis, gastric ulcers, and some forms of gastric cancer. This bacterium is highly prevalent in humans, with infection rates ranging from 25% to >75% of the population, depending on geographic location, sanitation levels, and other socioeconomic factors. For example, in year 2000, it was estimated that approximately 80 million inhabitants of the United States were infected with *H. pylori*. Since the discovery of the causal relationship between *H. pylori* infection and gastric disease, the standard of care for these ailments has shifted from palliative treatments to the use of antibiotics to cure the infection. However, treatment of *H. pylori* with broad-spectrum antibiotics has several drawbacks, including eradication of beneficial gastrointestinal flora and antibiotic resistance. Hence, there is great interest in selective antibiotics against this organism.

Cell proliferation requires a sufficient supply of purines and pyrimidines for, among other metabolic uses, the synthesis of nucleic acids. Mammals and most organisms have two sources of purines and pyrimidines: *de novo* biosynthesis and salvage pathways. However, in highly proliferating cells, such as human cancer cells, the pyrimidine salvage pathway cannot keep up with the requirements for rapid cell division and proliferation. Hence, these cells are uniquely dependent on the pyrimidine biosynthetic pathway. The rate-limiting enzyme in pyrimidine biosynthesis is DHODase, a mitochondrial enzyme that catalyzes the conversion of dihydroorotate to orotic acid, using a flavin mononucleotide (FMN) and an isoprenoid quinone (ubiquinone or coenzyme Q_n) co-substrate as redox cofactors. Hence, attempts have been made to develop potent, selective inhibitors of human DHODase as potential treatments for human cancers, and these efforts have resulted in the compounds brequinar sodium (brequinar) and the active metabolite of leflunomide, A771726 (Figure 3.1).

Copeland et al. (2000) brought together several observations to form a hypothesis that targeting of the *H. pylori* DHODase could result in selective antibiotics for this infectious species. First, they noted that whereas brequinar and A771726 were nanomolar inhibitors of the mammalian DHODases, they displayed no inhibition of the enzyme from the Gram-negative bacterium *E. coli* at concentrations up to 100 μM. Clardy and coworkers (Liu et al. 2000) had solved the crystal structures of complexes of human DHODase with brequinar and A771726 bound and showed unequivocally that both compounds bound within the ubiquinone-binding pocket. This suggested some structural divergence between the quinone-binding pocket of human and Gram-negative bacterial DHODase enzymes. Copeland et al. (2000) also noted that genomic analysis of two strains of

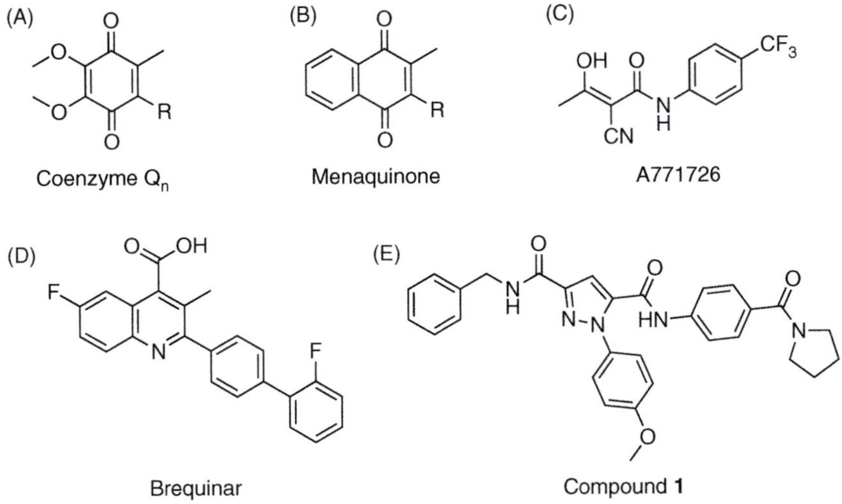

Figure 3.1 Ligands (substrates and inhibitors) of human and/or *H. pylori* dihydroorotate dehydrogenase (DHODase). (A) Coenzyme Q_n. R is an isoprenoid chain of varying length, depending on the specific molecule. (B) Menaquinone. R is an isoprenoid chain of varying length, depending on the specific molecule. (C) A771726, the active metabolite of leflunomide that acts as a potent inhibitor of human, but not *H. pylori* DHODase. (D) Brequinar, a potent inhibitor of human, but not *H. pylori* DHODase. (E) Compound **1** of Copeland et al. (2000), a potent inhibitor of *H. pylori*, but not human DHODase. *Source:* Adapted from Copeland et al. (2000).

H. pylori demonstrated that unlike most Gram-negative bacteria, *H. pylori* lacked the genes encoding enzymes of the pyrimidine salvage pathway, indicating that this bacterium would be uniquely dependent on *de novo* pyrimidine biosynthesis for proliferation. Finally, Copeland et al. (2000) noted that whereas the primary isoprenoid quinone in mammalian mitochondria was ubiquinone, the only major isoprenoid quinones in *H. pylori* cells were the naphthoquinones menaquinone-6 and menaquinone-4 (Figure 3.1). Steady-state enzyme kinetics demonstrated that whereas the *H. pylori* DHODase could utilize either ubiquinones or menaquinones as co-substrates, the human enzyme was only able to utilize ubiquinones (Copeland et al. (2000) and Table 3.1). With these target validation data in hand, Copeland et al. (2000) launched a hit-finding effort for selective inhibitors of *H. pylori* DHODase, resulting in the discovery of pyrazole-based compound **1** (Copeland et al. 2000; Haque et al. 2002; Figure 3.1). Compound **1** was a 26 nM inhibitor of the *H. pylori* DHODase but showed no inhibition of the human enzyme at concentrations up to 100 μM. They went on to demonstrate that compound **1** selectively killed cultures of *H. pylori* with no effect on proliferation of Gram-positive bacteria, other Gram-negative bacteria (presumably owing to

Table 3.1 Ligands of the human and *H. pylori* DHODase enzymes.

Ligand (substrate or inhibitor)	K_M or K_i Human DHODase	K_M or K_i *H. pylori* DHODase
Coenzyme Q_0 (substrate)	Not active	5.3 μM
Coenzyme Q_6 (substrate)	8–20 μM	19 μM
Menaquinone (substrate)	Not active	8.7 μM
Menadione (substrate)	Not active	5.9 μM
Brequinar (inhibitor)	≤6 nM	>100 μM
A771726 (inhibitor)	179 nM	>100 μM
Compound **1** (inhibitor)	>100 μM	26 nM

Source: Adapted from Copeland et al. (2000).

their competent pyrimidine salvage pathways), nor human lymphocytes or Jurkat cells (an immortalized human T-lymphocyte cell line). These data provide strong validation of DHODase as a target for *H. pylori*-selective antibiotics. This target has subsequently also been proposed for other infectious disease indications (Boschi et al. 2019).

3.1.1.2 Gene Amplification and Protein Overexpression

A second form of genetic change that is associated with some diseases is increased gene expression through, for example, gene amplification – a process in which the copy number of a particular gene or chromosomal segment is increased within a cell. Gene amplification leads to increased levels of the gene product and thus increased biological function associated with the protein.

A clear example of a gene amplification is that of the receptor tyrosine kinase ErbB2 in cancer. The ErbB2 (also known as HER2/neu) gene is amplified in certain breast cancers, leading to aberrantly high expression levels of the receptor and in turn aberrantly high tyrosine kinase activity. ErbB2 is the target of the biologic anticancer drug Herceptin and of the small-molecule kinase inhibitor Tykerb (lapatinib). As this receptor kinase activity is an initiator of a signal transduction cascade that results in cellular proliferation, the enhanced ErbB2 kinase activity seen in cells with ErbB2 gene amplification leads directly to tumorigenesis. Importantly, gene amplification is not the only mechanism for increased protein expression or function. In the case of ErbB2, some breast cancer cells have extra copies of the gene that lead to enhanced message levels and, in turn, to increased protein levels of ErbB2. However, studies have shown that other breast cancer cells display increased ErbB2 message and protein without gene amplification, and still other breast cancer cells show increased ErbB2 protein levels without accompanying increases in gene copy or message level (Slamon et al., 1987; Weinberg 2007).

3.1.1.3 Target Mutations in Disease

We have already encountered some examples of disease-associated target mutations in Chapters 1 and 2. A mutation or single-nucleotide polymorphism (SNP) occurs when a nucleotide substitution occurs within the DNA sequence of a gene. The substitution can have one of three outcomes with respect to the protein product of gene transcription and message translation: a silent SNP, a missense SNP, or a nonsense SNP (Copeland 2023).

Silent SNPs do not alter the amino acid sequence of the encoded proteins and are thus not relevant to our current discussion. A missense SNP results in a change in the amino acid encoded by the codon containing the SNP, and therefore a change in the amino acid sequence of the protein encoded by the gene. Missense SNPs can have one of three consequences on the biological function of a protein: they can have no effect on function; they can diminish or completely ablate function, in which case the SNP is referred to as a *loss-of-function (LoF) mutation*; or the SNP can augment the function of the protein, in which case the SNP is referred to as a *gain-of-function (GoF) mutation*; both LoF and GoF mutations can be disease-causing. Finally, a nonsense SNP transforms an amino acid–coding codon into a Stop codon, resulting in a premature termination of the protein sequence during translation. In most cases, the premature termination of the protein sequence results in an LoF due to the loss of key structural elements needed for function. However, it is also possible that the early termination of protein sequence eliminates an auto-inhibitory domain of the protein, thereby resulting in a GoF and constitutive activation.

LoF mutations can cause disease by reducing or eliminating a critical biological function of the protein. For example, LoF mutations in the low-density lipoprotein receptor (LDLR) are the most common cause of familial hypercholesterolemia, and LoF mutations in tumor suppressor proteins, such as p53, are one of the most common genetic alterations seen in human cancers. It is generally difficult to intervene in diseases caused by LoF mutations with small-molecule drugs. Restoration of biological function by binding of small molecules to LoF mutant proteins has proved to be extremely challenging in most cases. However, as we discussed in Chapter 2 (Section 2.3.5.1), some success has been realized in reactivating misfolded membrane-associated proteins with small-molecule correctors and potentiators. Also, as we shall see below, it is possible to indirectly target LoF mutations through synthetic lethal approaches (see below).

In contrast, GoF mutations can often serve as good targets for drug discovery efforts aimed at identifying and optimizing small-molecule inhibitors of the augmented activity engendered by GoF mutations. An excellent example of a GoF mutation that is causal in disease is the $BRAF^{V600E}$ mutation in melanoma and certain other cancers.

There are three isozymes of the rapidly accelerated fibrosarcoma (RAF) protein present in humans: ARAF, BRAF, and CRAF. Of these, oncogenic mutations in BRAF are the most abundant RAF alterations seen in cancer, including 70% of melanomas, 30% of thyroid cancers, and 15% of colorectal cancers (CRCs). The most common mutation of BRAF observed in cancer is the V600E mutation, which accounts for 90% of all RAF oncogenic mutations. The RAF proteins function as serine/threonine protein kinases and represent the first kinase of the RAS–RAF–MEK–ERK signal transduction cascade. In normal physiology, the kinase activity of the RAF proteins is highly regulated. However, the mutation of valine to glutamic acid at position 600 of BRAF induces a conformational change in the protein that leads to constitutive activity of the enzyme, resultant constitutive downstream signal transduction, and transcriptional activation that drives oncogenesis. The clear evidence of a causal role of $BRAF^{V600E}$ in melanoma and other cancers spurred drug discovery efforts leading to several $BRAF^{V600E}$ inhibitors entering the clinic. However, as is often the case with targeted cancer therapeutics, mutation-based resistance to these inhibitors rapidly emerged in patients treated with $BRAF^{V600E}$ inhibitors as monotherapy. Resistance could be greatly mitigated by combining a $BRAF^{V600E}$ inhibitor with an inhibitor of the downstream kinase MEK, and this "double hit" to the common signal transduction cascade also increased the objective response rate for the combination to 76%, as compared to 54% for the monotherapy (Flaherty et al. 2012). Today, three combinations of $BRAF^{V600E}$/MEK inhibitors are approved for the treatment of melanomas and other cancers harboring $BRAF^{V600E}$ mutations: dabrafenib/trametinib, vemurafenib/cobimetinib, and encorafenib/binimetinib.

In addition to GoF mutations, researchers have recently identified another potential consequence of SNPs, referred to as change-of-function (CoF) mutations, in certain cancer indications. In CoF mutations, the biological function of the normal gene product is replaced by a different function, so that these mutants are commonly first thought to be LoF mutations. Two examples of this mechanism of pathogenesis have been identified.

The first example is mutations in the cytosolic and mitochondrial isocitrate dehydrogenase enzymes, IDH1 and IDH2, respectively. The normal (wild-type [WT]) forms of these enzymes catalyze the oxidative decarboxylation of isocitrate to generate α-ketoglutarate. Point mutations in IDH1 and IDH2 have been identified in a broad spectrum of cancers, including acute myeloid leukemia (AML) and myelodysplastic syndromes (MDSs). When these mutants were tested for their ability to convert isocitrate to α-ketoglutarate they were found to be inactive and deemed LoF mutations. Subsequently, the mutant enzymes were found to be active in catalyzing a totally unexpected reaction, the conversion of α-ketoglutarate to 2-hydroxyglutarate (Dang et al. 2009). Three drugs have been approved for the treatment of AML patients based on their inhibition of mutant IDH1/2: vorasidenib, enasidenib, and olutasidenib.

The second example involves mutations of tyrosine 641 of the protein lysine methyltransferase EZH2 in subsets of patients with germinal center non-Hodgkin's lymphoma (NHL). The WT EZH2 catalyzes the mono-, di-, and tri-methylation of histone H3 on lysine 27 (H3K27). It is most active at catalyzing the mono-methylation reaction and its activity wanes with each subsequent methylation step. The NHL-associated mutant EZH2 enzymes were tested for their ability to methylate a peptide bracketing the amino acid sequence surrounding H3K27, using an antibody specific for tri-methylated H3K27 for detection of reaction product. The WT EZH2 was quite active in this assay, but all the NHL-associated mutants were not. Hence, these mutants were considered to be LoF. Subsequently, Sneeringer et al. (2010) demonstrated that the mutant EZH2 enzymes were effectively inactive in performing the first methylation reaction, but were quite active in performing the di-methylation reaction and far more active than the WT enzyme in performing the tri-methylation reaction. Therefore, Snerniger et al. (2010) concluded that the mutations observed in NHL represented the first example of a novel mechanism of pathogenesis, the requisite coupling of the activity of a WT and mutant enzyme for disease. In NHL patients bearing EZH2 mutations, the WT enzyme is required to perform the mono-methylation reaction; both WT and mutant enzyme can perform the di-methylation reaction, but the mutant enzyme greatly accelerates formation of the final tri-methylated product species. These data were a springboard for drug discovery efforts leading to the potent, selective EZH2 inhibitor tazemetostat, which is now approved for treatment of follicular lymphoma (Ho and Copeland 2024). A more complete description of the IDH1/2 and EZH2 CoF mutations, their roles in disease, and the successful discovery and development of potent, selective inhibitors of these enzymes as cancer therapeutics can be found in the reviews by Copeland (2021) and Ho and Copeland (2024).

3.1.1.4 Chromosomal Translocations in Disease

Chromosomal translocations were introduced in Chapter 1 and exemplified by the Bcr-Abl fusion protein. These types of chromosomal translocation are most often seen in cancer indications, due to the inherent genetic instability that attends that disease. Often the fusion proteins encoded by oncogenic chromosomal translocations result in loss of enzymatic regulation. For example, Table 3.2 provides some examples of oncogenic chromosomal translocations that lead to dysregulated kinase enzymatic activity.

Not all chromosomal translocations result directly in dysregulated enzymatic activity. Consider the example of DOT1L enzymatic activity as a driver

Table 3.2 Some examples of oncogenic chromosomal translocations leading to aberrant kinase activity.

Partner protein gene	Kinase gene	Cancer indication(s)
BCR	ABL	Chronic myelogenous leukemia, acute lymphocytic leukemia
NPM	ALK	Non-Hodgkin's lymphoma, anaplastic large-cell lymphoma
ETV6	NTTRK3	Breast carcinoma
EML4	ALK	Lung adenocarcinoma
PTC (1–11)	RET	Papillary thyroid carcinoma

Source: Adapted from Copeland et al. (2000).

of mixed lineage leukemia (MLL). Patients that present with MLL invariably display a chromosomal translocation involving the KMT2A gene. This gene normally encodes a protein lysine methyltransferase containing a catalytic Su(var)3-9, Enhancer-of-zeste, and Trithorx (SET) domain at its C-terminus and DNA-binding domains near its N-terminus. This enzyme catalyzes the methylation of histone protein H3 on lysine K4 (H3K4). However, in MLL, the KMT2A gene undergoes chromosomal translocations that result in fusion proteins between KMT2A and a variety of fusion partner proteins of the AF4 and ENL families. In all these fusion proteins, the KMT2A DNA-binding domains remain intact, allowing for protein binding to specific gene locations, but the C-terminal SET domain is lost, so that the resulting fusion proteins lack the lysine methylation activity. A common feature of the various fusion partner proteins is the ability to bind and therefore recruit another protein lysine methyltransferase enzyme, known as DOT1L. Unlike KMT2A, DOT1L does not methylate H3K4, but instead methylates another histone location, H3K79. Hence, as the result of the fusion protein:DOT1L complex, genes usually methylated at H3K4 are now methylated at H3K79, resulting in transcriptional activation of several leukemogenic genes, such as members of the HOXA gene family. Thus, oncogenesis in MLL is indirectly driven by the fusion proteins that are a diagnostic hallmark of this disease (Daigle et al. 2011).

3.1.1.5 Synthetic Lethality

The term synthetic lethality refers to the genetic situation illustrated in Figure 3.2. Suppose that in a specific organism, the LoF of the product of gene A is inconsequential to the organism and likewise the LoF of gene B is also inconsequential. However, when LoF occurs in both gene A and B simultaneously, it is lethal to the organism. Therefore, there is some functional or mechanistic relationship between

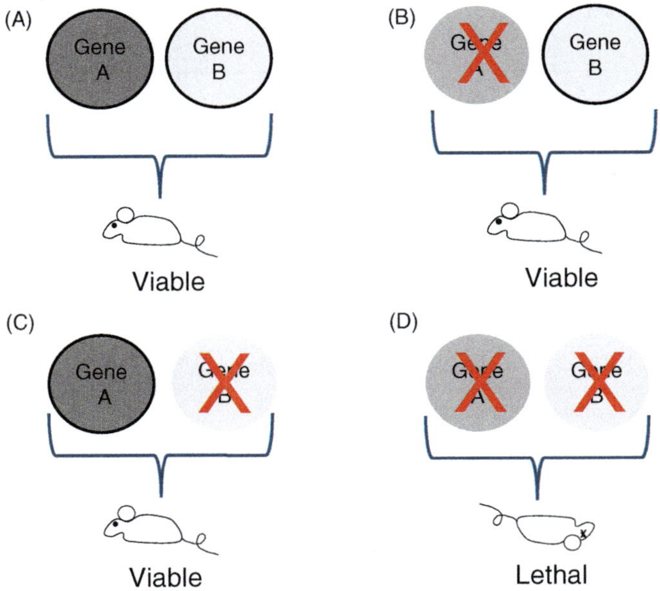

Figure 3.2 Schematic model of synthetic lethality. (**A**) When the products of gene A and B are both functional, the organism is viable. (**B**) When the product of gene A has lost function but that of gene B is still functional, the organism remains viable. (**C**) When the product of gene B has lost function but that of gene A is still functional, the organism remains viable. (**D**) When the products of both gene A and B have simultaneously lost function, the organism is no longer viable. Hence, the combined loss of function results in lethality.

the function of these two gene products, whether such a relationship is evident to the researchers. This situation was first encountered by the geneticist Calvin Bridges in studies of the fruit fly *Drosophila melanogaster* (Bridges, 1922). Later, Dobzhansky (1946) coined the term synthetic lethality for this phenomenon, based on the original ancient Greek meaning of synthetic: to combine two things to form something new (Nijilman, 2011).

Kaelin (1999, 2005) and others (e.g., Beijersbergen et al., 2017) have extended the concept of synthetic lethality to cancer drug targets. Suppose that in normal (non-cancer) cells, the products of both gene A and B are functional, but in certain cancers, gene A has a LoF mutation. In this genetic background, an inhibitor of gene B function (the pharmacological equivalent of a LoF mutation) would be tolerated by the non-cancer cells, but in those cancer cells with a gene A LoF mutation, the compound would be lethal.

Kaelin (2005) points out an additional potential advantage of targeting synthetic lethality in cancer cells. Most of the genetic changes observed in cancers are LoF changes that abrogate the activity of the gene product. Hence, to correct such a change, one would want a drug molecule to restore or enhance the activity of that gene product. However, this is generally a difficult pharmacologic outcome to achieve, as most drug molecules act as inhibitors of target function (but see Chapter 2 for exceptions to this generality). Yet, in biological networks, it is often the case that LoF of one protein is phenotypically equivalent to GoF of another protein. For example, p16INK4a is a natural inhibitor of the cyclin-dependent kinase CDK4, and loss of p16INK4 protein levels is a common signature of some cancers (e.g., melanoma). Pharmacologically restoring the protein levels of this tumor suppressor would be extremely challenging. However, inhibition of CDK4 kinase activity would, in principle, be phenotypically equivalent to restoration of the p16INK4a protein, and inhibition of such a kinase by a small-molecule drug is certainly a pharmacologically tractable approach (Hartwell et al. 1997). Furthermore, it is possible that within the context of synthetic lethality, LoF of one protein renders a cell abnormally sensitive to inhibition of a different protein.

Several pharmacologically relevant examples of synthetic lethality have been identified and, in some cases, used prospectively in drug discovery. One example is provided by the drug rapamycin, which targets the kinase mTOR (mammalian target of rapamycin), an important enzyme of the FRAP/RAFT/TOR signaling pathway. The sensitivity of cells to rapamycin depends on their having lost a specific phosphatase and tensin homolog (PTEN), or equivalently on their having gained activity of the kinase Pi3K; both act at the same point within a convergent signaling cascade, the Pi3K/AKT pathway (Mills et al. 2001). A second example of a pharmacologically relevant synthetic lethal interaction comes from the use of inhibitors of the poly-(ADP-ribose) polymerase (PARPs) enzymes in cancers bearing LoF mutations of BRCA1 or BRCA2 (Faraoni and Graziani 2018). A final example of a pharmacologically relevant synthetic lethal relationship is that between the protein methyltransferase EZH2 (see above) and LoF mutations in the switch/sucrose nonfermentable (SWI/SNF) chromatin remodeling complex. Conformational changes in chromatin act to control gene transcription and these conformational changes are mediated by several mechanisms, including histone protein methylation and remodeling by an ATP-dependent multi-subunit complex known as the SWI/SNF complex. The activities of EZH2 and SWI/SNF act in opposition to one another with respect to inducing chromatin conformational states during development and cellular differentiation. The H3K27 tri-methylated state formed by the catalytic activity of EZH2 is a transcriptionally repressive mark, whereas the action of the SWI/SNF complex leads to a chromatin state that activates gene transcription. Chromatin remodeling by SWI/SNF requires all the protein subunits of the complex to be present and properly assembled. In several cancers, inactivating mutations within various SWI/SNF complex subunits are known to occur. For example, in epithelioid sarcoma (ES), malignant rhabdoid tumors (MRTs), and several other soft tissue sarcomas, biallelic deletions or mutations of the INI1 subunit result in the complete loss of expression of this SWI/SNF subunit. In fact, INI1 negativity, measured by immunohistochemistry, is considered a diagnostic hallmark of these cancers. ES and MRT are aggressive cancers that are considered "chemotherapy refractory" because neither responds well to available therapeutics.

Based on this information, Wilson et al. (2010) investigated the role of SWI/SNF and EZH2 in oncogenic transformation. They found that INI1 negativity led to elevated EZH2 expression and broad H3K27 trimethylation–mediated transcriptional repression. INI1 loss was sufficient to induce tumors in conditional mouse models, and genetic ablation of EZH2 blocked tumor formation in this animal model. Based on these data, Knutson et al. (2013) tested the potent, selective EZH2 inhibitor tazemetostat for antiproliferative activity in cell lines representing INI1-negative MRT and in INI1-replete rhabdomyosarcoma cell lines. They found that the INI1-replete cancer cells were insensitive to tazemetostat ($EC_{50} > 10\,\mu M$) while the INI1-negative cancer cells showed an average antiproliferative EC_{50} of 100 nM. They further showed that tazemetostat caused dose-dependent tumor regression in a mouse xenograft model of MRT. Tazemetostat was subsequently approved by the FDA for the treatment of ES (Ho and Copeland 2024).

Another approach based on the synthetic lethal concept is that of combining two small-molecule drugs, each targeting a different protein of a synthetic lethal pair. Through such a combination drug approach, one may be able to achieve efficacy at lower doses of each individual drug and in this way avoid toxicities that might otherwise be efficacy-limiting for the drugs used in isolation. Essentially all cancer chemotherapies are used in combinations in the clinic, and there are often toxicities associated with the drug combinations, in addition to those associated with each individual drug. The clinical oncologist must thus understand these toxicities and manage them through judicious choices of combination strengths, dosing schedules and the like. In the case of synthetic lethal combinations, the objective is to identify two or more drugs that when dosed in combination have a selective lethal effect on the cancer cells due to an underlying synthetic lethal interaction between the genes for the two drug targets in the context of the tumor cells.

Yet another extension of the concept of synthetic lethality is selective vulnerability to drugs that inhibit the activity of a specific target that is conferred not by the inactivation of a specific, second protein, but instead by a general pathway alteration or pathological state of the cell (Ngoi et al. 2024). These researchers make the point that alterations at different points (e.g., proteins) in specific cellular pathways may confer a common indispensability to another specific protein, so that inhibition of the function of that protein becomes lethal exclusively in cells harboring such pathway alterations. Likewise, pathological cell states, such as deficient DNA damage repair (dDDR), genomic instability, or replication stress, may be sufficient to confer to the cell a unique vulnerability for inhibition of a specific protein. They exemplify this concept with the clinical use of inhibitors of PARPs. PARP inhibitors are most effective in cancer cells with a particular dDDR, homologous recombination deficiency (HRD), as typified by deleterious BRCA1 or BRCA2 mutations. They go on to exemplify several kinase targets for which synthetic lethality with cellular pathways or cell state confer unique vulnerabilities to inhibitors of these kinases. They further describe the synthetic lethal relationship between loss of Werner syndrome ATP-dependent helicase (WRN) function and the microsatellite instability-high (MSI-H) phenotype in certain cancer cells (e.g., CRC cells). Similarly, Castro et al. (2025) have recently shown that a small-molecule inhibitor of the RNA and RNA/DNA hybrid helicase, DHX9, selectively kills MSI-H CRC cells with defective mismatch repair (dMMR), *in vitro* and in mouse xenograft models but has minimal effect on microsatellite stable (MSS) CRC cells with proficient MMR. They hypothesize that DHX9 inhibitors may be effective more broadly in cancer cells characterized by high, baseline levels of replication stress and genome instability, with dMMR/MSI-H representing one subset of high-replication-stress cancers.

Thus, synthetic lethality – as broadly defined here – offers many opportunities to define unique vulnerabilities of disease cells that can be exploited as targets for drug discovery.

3.1.1.6 Disease Association Versus Disease Causality

For all the genetic alterations that we have discussed, an important caveat is that the association of a genetic alteration with disease does not necessarily define a causal relationship between the alteration and pathogenesis or disease maintenance. This is perhaps most apparent in cancer, where genomic instability is a key hallmark of the disease. Thus, cancer cells often carry myriad genetic alterations, some of which may play a role in oncogenesis, but many others are merely a consequence of the disease state (Kaelin 1999, 2017). These latter alterations are referred to as bystander or epiphenomenal (Kaelin, 1999) changes.

Early advances in genomics, transcriptomics, and proteomics spawned frenetic searches for disease-associated genetic alterations in cancer and other diseases, with disappointingly mixed results. Unfortunately, many years and millions of dollars were squandered by researchers not appreciating the difference between association and causality. This has been particularly true in the case of gene amplifications, which may or may not translate into elevated levels of their gene product. Even when the encoded protein of an amplified gene is expressed at higher-than-normal levels, this does not necessarily imply a causal or facilitator role in disease. While this has been a pervasive issue in cancer research, it is not limited to cancers. Many examples of genetic alterations in other diseases have also turned out to be epiphenomenal in, for example, neurodegeneration, psychiatric diseases, obesity, and osteoarthritis. Hence, it is not sufficient to demonstrate the occurrence of a genetic alteration in a disease state. Rather, one must also demonstrate that the genetic alteration is selectively essential to pathogenesis or disease state maintenance, as we shall discuss next.

3.2 Selective Essentiality and Therapeutic Index

To qualify as a drug target, one must demonstrate that the target is essential for pathogenesis or disease state maintenance, such that modulation of that target is likely to result in a desired phenotypic consequence that is disease-modifying. Ideally, the essentiality of the target should be unique to the disease state, to avoid adverse effects due to target engagement within normal tissue (what is referred to as on-target toxicity). As we shall discuss in a subsequent chapter, toxicity can also be the result of collateral modulation of proteins other than the intended target, in what is commonly referred to as off-target toxicity. Hence, as we shall see, this is why target selectivity is an important characteristic of small-molecule therapeutics.

To have clinical utility, a drug should elicit its desired pharmacological effect (efficacy) at a dose less than the dose at which the same drug may cause untoward side effects due to on-target and/or off-target toxicity. In clinical studies and preclinical animal models, one can determine the systemic drug exposure (i.e., the drug concentration in blood) that elicits a therapeutic response in 50% of the population tested. The dose administered to the organism that produces this level of systemic exposure is referred to as the ED_{50} (for **E**ffective **D**ose resulting in **50**% response). Likewise, one can also experimentally determine the dose and systemic exposure of drug that produces toxicity in 50% of the population tested, and this dose is referred to as the TD_{50} (for **T**oxic **D**ose resulting in **50**% toxicity). The *therapeutic index* (TI) is used as a relative measure of drug safety in the clinic, and this metric is defined as the ratio of the TD_{50} to ED_{50}.

$$\mathrm{TI} = \frac{\mathrm{TD}_{50}}{\mathrm{ED}_{50}} \tag{3.1}$$

Thus, the TI provides a measure of the safety margin between effective and toxic doses in a given organism (e.g., humans). A drug with a TI value of 10 or greater is generally deemed safe enough for many applications, whereas a drug with a TI of 1.0 might only be useful in very specific, acute applications, and a drug with a $\mathrm{TI} < 1.0$ might be very challenging to administer safely.

While the TI is a well-established clinical metric of safety, at the early stages of target selection and drug discovery, one cannot determine the value of TI. The best one can do at these early stages is to try to avoid, or minimize, the molecular predicates of on-target and off-target toxicities: target essentiality in normal tissues and target selectivity of compound(s), respectively.

We can define three broad categories of target essentiality as follows.

Pan-essentiality
Selective essentiality
Nonessential or inconsequential

A pan-essential target is one that is required by all cells, both diseased and normal. Hence, a pan-essential phenotype would generally disqualify a protein as an actionable drug target (except, perhaps, in the case of an ADC strategy; see above and Chapter 2). A nonessential target is one that is not required by any cells. Again, this phenotype would disqualify a protein as an actionable drug target. What is most desirable is a target that displays selective essentiality, such that the target is required for the disease state but dispensable in normal tissue.

How can one establish the scope of essentiality of a potential target protein? The most compelling data, for drug discovery purposes, would be the use of a potent, selective modulator of the target that could be used to determine the phenotypic consequences of target modulation across a broad collection of disease and normal cell lines. This statement may seem counterintuitive, as the point of validating a target is to then search for modulators as starting points for drug discovery. Yet, in some favorable cases, there may be existing tool compounds, peptides, or modulators of mechanistically related targets that may be applicable here. These cases notwithstanding, for most potential targets, essentiality is commonly assessed using molecular biology methods to modulate the expression of the target of interest. The gene for a putative target can be expressed in a disease and/or normal cell to assess the phenotypic consequence(s) of the introduction, or reintroduction, of target function. Alternatively, the message (i.e., mRNA) of a putative target can be ablated by RNA-silencing methods (e.g., clustered regularly interspaced short palindromic repeat [CRISPR], siRNA, and the like) in disease and/or normal cells to assess the phenotypic consequence(s) of knockout or knockdown of target function as a portent of the effects of target inhibition. Here, we will introduce the use of RNA-silencing methods as they apply to assessment of target essentiality. We focus on these methods because they are illustrative of the strategy one may take toward target validation and because a large portion of drug discovery efforts are focused on inhibition of specific targets as a molecular modality for therapeutics.

3.2.1 RNA-Silencing Methods for Target Validation

Today, the two most used approaches to RNA silencing for target validation are RNA interference (RNAi) and CRISPR together with the bacterial endonuclease, CAS9 (CRISPR-CAS9).

3.2.1.1 RNAi

RNAi is a natural system of eukaryotic cells that is designed as a defense against viral infection (Fire et al. 1998; Hannon 2002). During replication, viral double-stranded RNA (dsRNA) is produced that can be recognized by the eukaryotic RNA endonuclease, Dicer. Dicer acts as a hydrolytic enzyme, cleaving the dsRNA into short, ds fragments. One of the fragment strands, referred to as the guide strand (also referred to as a small, interfering RNA or siRNA), becomes associated with a multiprotein complex known as RISC (RNA-induced silencing complex), while the other, passenger strand is degraded as a substrate of the RISC. In the context of the RISC, the guide strand recognizes and binds to complementary mRNA molecules. Once bound, the mRNA molecule is cleaved and degraded by the enzyme argonaut, an RNase component of the RISC. In this manner, mRNA that is complementary to the siRNA strand is degraded, resulting in a substantial decrease in the level of protein translation for the affected message. This natural system can be exploited experimentally by designing and synthesizing siRNA strands that are complementary to the mRNA for a gene product of interest. The synthetic siRNA strand(s) is then transfected into cells of interest. Once the exogenous siRNA is delivered to the cell cytoplasm, it can be incorporated into RISC and perform in the same way as described above. In this manner, one can target a specific message for degradation, thus reducing the level of expressed protein from that message. The cellular biochemical and phenotypic impact of the message knockdown can then be systematically assessed by the researcher.

In principle, one could use this technology to look for desirable phenotypes that would result from RNAi-mediated gene silencing in disease cells, with minimal impact on normal cells. From such results, one might conclude that pharmacological inhibition of target function would result in a similar, disease-selective phenotypic impact. However, there are several caveats that must be borne in mind in interpreting the results of RNAi effects in cells. These caveats can largely be classified into two types: false-positive and false-negative results.

A false-positive result is one in which the phenotypic impact of the RNAi treatment is not due to LoF for the gene product of interest. Why might this be the case? Two reasons can be envisaged. First, the gene-silencing effects of RNAi are not as specific as one would like. Indeed, it has become clear in recent studies that there are considerable off-target effects of introducing a particular siRNA construct into cells. The impact of off-target effects can be diminished by using multiple, distinct siRNA constructs, targeting the same mRNA, in knockdown experiments – an approach referred to as redundancy (Echeverri et al. 2006). The multiple siRNAs all impact the same target mRNA but would be unlikely to share overlapping off-target effects. Nevertheless, even with the use of multiple siRNAs, there is some danger that phenotypic effects seen in RNAi experiments may not be uniquely due to knockdown of the target mRNA of interest. An additional approach to assessing off-target effects of siRNA is the use of a C911 guide as a control (Buehler et al., 2012). This is a mismatch design approach in which the middle three bases (bases 9 through 11; hence, the name C911) of an siRNA are replaced with their complementary bases. C911 has been found to reverse the phenotypic output of on-target siRNA treatment without impact on off-target-mediated effects (Buehler et al., 2012).

An additional way to disambiguate off-target from on-target effects is by rescue experiment, in which a functional version of the knocked-down gene is reintroduced in a form that is resistant to the RNAi reagent (Lassus et al. 2002). If the observed phenotype is indeed a result of silencing the message of interest, then it should be reversed by the reintroduction of the cognate gene. If reintroduction of the gene does not reverse the phenotype, then the observed phenotype is likely the result of some off-target effect of the RNAi treatment.

A more subtle form of false-positive effect with RNAi knockdown is a phenotypic effect due to reduction in protein levels of a target rather than due to loss of a specific target function per se. Many proteins play pleiotropic roles in cellular biology, interacting with other proteins and other cell components in myriad ways (Copeland 2023). Suppose, for example, that a target of interest is an enzyme, and our hypothesis for pharmacological intervention is to inhibit the enzymatic activity of the target protein. If a favorable phenotype results from RNAi-based knockdown of the target message, and the phenotype is reversed by rescue with an siRNA-resistant version of the cognate gene, one might conclude that the pharmacological intervention strategy is validated for this target. Yet, these data do not inform us on whether it is the enzymatic activity of the protein that is critical to the phenotype, or whether some other role of the protein is critical. In other words, RNAi-mediated knockdown of the protein disrupts all function of the target, whereas drug therapy might impact only one of these functions. How then can we distinguish between these possibilities? One common approach is to perform the gene rescue experiment twice; once with the WT protein and once with a point mutation that is designed to specifically abrogate enzymatic activity (commonly referred to as a catalytically dead mutant). If the enzymatic activity of the protein is what is phenotypically critical, then rescue with the WT enzyme should reverse the phenotype, but rescue with the catalytically dead enzyme should fail to reverse the phenotype. Of course, the interpretation of such an experiment rests on the assumptions that the mutation(s) introduced does indeed disrupt enzymatic activity while not impacting the overall structural integrity of the protein, assumptions that may be experimentally verified (Copeland 2013, 2023).

Let us consider the example of Werner syndrome protein (WRN). WRN is a multidomain protein (Figure 3.3A) that has been found to have a synthetic lethal relationship with the DNA mismatch repair protein, MutL homolog 1 (MLH1); loss of MLH1 is associated with high levels of microsatellite instability (MSI-H)

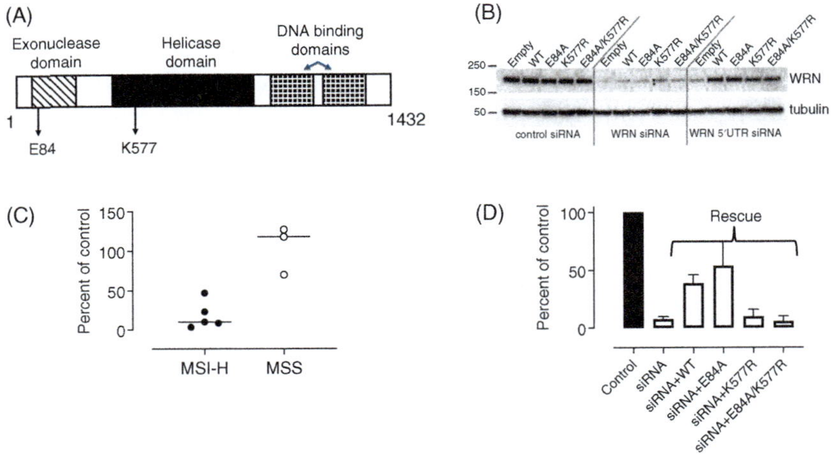

Figure 3.3 (A) Schematic illustration of the domain structure of human WRN. The approximate locations of amino acids E84 and K577 are highlighted. (B) Immunoblots of the effects of siRNA constructs on the protein level of WRN in HCT116 cells. The immunoblots of tubulin are presented as a control. The immunoblots shown in this panel were taken from figure 2B of Kategaya et al. 2019 / with permission of Elsevier. (C) Effect of WRN siRNA on cell proliferation, measured as percent of control, in five MSI-H cell lines (HCT116, LoVo, RKO, SW48, and LS174T) and three MSS cell lines (SW820, SW948, and T84). The horizontal lines represent the median value for each cell line type. (D) Effect of WRN 5′UTR siRNA on three MSI-H cell lines (HCT116, LoVo, and RKO) and with rescue by WT, E84A, K577R, or the double-mutant E84A/K577R WRN. *Source:* Data for panels C and D were estimated from data reported by Kategaya et al. (2019).

in certain cancers, such as CRC. Other CRCs, however, do not demonstrate high microsatellite instability and are said to be microsatellite stable (MSS). This suggests that inhibition of WRN function would be selectively lethal in MSI-H cancers but would be inconsequential in MSS cancers and normal cells. However, WRN contains two domains that confer distinct catalytic activities: a 3′-5′ exonuclease domain and a 3′-5′ helicase domain. Kategaya et al. (2019) set out to confirm a selective essentiality for WRN in MSI-H CRC and to determine if one or the other or both, or neither of WRN's catalytic activities were required for MSI-H cancer proliferation and survival. They noted that mutation of glutamic acid 84 to alanine (E84A), within the exonuclease domain, ablated exonuclease activity without affecting the helicase activity of WRN. Likewise, mutation of lysine 577 to arginine (K577R), within the helicase domain, ablated ATPase and helicase activities, without affecting the endonuclease activity of WRN. Simultaneous mutation of E84 to A and K577 to R rendered WRN devoid of both endonuclease and helicase activities. They then designed siRNA constructs that would selectively knockdown endogenous transcripts by targeting their 5′

UTR (WRN 5′UTR siRNA) and other internally targeting constructs that would knockdown endogenous and exogenous transcripts (WRN siRNA). Figure 3.3B illustrates the impact of these siRNA constructs on the level of expressed WRN forms in MSI-H HCT116 CRC cells, using immunoblots to assess protein level. The effect of siRNA-mediated WRN knockdown on cell proliferation was tested by transfecting five MSI-H cell lines (HCT116, LoVo, RKO, SW48, and LS14T) and three MSS cell lines (SW620, SW948, and T84) and measuring terminal cell counts in 7- to 10-day proliferation assays. As summarized in Figure 3.3C, the hypothesized selective essentiality of WRN in MSI-H cancers was borne out by the significantly greater reduction in median cell counts for MSI-H cells (median % of control = 10%) as compared to that for MSS cells (median % control = 118%). Kategaya et al. (2019) next looked at the ability of different exogenous WRN constructs to rescue cell proliferation in three MSI-H cancer cell lines (HCT116, LoVo, and RKO). The results of these experiments are summarized in Figure 3.3D, where the mean values of terminal cell counts at five days after transfection are shown as percentages of that for the control (non-transfected) cells (in Figure 3.3D, all the siRNA constructs used were WRN 5′UTR siRNAs). We see that the WRN 5′UTR siRNA reduced cell proliferation to about 8% of control. Rescue with WT WRN increased proliferation to about 39% of control and rescue with the E84A mutant increased proliferation to 54% of control. Hence both WT and E84A were able to partially rescue the effects of knockdown, suggesting that the exonuclease activity of WRN is not critical for MSI-H cancer cell proliferation. In stark contrast, no significant rescue was realized by expression of the K577R mutant nor the double-mutant E84A/K577R. These data provide strong evidence that the helicase activity of WRN is necessary and sufficient to drive proliferation in MSH-H cancers. These data allowed drug discovery researchers to focus their efforts on the discovery of small-molecule inhibitors of the helicase activity of WRN, leading to several such inhibitors entering clinical studies.

False-negative results of RNAi treatments most often result from insufficient knockdown of the protein of interest. For a given siRNA, or collection of targeted siRNA constructs, the degree of message knockdown can be quite variable from cell line to cell line, and even from experiment to experiment in the same cell line. A robust message knockdown mediated by RNAi methods may knockdown the target message by 70–90%. While that is an impressive change, the remaining 10–30% of message leak-through may be sufficient to drive the cellular processes required for disease, especially if the target of interest is an enzyme for which the cell may contain a considerable excess of catalytic capacity. Thus, insufficient knockdown could result in a failure to observe an induced phenotypic change. This potential source of false-negative result is exacerbated by the fact that a particular percent knockdown of message level does not necessarily translate into a quantitatively similar knockdown of protein level, and with respect to biological function, it is the protein, not message, level that is most critical. There are several examples in medicine of enzyme targets for which more than 90% inhibition of catalytic activity is required to elicit a pharmacological effect, and this is often well achieved by small-molecule drugs targeted against the enzyme and dosed at appropriate levels and with appropriate frequency. The aspartyl protease of HIV, the virus that causes acquired immunodeficiency syndrome (AIDS), is a well-known example of this phenomenon. AIDS drugs that work by inhibiting the HIV protease must be potent and safe enough to dose them at levels that essentially block all enzymatic activity of the viral proteases. In a knockdown experiment where the target level is reduced by only 70–80%, one might not observe a favorable phenotypic result in cases like this, where essentially complete loss of enzymatic activity is required. Hence, one runs the risk of eliminating or deprioritizing a target if care isn't taken in interpreting negative results from RNAi-mediated knockdown experiments alone.

3.2.1.2 CRISPR-Cas9

To achieve complete knockout of a potential target gene, researchers often turn to the CRISPR-Cas9 system – a bacterial system for viral defense – that permits targeted gene cleavage and editing in eukaryotic cells (Lu et al. 2017). The cleavage specificity of the endonuclease Cas9 is guided by the RNA sequence. Hence, editing can be directed to any genomic site by engineering a complementary guide RNA and delivering this, together with Cas9, to a target cell.

The guide RNA (gRNA) is a recombinant fusion of two elements. The first element is a target-specific CRISPR RNA (crRNA) that base pairs to a specific sequence within the target gene DNA (referred to as the protospacer), to thus localize the action of the Cas9 endonuclease. The second element is an auxiliary trans-activating CRISPR RNA (tracrRNA) that forms a hairpin structure. The tracrRNA element of the gRNA forms the recognition locus for Cas9 binding. In this way, the gRNA binds Cas9 and localizes the endonuclease to specific gene locations that are then cleaved by the enzyme. Cleavage of the gene also requires a three-nucleotide recognition motif of sequence NGG (where N is any nucleotide base followed by a pair of guanine [G] bases), that is immediately downstream on the 3′ end of the target protospacer region. The NGG motif is referred to as the protospacer-adjacent motif (PAM). The PAM must be present within the target DNA but is not a component of the crRNA. Cas9-mediated cleavage of the target DNA occurs three base pairs upstream of the PAM sequence on both DNA strands. The gRNA and the mRNA coding for Cas9 are typically delivered to cells of interest by viral transfection using, for example, lentivirus-mediated delivery.

Both siRNA and CRISPR-Cas9 can be used to investigate the phenotypic consequences of knockdown or knockout of a single gene in different disease and/or normal cell lines. However, lentiviral-mediated CRISPR-Cas9 can be used to deliver multiple gRNAs for different target genes in different cell lines. This has allowed researchers to evaluate the phenotypic consequences of knockout of large numbers of individual genes across a library of cell lines in what is referred to as CRISPR pooled screening (Hsu et al. 2016). For example, researchers at the Broad Institute of the Massachusetts Institute of Technology and Harvard are using CRISPR-Cas9 to knockout hundreds of individual genes in over 1,000 human cancer cell lines to determine the effects of knockout of specific genes on the proliferation of specific cancer cell lines. They have codified these data and have made them publicly available as a database portal known as DepMap (www.depmap.org). To compare the impact of CRISPR-Cas9–mediated gene knockout among genes or for a single gene across multiple cell lines, one needs a cogent scoring system for phenotypic impact. Several mathematical approaches have been suggested for this purpose, but the CERES scoring system of Meyers et al. (2017) is generally considered to be the most appropriate for this purpose. This scoring system accounts for the overestimation of essentiality for highly amplified genes, as well as confounding copy number and gene-independent antiproliferative effects of CRISPR-Cas9 knockdown. A CERES score of zero represents the median value for a gene that is nonessential in any cell line. A CERES score of −1 represents the median value for an essential gene, and any CERES value > −0.5 is assumed to represent a gene that is not essential. By plotting the CERES score for a particular gene across multiple cancer cell lines, one can assess the essentiality of the gene across cancer cells. Figure 3.4 illustrates such plots, across 341 cancer cell lines, for three genes: KRAS, serine- and arginine-rich splicing factor 3 (SRSF3), and protocadherin alpha subfamily C2 (PCDHAC2), representing a selectively essential, a pan-essential, and a nonessential gene, respectively. As we saw in Chapter 2, it makes sense that KRAS is a selectively essential gene, with cancer cells bearing KRAS GoF mutations being the most sensitive to knockout of this gene. Likewise, a similar plot for the BRAF gene would also illustrate a narrow set of cancer cell lines with CERES scores <−0.5, consistent with the expected selective essentiality of this gene in cancers bearing $BRAF^{V600}$ mutations (see Chapter 2). Although the data compiled in DepMap is limited to cancer cell lines, a logical starting assumption would be that if a gene demonstrates selective essentiality across cancer cell lines, it is unlikely to be essential in noncancer cell lines. This assumption is often invoked in the early stages of cancer target validation; however, this must eventually be experimentally tested *in vitro* and/or *in vivo*.

The development of RNA knockdown (i.e., siRNA) and knockout (i.e., CRISPR-Cas9) technologies has also been used *in vivo* to create knockdown or knockout mouse models. Many cancers recapitulate genetic and epigenetic aspects of embryonic development (Sperger et al. 2003), and therefore knockdown or knockout of such targets often proves to be embryonically lethal. The use of inducible promoter systems has allowed researchers to configure conditional transgenic mouse models wherein knockdown or knockout of a gene only occurs

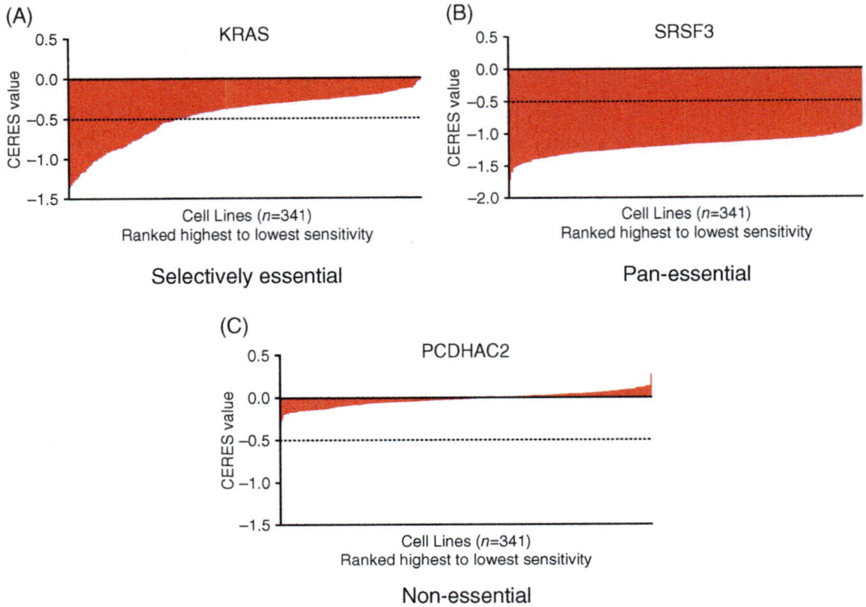

Figure 3.4 CERES scores across a library of 341 cancer cell lines for CRISPR-Cas9 knockout of three different genes, illustrating selectively essential, pan-essential, and nonessential gene profiles. (A) Plot for KRAS, a selectively essential gene. (B) Plot for SRSF3, a pan-essential gene. (C) Plot for PCDHAC2, a nonessential gene. Data for these plots provided by Scott Ribich and Jennifer Castro.

at a particular stage of development or only in the adult animal (Bockamp et al. 2008), thereby avoiding the complication of embryonic lethality. These models offer great opportunities to explore the phenotypic consequences of target knockdown/knockout in normal tissues, as part of the establishment of an estimated TI (see above). However, the creation of such models is time-consuming and costly. Hence, most drug discovery researchers do not invest directly in creating such models, but instead often rely on the academic community to develop and test these models for specific genes.

3.2.2 Genetic Versus Disease Vulnerability Target Validation Funnels

For most drug discovery teams, there is a clear focus on a specific disease or small set of specific diseases that define the strategic *sine qua non* of the laboratory or organization. This disease focus is based on a clear view that there is an unmet medical need among patients suffering from the disease, because of a lack of therapeutic options or because patient needs are underserved by existing therapies. Hence, the first step – albeit an often-unrecognized step – in target selection is

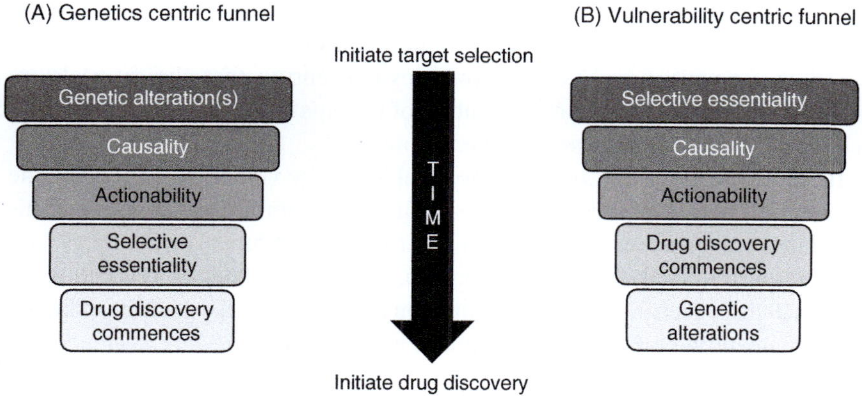

Figure 3.5 Two distinct sequences of target selection criteria and supporting experiments. (A) The genetics-centric target selection funnel and (B) the vulnerability-centric target selection funnel. The term "funnel" is used here as it is assumed that one starts with a large collection of putative target genes and with each successive stage of validation, the target pool is winnowed further as fewer targets pass the criteria for continuation to the next stage.

defining the disease in which the researcher wishes to intervene. Taking this as a given, we can ask what then becomes the first practical step in target selection.

In the post-genomic era, target selection for molecular targeted drug discovery has traditionally been initiated by identifying specific genetic alterations associated with disease. Criteria such as causality, selective essentiality, and actionability were usually assessed subsequently in what may be referred to as a *genetics-centric target selection funnel*, as illustrated in Figure 3.5A. With the introduction of large-scale CRISPR pooled screening, researchers can today take a different sequential path for target selection and validation: a *vulnerability-centric target selection funnel* (Figure 3.5B).

In this latter approach, researchers begin by analyzing the results of CRISPR pooled screening and like approaches to identify selectively essential targets that constitute unique vulnerabilities of the specific disease state of interest. Hypotheses of selective essentiality and causality in disease are then verified with siRNA studies and gene rescue studies across multiple disease and control cell lines (i.e., normal or otherwise invulnerable cells). Target actionability is typically defined next. Only after these criteria have been met do researchers then expend the time, effort, and funding to identify the specific genomic, epigenomic, transcriptomic, proteomic, or pathway alterations that may form the genetic/physiological basis for target vulnerability. In fact, many times a target may be selected and drug discovery efforts commenced, with exploration of the genetic/physiological basis of target vulnerability being performed in parallel with hit-finding and lead optimization, rather than as a prerequisite to these activities. That is not to say that these activities are unimportant; on the contrary, these studies often provide the basis for patient stratification strategies that ensure that clinical trials are designed to select or enrich for the subset of patients most likely to benefit from a targeted therapy, as discussed further below.

Because one is starting with a demonstrable disease vulnerability in using the vulnerability-centric funnel, many researchers (this author included) view this as a more pragmatic and cost-effective (cost in terms of funds, time, energy, and other resources) approach to target selection. Other researchers, however, find this approach disconcerting, as they cannot seem to divest from the doctrine of knowing the genetic alterations as a prerequisite to target selection.

3.3 Actionability

As we discussed in Chapter 2, for a target to be actionable toward modulation by small-molecule drugs, it must contain a binding pocket(s) that conform to specific physicochemical properties that are complementary to the physicochemical properties required of small-molecule drugs (as will be further discussed in subsequent chapters). We saw that such pockets are largely hydrophobic, of limited volume, and display a degree of surface roughness that confers three-dimensional complexity and uniqueness to the pocket. As part of our assessment of target validation, it is important to determine whether a putative target contains at least one such druggable binding pocket. How might one make this determination at the early stage of target selection? Below we shall discuss several approaches that are commonly used to mitigate risk in terms of target druggability.

3.3.1 Existing Structures and/or Ligands

The most definitive approach to assessing druggability is to have either (1) an existing structure (from, for example, crystallography, nuclear magnetic resonance [NMR], or cryo-electron microscopy [EM]; Copeland 2023) of the putative target from the literature that clearly identifies an appropriate binding pocket or (2) an existing ligand molecule that binds this target from previous efforts, or both. The availability of a preexisting structure for a putative target is not as unlikely as one might think. In favorable cases, academic groups may have already devoted resources to a target, not for the purpose of drug discovery, but instead to understand the role of the target in cellular biology or pathobiology. Perhaps less commonly, a protein may have been previously targeted for drug discovery by other researchers, so that small-molecule ligands of the target have been reported in the scientific and/or patent literature; recently, Carpenter and Altman (2024) have provided a comprehensive overview of available databases of ligand-binding pockets and protein–ligand interactions. It may be that these existing modulators lacked key features to progress them to the clinic but nonetheless are sufficient to credential the target as druggable. In other cases, a clinical-stage drug against the target may exist, but the researcher believes that there is still value in targeting the protein to create improved drugs that mitigate some deficiencies of the existing drug (such an approach is commonly referred to as a fast follower or best-in-class approach).

3.3.2 Target Classes

If an existing structure or ligand does not exist for a putative target, one may be able to infer the existence of a druggable binding pocket from the biochemical/biological function of the target. Today, the gene sequence for virtually any putative target will be known and available from public data bases (e.g., GeneCards; www.genecards.org). From this, the protein sequence of the gene product can be determined, and this can often define the protein as a member of a specific functional target class. The protein structure may, for example, reveal the existence of a known functional domain, such as a nucleotide triphosphate (e.g., ATP)-binding domain, kinase domain, a protease domain, a hormone-binding domain, etc., within the protein sequence. Proteins containing such domains are well precedented as drug targets (Barnash et al. 2017; Copeland 2023). As we saw in Chapter 2, other examples of druggable target classes include ion channels, G-protein–coupled receptors (GPCRs), protein, DNA and RNA methyltransferases (PMTs, DMTs, and RMTs), nucleotide triphosphate (NTP)-coupled motor proteins (e.g., helicases, kinesins), and the like.

Essentially, any member of a protein target class known to bind small-molecule cofactors, metabolites, nucleic acids, or peptides can be inferred to contain at least one druggable binding pocket. However, the history of drug discovery suggests that greater success in developing clinical drugs has been associated with two broad target classes: enzymes and receptors. For example, a survey of recent FDA-approved anticancer drugs found that 60% targeted enzymes as targets and another 25% targeted receptors (Sun et al. 2017). Thus, we can account for 85% of recent FDA-approved cancer therapeutics by considering only these two broad categories of targets. Some of the physicochemical reasons for the predominance of enzymes and receptors as drug targets have been presented by Copeland (2013, 2023).

3.3.3 Sequence Alignment and Homology Modeling

Related to the target class approach, one may be able to infer the structure of a putative target by homology to related protein sequences (for example, orthologues from other organisms, or homologues from the same organism) for which a crystal (or other) structure is available. In this context, the related protein is referred to as the template. The availability of appropriate template proteins has been greatly enhanced by the ever-growing number of protein structures that continue to be deposited in the Protein Data Bank (PDB; wwPDB Consortium 2018).

Homology modeling rests on several well-established assumptions about how the physicochemical properties of proteins, conferred by their amino acid sequences, dictate their 3-dimensional structures under specific solution conditions: the protein folding problem (Dill et al. 2008). The salient assumptions for our purposes are as follows.

- Amino acid sequence determines secondary and tertiary structure of proteins under specific solution or cellular conditions.
- Amino acid sequences compatible with specific elements of structure (e.g., alpha helices, beta sheets, and the like) are not unique; multiple sequences can adopt the same secondary structures.
- Proteins within families share generalized homologous structure and simple amino acid sequence variation can determine functional details, such as ligand specificity.
- Evolutionarily related proteins (e.g., families) have similar three-dimensional structures and the three-dimensional structure is more highly conserved than are amino acid sequences. Hence, small or medium changes in sequence typically result in minimal changes in three-dimensional structure.
- Functional sites maintain nearly identical structural folds so that proteins of the same (or similar) function and/or family have similar structures, at least within functional domains.

The first step in homology modeling is to identify the template protein(s) of known structure(s). These are selected based on functional and/or family association with the target. Next, the amino acid sequences of the target and template(s) are aligned, based on sequence identity or similarity (i.e., identical or conservative substitution of amino acid residues; for example, a replacement of a glutamic acid by an aspartic acid residue may be considered a conservative substitution) among structurally conserved regions (e.g., domains) of the proteins. This process must account for variable regions, such as insertions, missing N- or C- termini, and other deletions, to align the amino acid sequences of the target protein and the template protein(s) properly. This process may take multiple rounds of alignment correction. Based on the optimized sequence alignment, model building begins, which today is done largely through established modeling programs (e.g., SWISS-MODEL; Waterhouse et al. 2018) and machine learning (Muhammed and Aki-Yalcin 2018; Hameduh et al. 2020). Model building begins with backbone modeling, followed by attempts to model loop regions and the addition of amino acid side chains. The initial model is then subjected to multiple rounds of iterative optimization, refinement, and validation. The quality of the final model will vary depending on the similarity of amino acid sequence between target and template(s). A homology model may not provide sufficient atomic detail to allow for *de novo* design of high-potency ligands (a topic we shall encounter in Chapter 4) but may be sufficient for initial ligand design approaches. However, for the current purpose of target validation, homology modeling is often sufficient to infer the presence of druggable binding pockets within a target protein, based on the presence of such binding pockets within the template protein(s).

3.3.4 Other Computational Methods

As computational hardware and software have evolved with time, so too has the sophistication of approaches for predicting the three-dimensional structure of proteins, based on sequence relatedness to other protein for which structural data are available. With recent advances in artificial intelligence (AI) applied to protein structural predictions, the community has seen a vast improvement in the predictive power of computation methods. These efforts have culminated in a highly accurate neural network–based modeling system referred to as AlphaFold 2 (Jumper et al. 2021). AlphaFold 2 uses a novel training process based on the evolutionary, physical, and geometric constraints of known protein structures and introduces a new architecture that jointly embeds multiple sequence alignments (MSAs) and pairwise features to produce accurate end-to-end structure predictions. Using this system, Jumper et al. (2021) were able to predict the three-dimensional structures of a set of 87 proteins that were selected to form the 14th Critical Assessment of Protein Structure Prediction (CASP14; Moult et al. 2020). Across this set of 87 proteins, the root mean square deviation (rmsd) of backbone (C_α) structure at 95% residue coverage was less than 0.96 Å. This is a substantial improvement in predictive power, relative to the next best prediction program (rmsd = 2.8 Å at 95% residue coverage; Jumper et al., 2021). Subsequently, Abramson et al. (2024) reported the use of AlphaFold 3 for accurate structure predictions of biomolecular interactions, including protein–ligand interactions. It will be interesting to see how these new approaches will be applied to target validation for drug discovery. Other computational methods for predicting protein–ligand binding have been reviewed by Zhao et al. (2020).

3.3.5 Target Vitality and Competition with Endogenous Ligands

The term target vitality was coined to quantify the threshold of residual target function below which a desired phenotypic outcome is realized. For example, suppose that one's goal is to kill specific cancer cells. The question becomes as follows: To what extent does one need to inhibit target function to elicit an antiproliferative effect? Is 50% inhibition sufficient, or does one need to inhibit the target by 70, 80, 90, or even 99% to affect an antiproliferative phenotype? This is not a consideration of target validation, per se, but it is nevertheless of practical importance to assess this, as far as possible at this early stage, to calibrate expectations for how challenging a particular target may be to modulate to a biologically meaningful extent. How might one assess this concern?

In favorable cases, a review of the existing literature on the disease may be informative. For example, in some diseases, a patient must be homozygous for a genetic alteration to present with symptoms, whereas others who are heterozygous for the alteration may be asymptomatic (i.e., are carriers of disease). This might suggest that a lower percent modulation of the genetically altered target might suffice to bring the patient to the pharmacological equivalent of asymptomatic heterozygosity. In other cases, one may be able to compare the phenotypic consequences of cellular target loss (by CRISPR, see above) versus target knockdown (by siRNA) to get some broad sense of target function vitality.

However, what would be optimal is a method by which one could systematically vary the degree of target protein expression and quantify the resulting functional output and phenotypic consequences. For example, Michaels et al. (2019) have reported a method for precise tuning of gene expression levels in mammalian cells using a library of microRNA (miRNA) response elements (MREs) with single-nucleotide resolution. The changes in nucleotide composition of the synthetic MREs result in variation in the repressive strength of the MRE. They then developed gene-selective panels of miRNA-silencing–mediated fine-tuners (miSFITs) with which they could precisely modulate the expression levels of multiple genes, including the T-cell co-inhibitory receptor programmed cell death 1 (PD-1). By titrating the level of PD-1 expression with the miSFIT panel, they were able to demonstrate a quantitative, Hill–Langmuir relationship between PD-1 expression level and T-cell response both *in vitro* and *in vivo*. A similar approach has also been reported by Mathis et al. (2021). To date, the application of these and related methods (e.g., auxin-inducible degron 2 technologies; Yesbolatova et al. 2020) for titration of target protein expression for the purpose of target validation has been quite limited. However, these methods provide a strong complement to current all-or-none approaches to regulation of target expression (e.g., siRNA and CRISPR) with which one could better assess the degree of target modulation required for phenotypic outcome.

As we shall see in subsequent chapters, there are many physiochemical properties that can affect the cellular permeability, solubility, and other aspects of cellular action of compounds. Hence, the quantitative relationship between cell-free biochemical/biophysical potency and cellular potency can be affected by several properties of the compound. Two target-related properties that also affect

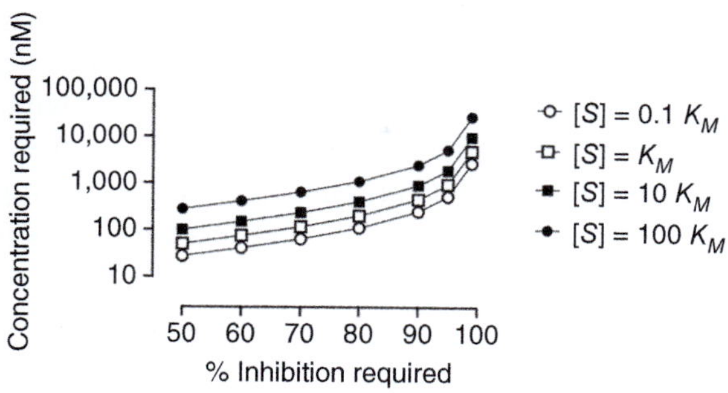

Figure 3.6 Percent inhibition required to reach a phenotypic threshold and the effect of intracellular substrate concentration ($[S]$), relative to K_M, on the concentration of compound required to achieve that phenotypic threshold.

this relationship are target vitality and target affinity for a competing endogenous ligand. Obviously, the greater percent modulation required to reach a phenotypic threshold, the greater the concentration of compound required to achieve that threshold. The concentration of compound required to achieve a specific level of target function modulation can be determined by rearrangement of the Hill–Langmuir equation that was presented in Chapter 2. Figure 3.6 illustrates this trend for a hypothetical enzyme inhibitor that displays a K_i of 25 nM in a cell-free assay of enzymatic activity. If this inhibitor happens to be competitive with an endogenous ligand, such as the substrate of an enzyme, the amount of compound required to reach the phenotypic threshold will then also depend on the intracellular concentration of substrate relative to the affinity of the enzyme target for that substrate (K_M, as described in Chapter 2). The effect of intracellular substrate concentration ($[S]$) and K_M value on the amount of compound required to reach the phenotypic threshold can be calculated by rearranging the Cheng–Prusoff equation for competitive inhibition (Equation 2.3). This trend is also captured in Figure 3.6 as the intracellular value of $[S]/K_M$ increases from 0.1 to 100.

As summarized in Figure 3.6, both the degree of target modulation required and any competition with endogenous ligands can lead to a requirement for higher concentrations of compound to reach a phenotypic threshold. In many cases, one may not have much information with which to assess either of these points at the early stage of target selection. However, where there is available information from the literature, these factors should be considered in prioritizing targets for further interest. Recall from Chapter 2 that this was a major hurdle in the discovery of GTP-competitive inhibitors of $KRAS^{G12C}$, which displays a K_M value of ca. 20 pM for GTP, whereas the intracellular concentration of GTP is about 500 μM ($[S]/K_M = 2.5 \times 10^7$). In contrast, there are many examples of successful drug discovery efforts targeting protein kinases with ATP-competitive inhibitors. In this case, the cellular concentration of ATP in human cells has been determined to be in the range of 700–1000 μM (see, for example, Copeland et al. 2005), but the K_M values for ATP among protein kinases are typically in the range of 10–100 μM ($[S]/K_M = 8$–100). Suppose that one was considering a protein as a potential target that is known to bind ATP as a critical part of its function. Based on the success of ATP-competitive inhibitors of protein kinases, we might be tempted to deem this target actionable, due to the presence of an ATP-binding pocket. If, however, the affinity of this target for ATP were much greater than a typical kinase, a focus on ATP-competitive inhibitors might prove more challenging than one might expect. For example, if the K_M for ATP of this target were 0.1 μM, the cellular $[S]/K_M$ would be about 8,000–10 000.

Neither required % inhibition nor high affinity for an endogenous ligand is a reason to disqualify a target for drug discovery. However, these properties may suggest that the researcher should consider different approaches to targeting such a protein. As described in Chapter 2, these could include a focus on noncompetitive or uncompetitive inhibitors, covalent inhibitors, PROTAC/glue approaches, and the like.

3.4 Unmet Medical Need

Usually, the researcher will have some general idea of their disease focus, prior to beginning target discovery. Hence, they are also likely to have some sense of the needs of the patient population for new medicines. This need may be because existing therapeutics are suboptimal with respect to, for example, efficacy, safety, route of administration, dosing interval, patient's refractory response to treatment, or the onset of resistance to treatment. A clear understanding of the benefits and limitations of existing medicines for a specific disease should be a critical part of the diligence performed prior to initiating target selection activities.

A secondary consideration is the size of the patient population that is impacted by this unmet need. Statistics on disease incidence and prevalence in developed countries are often available from multiple government or private organizations. For example, the American Cancer Society provides detailed statistics for cancer incidence and prevalence in the United States for most major cancer indications. Understanding what the addressable patient population is likely to be is important in making decisions on commercial value, time horizon for recouping research, development and manufacturing costs, and the potential relevance to special incentive programs from the FDA to encourage drug discovery and development in orphan and rare diseases.

As we have already seen, many therapeutic areas, such as cardiovascular disease, inflammatory disease, or cancer type (e.g., breast cancer, prostate cancer), reflect umbrella terms, representing a collection of diseases, rather than a single, genetically, or molecularly homogeneous disease. Hence, it is often the case that a target will be relevant to only a subset of patients, as potentially reflected by a selectively essential profile when the target is knockout across a broad collection of disease-relevant cell lines (e.g., as illustrated in Figure 3.4). Ultimately, the question must be addressed as to which patients are most likely to benefit from a treatment based on modulation of the specific target. This becomes a critical question later, as programs transition to clinical development. Generally, however, this question reduces to understanding the molecular and/or genetic determinants of cell sensitivity to target knockout, knockdown, or inhibition. The better one can define specific determinants of cell sensitivity to target modulation, the better one can design methods to identify patients that are likely to benefit from a molecular targeted therapy. This is the basis of the fields of translational biology and translational medicine, which we will touch upon in subsequent chapters.

At the early stage of target selection, we need not have a fully validated strategy for stratifying patients for clinical studies; this will be an evolving topic of research throughout the drug discovery process. Rather, what is needed with respect to target validation is to have cogent and testable hypotheses that can be further refined and modified during the drug discovery campaign. Let us revisit two examples from above to illustrate this concept.

As discussed above, heterozygous mutations of EZH2 are observed in about 22–30% of germinal center NHL patients. Sneeringer et al. (2010) demonstrated that the mutant EZH2 proteins, in concert with the WT EZH2, were drivers of oncogenesis. From these data, the foundational hypothesis for drug discovery was that NHL patients presenting with heterozygous mutations in EZH2 would be most likely to benefit from a therapy based on inhibition of EZH2. As the team began to identify potent, selective inhibitors of the enzyme, this hypothesis was confirmed in cell culture studies, comparing the antiproliferative effect of EZH2 inhibitors in NHL cells bearing heterozygous EZH2 mutations to other NHL cells that were homogeneous for the WT enzyme (Figure 3.7A). These studies ultimately led to the discovery of tazemetostat, a drug that was approved by FDA in 2020 for treatment of EZH2 mutant-bearing follicular lymphoma patients (Ho and Copeland 2024).

Earlier in this chapter, we saw that knockdown of WRN led to selective antiproliferative effects in MSI-H but not MSS cancers, such as CRC. In this case, we do not have a single, unique genetic determinant of drug sensitivity, as we did in the case of EZH2 mutations in NHL. Rather, here the foundational hypothesis is that patients that present with MSI-H cancers are most likely to benefit from a therapy based on inhibition of WRN. Based on this hypothesis, Ferretti et al. (2024) began drug discovery efforts toward this enzyme, resulting in compound HRO761. HRO761 was found to be a potent, selective, allosteric inhibitor of WRN. The effects of treatment with HRO761 on colony formation was tested in a collection of MSI-H ($n = 20$) and MSS ($n = 8$) cell lines. As shown in Figure 3.7B, the median EC_{50} for MSH-h cell lines was about 100 nM, although there was considerable variation across MSI-H cell lines, suggesting that other factors

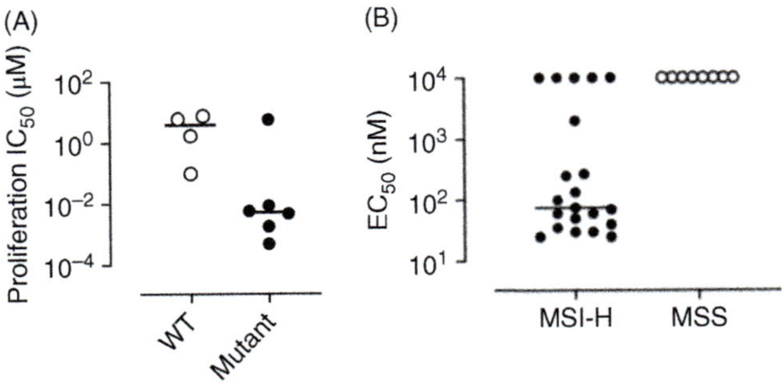

Figure 3.7 (A) Comparison of the IC_{50}s for inhibition of cell proliferation among NHL cells that are wild type for EZH2 (WT) or bear heterozygous mutations at tyrosine 641 (mutant) treated with the EZH2 inhibitor tazemetostat. The median value for each group is displayed as a horizontal line. *Source:* Adapted from Knutson et al. (2014). (B) Comparison of the EC_{50}s for inhibition of colony formation among MSI-H and MSS cell lines treated with the WRN inhibitor HRO761. Again, the median value for each group is displayed as a horizontal line. *Source:* Adapted from Ferretti et al. (2024).

besides MSI-H status may influence cell sensitivity to WRN inhibition. However, in stark contrast, none of the MSS cell lines responded to compound treatment at concentrations as high as 10 μM (i.e., 10 000 nM). Hence, as with the example of tazemetostat inhibition of EZH2, the cellular data with compound here support the foundational hypothesis for selective essentiality of WRN inhibitors for MSI-H tumors.

3.5 DHX9: An Illustrative Example of Target Validation

In this chapter, we have described various criteria by which to judge whether a target is appropriate for drug discovery. Let us now look at an example from the recent literature that illustrates how these various criteria may be integrated into a cogent target validation process. We shall illustrate this with the example of the enzyme DHX9 as a target for MSI-H cancers with dMMR, as recently reported by Castro et al. (2025).

The team at Accent Therapeutics was focused on a large collection of proteins involved in RNA modification (the RNA-modifying proteins or RMPs) as potential targets for small-molecule cancer therapeutics. Early in the history of this company, the team surveyed a publicly available database (DepMap) for the effects of individual CRISPR knockout of each of over 400 RMPs across about 450 different cancer cell lines, in a search for selectively essential targets among the RMPs. Among the RMPs demonstrating a selectively essential profile was the DExH-box

RNA helicase, DHX9 (Figure 3.8A). DHX9 is known to play important roles in regulating transcription, translation, and maintenance of genome stability, and DHX9 levels had been shown to be elevated in several cancer types, including CRC. DHX9 plays a critical role in resolving R-loops, DNA/RNA hybrid structures that are formed during nascent RNA hybridization with the DNA template strand in the process of gene transcription. The formation of R-loops leaves the non-template strand of the DNA single-stranded and therefore susceptible to DNA damage, transcription-associated mutagenesis, or transcription-associated recombination (Skourti-Stathaki and Proudfoot 2014). Left unresolved, these R-loops can also stall further RNA polymerase processing, leading to replication stress in the cell. If the level of such replication stress exceeds a critical threshold, the cell cannot survive and will enter the programmed cell death process (apoptosis).

In addition to displaying a selectively essential profile, DHX9 had previously been conditionally knocked down in adult mice (for six months) without any changes in animal health, body weight, blood biochemistry, or histology (Lee et al. 2016). Because DHX9 was known to be a helicase, the team knew that it contained an ATP-binding pocket as well as a channel for binding single-stranded RNA as part of its RNA unwinding (helicase) function. These data suggested that DHX9 was actionable as a small-molecule drug target and that inhibitors of the enzyme would likely demonstrate a favorable TI.

When Castro et al. (2025) reviewed the CRISPR knockout profile for this enzyme (Figure 3.8A), they noted that only a small subset of cancer cells were significantly affected by the knockout, and this subset was enriched for MSI-H CRC or endometrial cancer cells (colored blue in the waterfall plot of Figure 3.8A). This suggested that like WRN (see above), DHX9 inhibition might be selectively antiproliferative in MSI-H cancers. To test this hypothesis, Castro et al. used two siRNA constructs targeting DHX9 (constructs s4020 and s4021) and the corresponding C911 versions (see above) of these as negative controls (s4020C911 and s4021C911). They demonstrated the ability of each of the positive constructs to significantly knockdown DHX9 protein by Western blot analysis, as well as the lack of knockdown with the two corresponding C911-negative controls, in two MSI-H CRC cell lines (SNU407 and LS411N), and two MSS CRC cell lines (HT29 and NCI-H747), as shown in Figure 3.8B (left panel). Next, they treated these cells with the positive and negative siRNA constructs and looked at the effects of DHX9 knockdown on colony formation. As shown in the right panel of Figure 3.8B, knockdown of DHX9 ablated colony formation, but the C911-negative control constructs had no significant effect on colony formation, supporting the on-target effects of the two DHX9 siRNA constructs. They next tested the effects of DHX9 knockdown by s24021, on proliferation in cell culture (14 days) for a larger panel of four MSI-H (SNU407, LS411N, HCT116, and LS180) and four MSS (HT29, NCI-H747, Colo205,

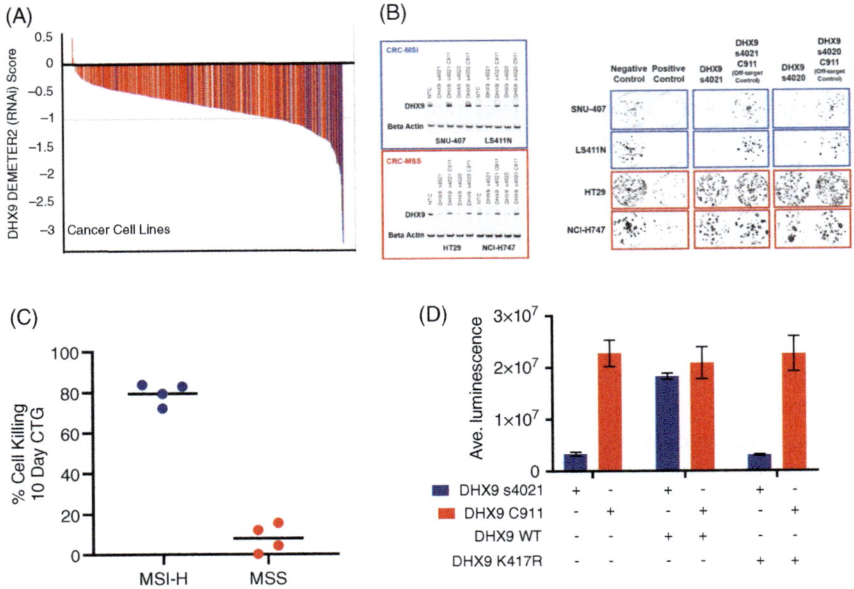

Figure 3.8 Target validation data for DHX9 as a selectively essential target for MSI-H cancers. (A) Waterfall plot demonstrating DHX9-dependency RNAi score as reported in DepMap across all cancer cell lines and colored by MSI-H and MSS status (MSI-H, blue; MSS, red; undetermined, gray). (B) Left: DHX9 expression in MSI-H (SNU407, LS411N, blue box) and MSS (HT29, NCI-H747, red box) CRC cells after exposure to two different DHX9 siRNA constructs (s4020 and s4021), a nontargeted control (NTC) or an off-target control (C911) ($n > 3$). Right: Colony formation assay demonstrating the sensitivity of selected MSI-H (blue boxes) and MSS (red boxes) CRC cell lines to DHX9 knockdown. ($n \geq 3$). (C) Quantification of percent cell killing as assessed by CellTiter-Glo (CTG) induced by DHX9 siRNA treated (s4021) in MSI-H (blue) and MSS (red) CRC cell lines ($n \geq 3$). (D) Quantification of proliferation effects of DHX9 knockdown followed by overexpression of WT DHX9 or DHX9-K417R in HCT116 cells. Error bars represent standard deviation ($n = 2$). *Source:* Castro et al. (2024) / with permission of American Association for Cancer Research.

and MDST8) CRC cell lines. Knockdown of DHX9 resulted in killing of about 80% of the MSI-H cells but less than 10% of the MSS cells were affected by the knockdown (Figure 3.8C). In aggregate, these data clearly support a critical role for DHX9 in the proliferation and survival of MSI-H cancer cells. The next question Castro et al. (2025) sought to answer was whether antiproliferative effects of DHX9 knockout/knockdown were the result of the loss of DHX9 enzymatic activity (helicase activity) or some other role of the protein in cell biology. To address this question, they knockdown endogenous DHX9 with s4021 in HTC116 cells and then attempted to rescue proliferation by overexpression of WT DHX9 or with a point mutant in the ATP-binding site of the enzyme, K417R, that would ablate the helicase activity of the protein. As shown in Figure 3.8D, relative to the C911 control, the siRNA treatment reduced proliferation to about 15% residual proliferation. Rescue with the WT DHX9 restored about 90% of control-level proliferation, whereas rescue with the K418R mutant was not able to restore proliferation at all. Thus, these data strongly implicate the helicase activity of DHX9 as critical for proliferation of MSI-H cancer cells. Finally, Castro et al. (2025) went on to test whether DHX9 knockdown would increase cellular levels of R-loops and thereby drive cells into apoptosis. Figure 3.9 illustrates immunofluorescence imaging of two cell lines (HTC116, MSI-H; and NCI-H747, MSS) cells with siRNA s4021 or the control C911 construct (labeled Negative Control in Figure 3.9) and labeling with 4′, 6-diamidino-2-phenylindole (DAPI) to visualize cell nuclei, with an anti-DHX9 antibody to visualize DHX9 protein, and with an antibody (S9.6) that recognizes RNA/DNA hybrids to visualize R-loop induction. The data clearly show induction of R-loops when DHX9 is knockdown in MSI-H cells, but not in MSS cells. Castro et al. (2025) additionally showed induction of several known markers of replication stress and DNA damage in MSI-H but not MSS cells, upon DHX9 knockdown. This induction of replication stress was followed by cell cycle arrest and subsequent apoptotic cell death for MSI-H cells (data not shown).

Based on the compelling target validation just described, Castro et al. (2025) initiated a drug discovery program against DHX9, including building a suite of biochemical and biophysical assays for DHX9 (Gotur et al. 2023) and solving the first crystal structures of mammalian forms of DHX9, including the cat and

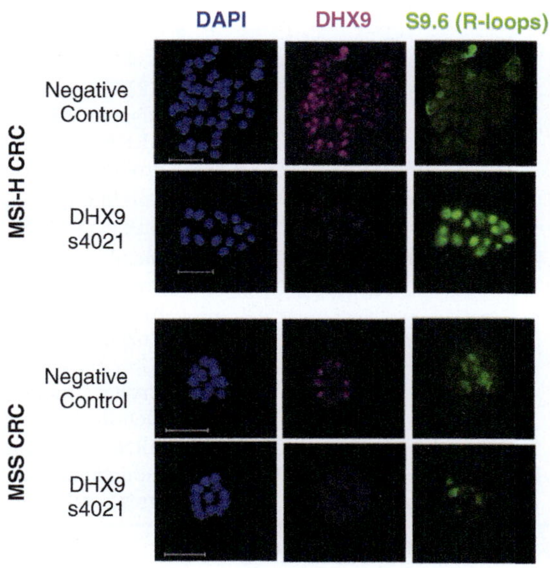

Figure 3.9 Immuno fluorescence imaging of HCT116 MSI-H (top) and NCI-H747 MSS (bottom) CRC cells treated with DHX9 siRNA (s4021) or negative (off-target s4021C911) control for three days (DAPI [nuclei], blue; DHX9, pink; S9.6 [R-loops], green. *Source:* Castro et al. (2025) / American Association for Cancer Research / CC BY-NC-ND 4.0.

human enzymes, with ADP bound (Lee et al. 2023). In their paper (Castro et al. 2025), they also reported the discovery of tool compound ATX968, a potent and selective inhibitor of DHX9 that recapitulated all the effects described for DHX9 knockdown. ATX968 selectively killed MSI-H cancer cells *in vitro* and demonstrated robust tumor growth inhibition in mice *in vivo* xenograft studies of MSI-H CRC models. Subsequent cell line screening with ATX968 demonstrated that in addition to MSI-H cancers, the DHX9 inhibitor demonstrated potent antiproliferative activity in other cancer indications exhibiting increased replication stress and genomic instability, including BRCA1/2-altered breast cancer (Copeland et al. 2024). Thus, unlike WRN inhibitors, DHX9 inhibitors may find utility in several cancer indications beyond MSI-H tumors. Ultimately, the DHX9 program identified a more potent and pharmacologically tractable novel DHX9 inhibitor which has recently entered phase 1 clinical testing (clinicaltrials.gov, NCT06625515).

Table 3.3 summarizes the target validation data presented by Castro et al. (2025) or referenced in their paper, to illustrate how each of the target validation criteria we have discussed in this chapter was systematically addressed in this example.

Table 3.3 Summary of the target validation for DHX9.

Target Validation Criterion	Supporting Data
Disease relevance	• Amplification in certain cancers. • CRISPR knockout shows selective essential profile.
Selective essentiality	• CRISPR knockout shows selective phenotypic effect in MSI-H. • siRNA effects colony formation and cell proliferation selectively in MSI-H CRC.
Therapeutic index	• Conditional knockdown of DHX9 in adult mice (Lee et al. 2016).
Actionable target	• Member of helicase target class. • Phenotypic consequence of knockdown rescued by WT but not catalytically inactive enzyme. • Enzymatic activity dependent on ATP hydrolysis (Gotur et al. 2023). • Crystal structure solved with ADP-binding site identified (Lee et al. 2023).
Target modulation associated with biological impact associated with phenotypic outcome	• siRNA knockdown leads sequentially to R-loop induction, induction of markers of replication stress, cell cycle arrest, and apoptosis.

Source: Adapted from Castro et al. (2024).

3.6 Summary

In this chapter, we have defined a set of four key elements of target validation that must be met to qualify a particular target as appropriate for drug discovery. These broad elements can be categorized as follows: (1) disease relevance, (2) selectivity or differential essentiality to the disease state, (3) actionability by specific therapeutic modalities (e.g., small-molecule drugs), and (4) fulfillment of a meaningful unmet need of patients. For pedagogic purposes, we described methods for assessing each of these criteria separately, as if they were independent and discrete criteria. However, as we have seen, in practice target validation is an integrated process and clear demarcation between criteria may be lacking for some targets. After discussing each criterion in turn, we ended the chapter with an example of how these are integrated to form a cogent target validation process, using the recent studies of the cancer target DHX9 as an example (Castro et al., 2024).

As we have discussed, target validation is an ongoing process that does not end until the pharmacological action and safety of a drug has been demonstrated in human patients. Nevertheless, the integrated criteria that form our target validation process are sufficient to qualify a specific target for drug discovery and to justify the expenditure of time, money, and effort required to launch a drug discovery program. Once a target has qualified as a valid target for drug discovery, the next step is to identify compounds that modulate the function of the target as a starting point for medicinal chemistry. These starting points are commonly referred to as "hits". Hence, we next turn our attention to the process of hit-finding and hit-validation in Chapter 4.

References

Abramson, J., Adler, J., Dunger, J., Evans, R. et al. (2024) Accurate structure prediction of biomolecular interactions with AlphaFold3. *Nature* **630**, 493–500.

Barnash, K.D., James, L.I. and Frye, S.V. (2017) Target class drug discovery. *Nat. Chem. Biol.* **13**, 1053–1066.

Beijersbergen, R.L., Wessels, L.F.A. and Bernards, R. (2017) Synthetic lethality in cancer therapeutics. *Ann. Rev. Cancer Biol.* **1**, 141–161.

Bockamp, E., Sprengel, R., Eshkind, L., Lehman, T., Braun, J.M., Emmrich, F. and Hengstler, J.G. (2008) Conditional transgenic mouse models: from basics to genome-wide sets of knockouts and current studies of tissue regeneration. *Regenerative Med.* **3**, 217–235.

Boschi, D., Pippione, A.C., Sainas, S. and Lolli, M.L. (2019) Dihydroorotate dehydrogenase inhibitors in anti-infective drug research. *Eur. J. Med. Chem.* **183**, 111681.

Bridges, C.B. (1922) The origin of variation. *Am. Nat.* **56**, 51–63.

Buehler, E., Khan, A.A., Marine, S., Rajaram, M. et al. (2012) siRNA off-target effects in genome-wide screens identify signaling pathway members. *Sci. Rep.* **2**, 428.

Cabral, K.P. and Ansell, J.E. (2015) The role of factor Xa inhibitors in venous thromboembolism treatment. *Vasc. Healt Risk Manage.* **30**, 117–123.

Carpenter, K.A. and Altman, R.B. (2024) Databases of ligand-binding pockets and protein-ligand interactions. *Comput. Struct. Biotechnol. J.* **23**, 1320–1338.

Castro, J., Daniels, M.H., Brennan, D., Johnston, B. et al. (2025) A potent, selective small-molecule inhibitor of DHX9 abrogates proliferation of microsatellite instable cancers with deficient mismatch repair. *Cancer Res.* **85**, 758–776.

Claussnitzer, M., Cho, J.H., Collins, R., Cox, N.J. et al. (2020) A brief history of human disease genetics. *Nature* **577**, 179–189.

Copeland, R.A. (2013) *Evaluation of Enzyme Inhibitors in Drug Discovery: A Guide for Medicinal Chemists and Pharmacologists.* 2nd Edition. Wiley, Hoboken, NJ.

Copeland, R.A. (2021) Chance favors the perplexed mind: the critical role of mechanistic biochemistry in drug discovery. *Biochemistry* **60**, 2275–2284.

Copeland, R.A. (2023) *Enzymes: A Practical Introduction to Structure, Mechanism and Data Analysis.* 3rd Edition. Wiley, Hoboken, NJ.

Copeland, R.A., Marcinkeviciene, J., Haque, T.S., Kopcho, L.M. et al. (2000) Helicobacter pylori-selective antibacterials based on inhibition of pyrimidine biosynthesis. *J. Biolumin. Chemilumin.* **275**, 33373–33378.

Copeland, R.A., Luo, L., Auger, K.R. and Huang, P.S. (2005) A method for determining intracellular concentrations of enzyme substrates from a combination of competitive inhibition and mutagenesis studies. *Anal. Biochem.* **337**, 351–353.

Copeland, R. A., Castro, J., Daniels, M. H., Gotur, D. Lee, Y.T., Yao, S., Brennan, D., Johnston, B.T., Laidlaw, M., Pai, S. and Wu, J. (2024) Small molecule inhibitors of RNA modifying enzymes as precision cancer therapeutics. *115th AACR Annual Meeting*, Abstract 3908.

Daigle, S.R., Olhava, E.J., Therkelsen, C.A., Majer, C.R. et al. (2011) Selective killing of mixed lineage leukemia cells by a potent small-molecule DOT1L inhibitor. *Cancer Cell* **20**, 53–65.

Dang, L., White, D.W., Gross, S., Bennett, B.D. et al. (2009) Cancer-associated IDH1 mutations produce 2-hydroxyglutamate. *Nature* **462**, 739–744.

Dill, K.A., Ozkan, S.B., Shell, M.S. and Weikl, T.R. (2008) The protein folding problem. *Annu. Rev. Biophys.* **37**, 289–316.

Dobzhansky, T. (1946) Genetics of natural populations. XIII. recombination and variability in populations of *Drosphila pseudoobscura*. *Genetics* **31**, 269–290.

Echeverri, C.J., Beachy, P.A., Baum, B., Boutros, M. et al. (2006) Minimizing the risk of reporting false positives in large-scale RNAi screens. *Nat. Methods* **3**, 777–779.

Faraoni, I. and Graziani, G. (2018) Role of BRCA mutations in cancer treatment with poly(ADP-ribose) polymerase (PARP) inhibitors. *Cancers* **10**, 487.

Ferretti, S., Hamon, J., de Kanter, R., Scheufler, C. et al. (2024) Discovery of WRN inhibitor HRO761 with synthetic lethality in MSI cancers. *Nature* **629**, 443–449.

Fire, A., Xu, S., Montgomery, M.K., Kostas, S.A., Driver, S.E. and Mello, C.C. (1998) Potent and specific genetic interference by double-stranded RNA in C elegans. *Nature* **391**, 806–811.

Flaherty, K.T., Infante, J.R., Daud, A., Gonzalez, R. et al. (2012) Combined BRAF and MEK inhibition in melanoma with BRAF V600 mutations. *N. Engl. J. Med.* **367**, 1694–1703.

Gotur, D., Case, A., Liu, J., Sickmier, E.A., Holt, N., Knockenhauer, K.E., Yao, S., Lee, Y.T., Copeland, R.A., Buker, S.M. and Boriack-Sjodin, P.A. (2023) Development of assays to support identification and characterization of modulators of DExH-box helicase DHX9. *SLAS Discov.* **28**, 376–384.

Hameduh, T., Haddad, Y., Adam, V. and Heger, Z. (2020) Homology modeling in the time of collective and artificial intelligence. *Comput. Struct. Biotechnol. J.* **18**, 3494–3506.

Hannon, G.J. (2002) RNA interference. *Nature* **418**, 244–251.

Haque, T.S., Tadesse, S., Marcinkeviciene, J., Rogers, M.J., Sizemore, C., Kopcho, L.M., Amsler, K., Ecret, L.D., Zhan, D.L., Hobbs, F., Slee, A., Trainor, G.L., Stern, A.M., Copeland, R.A. and Combs, A.P. (2002) Parallel synthesis of potent, pyrazole-based inhibitors of Helicobacter pylori dihydroorotate dehydrogenase. *J. Med. Chem.* **45**, 4669–4678.

Hartwell, L.H., Szankasi, P., Roberts, C.J., Murraay, A.W. and Friend, S.H. (1997) Integrating genetic approaches into the discovery of anticancer drugs. *Science* **278**, 1064–1068.

Ho, P.T.C., Copeland, R.A., Chan, H.M. and Arrowsmith, C.H. (2024) Discovery and development of tazemetostat. In *Epigenetic Drug Discovery* (H.M. Chan and C.H. Arrowsmith, Eds.). Royal Society of Chemistry, London, UK, pp. 563–595.

Hsu, P.D., Lander, E.S. and Zhang, F. (2016) Development and application of CRISPR-Cas9 for genome engineering. *Cell* **157**, 1262–1278.

Jumper, J., Evans, R., Pritzel, A., Green, T. et al. (2021) Highly accurate protein structure predictions with AlphaFold. *Nature* **596**, 583–589.

Kaelin, W.G. Jr., (1999) Choosing anticancer drug targets in the postgenomic era. *J. Clin. Invest.* **104**, 1503–1506.

Kaelin, W.G. Jr., (2005) The concept of synthetic lethality in the context of anticancer therapy. *Nat. Rev. Cancer* **5**, 689–698.

Kaelin, W.G. Jr., (2017) Common pitfalls in preclinical cancer target validation. *Nat. Rev. Cancer* **17**, 441–450.

Kategaya, L., Perumal, S.K., Hager, J.H. and Belmont, L.D. (2019) Werner syndrome helicase is required for the survival of cancer cells with microsatellite instability. *iScience* **13**, 488–497.

Knutson, S.K., Warholic, N.M., Wigle, T.J., Klaus, C.R., Allain, C.J., Raimondi, A., Porter, S.M., Chesworth, R., Moyer, M.P., Copeland, R.A., Richon, V.M., Pollock, R.M., Kuntz, K.W. and Keilhack, H. (2013) Durable tumor regression in genetically altered malignant rhabdoid tumors by inhibition of methyltransferase EZH2. *PNAS* **110**, 7922–7927.

Knutson, S.K., Kawano, S., Minoshima, Y., Warholic, N.M., Huang, K.C., Xiao, Y., Kadowaki, T., Uesugi, M., Kuznetsov, G., Kumar, N. and Wigle, T.J. (2014) Selective inhibition of EZH2 by EPZ-6438 leads to potent antitumor activity in EZH2-mutant non-Hodgkin lymphoma. *Mol. Cancer Res.* **13**, 842–854.

Lassus, P., Rodriguez, J. and Lazbnik, Y. (2002) Confirming specificity of RNAi in mammalian cells. *Science STKE* **147**, pI13.

Lee, T., Paquet, M., Larsson, O. and Pelletier, J. (2016) Tumor cell survival dependence on the DHX9 DExH-box helicase. *Oncogene* **35**, 5093–5105.

Lee, Y.T., Sickmier, E.A., Grigoriu, S., Castro, J. and Boriack-Sjodin, P.A. (2023) Crystal structures of the DExH-box RNA helicase DHX9. *Acta Crystallogr. D Struct. Biol.* **79**, 980–991.

Liu, S., Neidhardt, E.A., Grossman, T.H., Ocain, T. and Clardy, J. (2000) Structures of human dihydroorotate dehydrogenase in complex with antiproliferative agents. *Structure* **8**, 25–33.

Lu, Q., Livi, G.P., Modha, S., Yusa, K., Macarrón, R. and Dow, D.J. (2017) Applications of CRISPR genome editing technology in drug target identification and validation. *Expert Opin. Drug Discovery* **12**, 541–552.

Mathis, A.D., Otto, R.M. and Reynolds, K.A. (2021) A simplified strategy for titrating gene expression reveals new relationships between genotype, environment, and bacterial growth. *Nucleic Acids Res.* **49**, e6.

Meyers, R.M., Bryan, J.G., McFarland, J.M., Weir, B.A., Sizemore, A.E., Xu, H., Dharia, N.V., Montgomery, P.G., Cowley, G.S., Pantel, S. and Goodale, A. (2017) Computational correction of copy number effect improves specificity of CRISPR-Cas9 essentiality screens in cancer cells. *Nat. Genet.* **49**, 779–784.

Michaels, Y.S., Barnkob, M.B., Barbosa, H., Baeumler, T.A., Thompson, M.K., Andre, V., Colin-York, H., Fritzsche, M., Gileadi, U., Sheppard, H.M. and Knapp, D.J. (2019) Precise tuning of gene expression levels in mammalian cells. *Nat. Commun.* DOI: 10.1038/s41467-019-0877-y.

Mills, G.B., Lu, Y., Fang, X., Wang, H., Eder, A., Mao, M., Swaby, R., Cheng, K.W., Stokoe, D., Siminovitch, K., Jaffe, R. and Gray, J. (2001) The role of genetic abnormalities of PTEN and the phosphatidylinositol 3-kinase pathway in breast and ovarian tumorigenesis, prognosis, and therapy. *Semin. Oncol.* **28**, 125–141.

Moult, J., Fidelis, K., Kryshtafovych, A., Schwede, T. & Topf, M. (2020) Critical assessment of techniques for protein structure prediction, fourteenth round. CASP 14 Abstract Book https://www.predictioncenter.org/casp14/doc/CASP14_Abstracts.pdf.

Muhammed, M.T. and Aki-Yalcin, E. (2018) Homology modeling in drug discovery: overview, current applications, and future perspectives. *Chem. Biol. Drug Discov.* **93**, 12–20.

Ngoi, N.Y.L., Gallo, D., Torrado, C., Nardo, M., Durocher, D. and Yap, T.A. (2024) Synthetic lethal strategies for the development of cancer therapeutics. *Nat. Rev. Clin. Oncol.* DOI: 10.1038/s41571-024-00966-z.

Nijilman, S.M.B. (2011) Synthetic lethality: general principles, utility and detection using genetic screens in human cells. *FEBS Lett.* **585**, 1–6.

Skourti-Stathaki, K. and Proudfoot, N. (2014) A double-edged sword: R loops as threats to genome integrity and powerful regulators of gene expression. *Genes Dev.* **28**, 1384–1396.

Slamon, D.J., Clark, G.M., Wong, S.G., Levin, W.J., Ullrich, A. and McGuire, W. (1987) Human breast cancer: correlation of relapse and survival with amplification of the HER-2/neu oncogene. *Science* **235**, 177–182.

Sneeringer, C.J., Scott, M.P., Kuntz, K.W., Knutson, S.K., Pollock, R.M., Richon, V.M. and Copeland, R.A. (2010) Coordinated activities of wild-type plus mutant EZH2 drive tumor-associated hypertrimethylation of lysine 27 on histone H3 (H3K27) in human B-cell lymphomas. *PNAS* **107**, 20980–20985.

Sperger, J.M., Chen, X., Darper, J.S., Antosiewicz, J.E., Chon, C.H., Jones, S.B., Brooks, J.D., Andrews, P.W., Brown, P.O. and Thomson, J.A. (2003) Gene expression patterns in human embryonic stem cells and human pluripotent germ cell tumors. *PNAS* **100**, 13350–13355.

Sun, J., Wei, Q., Zhou, Y., Wang, J., Liu, Q. and Xu, H. (2017) A systematic analysis of FDA-approved anticancer drugs. *BMC Syst. Biol.* **11**, 87.

Waterhouse, A., Bertoni, M., Bienert, S., Studer, G., Tauriello, G., Gumienny, R., Heer, F.T., de Beer, T.A.P., Rempfer, C., Bordoli, L. and Lepore, R. (2018) SWISS-MODEL: homology modeling of protein structures and complexes. *Nucleic Acids Res.* **46**, W296–W303.

Weinberg, R.A. (2007) *The Biology of Cancer*. Garland Science, New York.

Wilson, B.G., Wang, X., Shen, X., McKenna, E.S., Lemieux, M.E., Cho, Y.J., Koellhoffer, E.C., Pomeroy, S.L., Orkin, S.H. and Roberts, C.W. (2010) Epigenetic antagonism between polycomb and SWI/SNF complexes during oncogenic transformation. *Cancer Cell* **18**, 316–328.

Wong, P.C., Pinto, D.J.P. and Zhang, D. (2011) Preclinical discovery of apixaban, a direct and orally bioavailable factor Xa inhibitor. *J. Thromb. Thrombolysis.* **31**, 478–492.

wwPDB Consortium (2018) Protein data bank: the single global archive for 3D macromolecular structure data. *Nucleic Acids Res.* **47**, D520–D528.

Yesbolatova, A., Saito, Y., Kitamoto, N., Makino-Itou, H., Ajima, R., Nakano, R., Nakaoka, H., Fukui, K., Gamo, K., Tominari, Y. and Takeuchi, H. (2020) The auxin-inducible degron 2 technology provides sharp degradation control in yeast, mammalian cells, and mice. *Nat. Commun.* **11**, 5701.

Zhao, J., Cao, Y. and Zhang, L. (2020) Exploring the computational methods for protein-ligand binding site prediction. *Comput. Struct. Biotechnol. J.* **18**, 417–428.

4

Hit-Finding and Hit-Validation

Key Learning Points

- The term "hit" refers to a compound, identified by screening against a target, that demonstrates some potency for binding to and/or modulating function of the target; hits thus constitute potential starting points for drug discovery.
- Several approaches to hit-finding are used in modern drug discovery. Of these, the most common approach is diversity library screening, in which a large collection (typically 10^6–10^9 compounds) of structurally diverse small molecules are tested. The quality of the library, the recombinant protein construct, and the assay format is critical to the success of the screening effort.
- Primary hits are tested in a battery of assays to assess:
  - target affinity
  - reproducibility in the primary and orthogonal assays
  - lack of reactive moieties or otherwise undesirable properties
  - verification of chemical structure and purity
  - pharmacological tractability
  - preliminary SAR
  - preliminary compound stability

  Only compounds that pass this battery of tests are considered validated hits.
- Most often validated hits display modest target affinity, limited selectivity, and limited pharmacological properties. Optimization of these and other properties of the hit(s) becomes the primary focus of medicinal chemistry efforts in subsequent hit-to-lead and lead-optimization campaigns.

Molecular Targeted Drug Discovery: A Guide to How Modern Medicines are Created,
First Edition. Robert A. Copeland.

- Although cellular screening has been successful in isolated cases, the vast majority of modern hit-finding efforts are performed with purified, recombinant target proteins, using primary assays of compound binding or modulation of target function.

Small-molecule targeted drug discovery begins with the identification of compounds that bind to and modulate the function of specific targets through a pharmacologically tractable mechanism and whose activity can be optimized through medicinal chemistry studies of structure–activity relationship (SAR). In this chapter, we review hit-finding strategies and systematic studies aimed at validating hits as true chemical starting points for drug discovery. To begin hit-finding, we must first understand and define properties that need to be incorporated into an optimized drug that is destined for the clinic. These integrated properties are rarely found in initial hit compounds. Hence, we must also define the minimum criteria needed to declare a compound as a validated hit, worthy of further drug discovery efforts. We begin our exploration of hit-finding and hit-validation with these two definitions. As a reminder, in this and subsequent chapters, we shall exemplify criteria and methods for small-molecule, reversible target modulators with the understanding that some of these criteria may be relaxed for other targeted modalities, as described previously in Chapter 2.

4.1 Properties of Small-Molecule Therapeutics

We have already discussed the physicochemical properties of drug-binding pockets on target proteins, and these properties have been reviewed in greater detail elsewhere (Copeland, 2013, 2023 and references therein). One can next logically ask the question, what are the physicochemical properties of therapeutics that would bind to such pockets. For many small-molecule therapeutics, the optimal route of administration is oral ingestion, and to be effective, the drug must be stable in systemic circulation for some reasonable amount of time. Combining these requirements, several authors have surveyed successful small-molecule drugs and developed the property guidelines summarized in Table 4.1 (Lipinski et al., 2001; Veber et al., 2002; Keller et al., 2006).

The values in Table 4.1 are generally considered to be reasonable guidelines for medicinal chemists, but they should not be viewed as inviolate rules. Many

Table 4.1 Physicochemical properties of drug-like small molecules.

Molecular property	Typical value
Molecular weight	<500 Da
cLog(*P*)[a]	<5
Number of H-bond donors	<5
Sum of N and O atoms	<10
Polar surface area	<140 Å
Rotatable bonds	<10

a) cLog(*P*) is the calculated base-10 logarithm of *P*, where *P* is the partition coefficient determined as the ratio of the equilibrium concentrations of a compound in 1-octanol over that in water.
Source: Copeland (2013) / with permission of John Wiley & Sons.

FDA-approved drugs deviated from one or more of these guidelines. Nevertheless, once validated hits have been identified and chemists begin the process of SAR-based optimization, these guidelines are generally referred to in establishing some of the criteria for declaring a lead compound (Chapter 5) and a development candidate compound (Chapter 6). At the hit-finding stage, one need not be too rigidly concerned about these guidelines, but it does make sense to consider these as part of hit prioritization (see below).

4.2 Hit-Validation Criteria

Prior to beginning any hit-finding campaign, it is important for the research team to agree prospectively to a minimum set of criteria for declaring hits as validated. The set of criteria and their minimum value vary from group to group, and there are no generally accepted standards for this. However, as an example, Table 4.2 provides hit-validation criteria that have proved useful at several biotechnology companies with which the author has been involved as a research leader or scientific advisor. We shall discuss each of these criteria further throughout the remainder of this chapter. They are presented here to give some sense of the appropriate stringency for hits to be considered validated for further use in drug discovery.

Table 4.2 Some typical hit-validation criteria.

General category	Specific property	Minimum criteria
Target interactions		
	Reproducible target modulation in primary assay	• Reproducible % modulation at screening concentration ($n > 2$)
	Biochemical or biophysical IC_{50} or K_d	• <10 μM with reproducibility within 3-fold
	Ligand efficiency[a)]	• >0.3
	Activity reproduced in orthogonal assay(s)	• Activity confirmed in ≥1 orthogonal biochemical or biophysical assay supporting specific, on-target binding/modulation
	Mechanism of action	• Tractable mechanism of action verified experimentally for at least 1 compound per series
	Structural biology	• Structure solved of ≥1 compound per series bound to target and binding site thereby identified (preferred but not required) • Absent structure, experimentally supported hypothesis of compound-binding site
False-positive tests		
	Interference/filtering assays	• No or minimal (10X differential) activity in assays of interference specific to assay detection method • No activity in assay of redox activity
	Chemical alerts	• No PAINS[b)]structures • No evidence of bad actors or promiscuous activity[c)] • No evidence of compound instability in assay • No heavy metal content
Chemistry		
	Structure and purity confirmed	• Initial structure and purity confirmed by liquid chromatography/mass spectroscopy (LC/MS) • Chemical structure of ≥1 compound per series confirmed by resynthesis
	Pharmacophore identified through initial SAR	• Hypothesis of pharmacophore developed and supported by preliminary SAR with >10-fold change in potency from founder compound

Table 4.2 (Continued)

General category	Specific property	Minimum criteria
	Intellectual property	• Preliminary patent search suggests freedom to operate
	Synthetic route	• Clear approach to compound synthesis to support SAR
Absorption, distribution, metabolism, excretion (ADME) and physical properties		
	Solubility	• Initial measurement indicates solubility >10-fold greater than EC_{50} and thought to be optimizable through SAR
	in vitro microsomal stability	• Measured and thought to be optimizable through SAR (preferred but not required)
	in vitro Caco-2 permeability	• Measured and thought to be optimizable through SAR (preferred but not required)

a) Ligand efficiency (L.E.) is defined as the free energy of ligand binding divided by the number of non-hydrogen atoms (N_{nonH}) in the compound: L. E. = $\Delta G_{\text{binding}}/N_{\text{nonH}}$. Where $\Delta G_{\text{binding}} = RT\ln(K_d)$, where R is the ideal gas constant, T is expressed in Kelvin (K) and K_d is expressed in molar units. The free energy of binding is typically expressed in units of kcal/mol·K, or in the S.I. unit of J/mol·K, (where K is Kelvin).

b) PAINS = pan-assay interference compounds; see below.

c) see below.

4.3 Screening Strategies

Many strategies for identifying pharmacologically active compounds have been attempted over the history of drug discovery. As described in Chapter 1, one of the first attempts to systematically screen compounds for pharmacological activity was made by Ehrlich and Hata (1910) by injecting individual compounds into syphilis-infected rabbits. The 606th compound to be thus tested proved effective and was subsequently marketed as Salvarsan to treat syphilis in humans. As one can imagine, this was a heroic and laborious effort. Modern, systematic drug discovery generally requires assay systems that can typically screen thousands to millions of compounds under highly controlled conditions, which can be rapidly processed and are generally amenable to robotic assay performance.

Attempts have been made to adapt cellular phenotypic (Zheng et al., 2013) and cellular biochemical assays (Seiffert et al., 2000) for compound screening, with some success in isolated instances. These methods, however, have several drawbacks. First, only hits with high levels of cell permeability will score in such assays, and this has the potential for missing many reasonable starting-point compounds that could be subsequently optimized for cell permeability. Second, it is often difficult in phenotypic assays to know if one is affecting a single target or multiple targets that effect the same phenotypic response. This can be quite confounding for developing cogent SAR hypotheses. For these and other reasons, the vast majority of novel, targeted drugs (i.e., first-in-class drugs) have their genesis in screening with cell-free, *in vitro* assays of target binding or modulation of target function (see, e.g., Eder et al., 2014; Southey and Brunavs, 2023). Hence, we will focus our attention for the remainder of this chapter on cell-free biochemical and biophysical assay formats using recombinant target proteins.

4.4 Protein Constructs for Screening

The goal of compound screening is to identify starting points for drug discovery that will ultimately demonstrate modulation of pathophysiological (e.g., intracellular) target function. Hence, the choice of protein construct to be used for screening is critical. In a perfect world, one would wish to screen with a full-length version of the target protein in the presence of any cell-relevant auxiliary proteins, cofactors, and the like, to recapitulate the intracellular form and environment of the target. This ideal, however, must be balanced with practical considerations of recombinant protein yield, stability, functionality, and the like.

Whenever practical, one should use the most full-length protein construct possible for screening (Copeland, 2013). When this proves impossible, one may need to experiment with different constructs of varying length to find the optimum construct for target function, stability, yield, and other factors that recapitulate, as best as possible, the pathophysiological form of the target. Let us say that the target of interest is a protein kinase. In some cases, just expressing the kinase domain of the target may be sufficient. In other cases, however, one may need to express constructs that contain other domains or loops of the proteins for optimal activity. In yet other cases, the full-length kinase may contain an auto-inhibitory domain that would need to be removed from the screening construct to reveal the cryptic catalytic activity. In still other cases, the protein may require specific posttranslational modifications to realize full activity. Hence, one may need to compare the specific functional activity of a target protein expressed in different hosts (e.g., *Escherichia coli*, insect cells, and mammalian cells) that provide different endogenous systems for posttranslational modification of expressed proteins. In favorable cases, specific posttranslational modifications may be mimicked by amino acid substitutions within the target protein. For example, glutamic acid can sometimes act to mimic phospho-serine, and aspartic acid can sometimes act to mimic phospho-threonine. Hence, replacement of a serine or threonine with glutamic or aspartic acid, respectively, provides a negative charge that often eliminates the need for posttranslational phosphorylation of these amino acid sites. All these factors must be taken into consideration in designing active protein constructs for screening, and often one will need to experiment with multiple constructs to identify the optimal form for screening.

As we discussed in Chapter 2, in some cases, the function of a target protein may depend on forming dimers or higher-order oligomers with itself (homo-oligomer) or other proteins (hetero-oligomer). This must also be explored in optimizing protein constructs and constituents for optimal activity and screening conditions. The state of oligomerization can be influenced by the protein concentration, and this must therefore be explored experimentally (see below) for optimal screening conditions.

With respect to recapitulating the pathophysiological form of the target, how would one know that they have identified an optimal recombinant protein construct? Unfortunately, one usually only knows this retrospectively by comparing the SAR trends of a large series of compounds, in the cell-free target assay and in a cellular assay of target function (pharmacodynamics), and/or phenotypic consequence (we shall discuss this point further in Chapters 5 and 6). However, there may be existing data (e.g., from the literature) on the pathophysiological form of the target and its interactions with natural ligands. Again, using the example of a protein kinase target, there may be existing information from intracellular or cell-extract studies that quantify the K_M for ATP, the K_M for the protein (or peptide, see below) substrate, the K_i for ADP, and the k_{cat} value of the kinase target. In such a case, one could compare these values to the corresponding values obtained with different recombinant constructs of the target kinase, as measures of similarity with the pathophysiological target. In a like manner, in favorable cases, there may be existing tool compounds available that can be tested in cells or cell extracts, and the EC_{50} or K_d values thus obtained can be compared with those of the recombinant target constructs.

The considerations just discussed for target proteins also apply to any auxiliary proteins needed for activity, and for protein-based ligands to be used in screening. Again, using protein kinases as an example, the two physiological substrates for these enzymes are ATP and a specific protein(s) that is the acceptor(s) of phosphorylation. When practical, one should use the full-length version of a protein ligand or auxiliary protein. For many kinases, proteases, and other enzymes that act on protein substrates, it is often more convenient to use a smaller peptide as a surrogate for the full-length protein. However, in such cases, it is critical to demonstrate that the peptide does indeed capture the relevant features of the full-length substrate/ligand. Often one finds that a peptide will bind the target protein with lower affinity (higher K_d) than the full-length protein ligand. This does not necessarily disqualify the peptide as a useful surrogate, but one must take this difference into account, especially later in hit-to-lead (Chapter 5) and lead optimization (Chapter 6), where one may be quantitatively comparing compound effects between cell-free biochemical/biophysical, and cellular assays. A useful litmus test for the appropriateness of peptide or truncated protein ligands is to ensure that these smaller ligands can displace the full-length protein ligand from the target, albeit at much higher concentrations of the former due to any differences in K_d.

4.5 Compound Sources for Screening

Multiple sources of compounds for screening have been envisaged over the history of drug discovery. Here we shall introduce some of the more common compound sources that are used in modern drug discovery. Depending on resources, timelines, and success in finding validated hits, a single compound source may be used for hit-finding, but in other cases, several of these sources may be combined or sequentially screened.

4.5.1 Diversity Libraries

Diversity libraries represent the most common source of compounds for screening. Today, commercially available and private diversity libraries typically contain from 100,000 to over 1 million compounds of diverse chemical structures. In the early days of high-throughput screening (HTS) in drug discovery, diversity libraries were assembled from organic molecules of chemistry industry sources indiscriminately – with little regard to drug-likeness, reactive moieties (Rishton, 1997), and other problematic structural elements (referred to collectively as "bad actors", see below). Over time, however, libraries have become better curated to favor more drug-like properties (i.e., Table 3.1), and to minimize the inclusion of reactive species, protein aggregators, nonspecific compounds (referred to as frequent hitters, pan-assay interference compounds [PAINS], or promiscuous inhibitors), and the like (Baell and Walters, 2014).

The goal of designing a diversity library for HTS is to balance diversity of chemical structure and diversity of bioactivity (i.e., a library not disproportionately composed of compound structures known to hit against a specific target class or classes; for example, kinase inhibitors) with physicochemical properties that are likely to provide chemical starting points for medicinal chemistry (Arrowsmith et al., 2015). For example, Schuffenhauer et al. (2020) have provided an in-depth description of the evolution of the private small-molecule screening deck used by the Novartis Institute of Biomedical Research (NIBR). Their diversity screening library at one time contained over 2 million compounds. Over the course of 2015–2019, they iteratively took learnings from successful screens to enrich their library for diverse compound subsets with optimized properties. Chemical diversity was assessed using two-dimensional and three-dimensional metrics of structural similarity, such as molecular fingerprinting, Tanimoto coefficients, and the like (Koutsoukas et al., 2014; Bajusz et al., 2015; Capecchi et al., 2020). Optimized properties that were considered included molecular weight, cLog(P), solubility, and cell permeability (either experimentally measured or predicted). At the same time, they aimed to cull the library of problematic structures such as reactive compounds (e.g., acylating agents, metal chelators, and metals), compounds leading to protein precipitation, PAINS compounds (Baell and Walters,

2014), and the like. This paper is rich in information regarding the multiple parameters considered and the iterative improvements gained by this analysis.

Diversity library components are typically stored as individual dry powders and/or in tubes (so that they can be robotically selected) containing high-concentration (i.e., 1–4 mM) solutions in dimethyl sulfoxide (DMSO). However, because of the size of these libraries, screening assays must be configured to run in microtiter plates of 96- or more typically 384- or 1536-well design. Hence, DMSO stocks of library components are also typically dispensed into microtiter plates, referred to as "mother plates" from which defined volumes of these stock solutions can be in turn dispensed into "daughter plates" in a defined well pattern (a plate map) for use in assay execution. Liquid dispensing into the wells of the plate and assay readout are today all done through automation, so that the full libraries can be screened over a reasonable period. Typically, each well of the screening plate contains an individual compound of the library, or a positive or negative assay control (see below), arranged in a specific plate map pattern (Malo et al., 2006). Library compounds are screened at a single concentration of typically 10 or 30 μM. After a fixed amount of assay time (after mixing the target, the compound, and any other assay components), the plate is read to measure a specific signal, and this is used to calculate a percent binding or percent modulation of the target. The expectation is that most library compounds will not bind or modulate the target, so that the percent of compounds scoring as hits (as defined below) will be relatively low; for diversity, library screening a hit rate of 0.01–0.1% is typical.

4.5.2 Fragment-Based Libraries

Fragment-based libraries are another form of structural diversity screening. Here, however, the library components are not drug-like molecules, but instead smaller fragments that may be combinable to form novel ligands for specific targets. The fragments generally conform to what is referred to as the "rule of 3" (Congreve et al., 2023), such that the molecular weight of fragments is <300 Da, the cLog(P) is <3, and the number of hydrogen bond donors and acceptors are each <3. The concept of fragment-based screening is based on the additivity of free energy terms for two substrates binding simultaneously to an enzyme active site (Jencks, 1981; Broom, 1989; see also Hajduk and Greer, 2007 and Copeland, 2013). Suppose that an enzyme bound two substrates, A and B, and structurally related inhibitors of each could be identified, A' and B'. If A' and B' each binds the enzyme with $K_d = 1$ mM (i.e., 1×10^{-3} M), then each has a binding free energy at 25 °C of

$$\Delta G_{\text{binding}} = RT\ln(K_d) = -4.08\ \text{kcal/mol} \tag{4.1}$$

If one could covalently link A' and B' in an optimal orientation, without any energetic losses, the free energy of the covalent compound A'–B' should be the sum of the free energies of A' and B':

$$\Delta G_{A'-B'} = -8.16\ \text{kcal/mol} \tag{4.2}$$

We can rearrange the equation for free energy as follows.

$$K_{A'-B'} = e^{\Delta G/RT} = 1 \times 10^{-6} = 1\ \mu\text{M} \tag{4.3}$$

Thus, while the free energies are additive for the covalently linked compound, A'–B′, the K_d values are multiplicative, such that from two millimolar fragments, we obtain a micromolar compound. In practice, one seldom realizes the full additivity of free energies in covalently combining fragments, as there are usually entropic, enthalpic, and heat capacity losses. Nevertheless, by combining two fragments in this fashion, one can achieve orders of magnitude increases in binding affinity.

To be successful, fragment-based screening must be combined with structural biology methods, such as crystallography or NMR spectroscopy, so that one can visualize the proximity and orientation of fragments relative to each other, and thereby plan optimal approaches to covalently linking them. An excellent review of the basis of fragment-based screening, with examples of its success, has been provided by Rees et al. (2004) and by Hajduk and Greer (2007).

4.5.3 DNA-Encoded Libraries (DEL)

As we discussed in Chapter 2, the use of phage display to generate libraries of peptides or proteins has revolutionized screening aimed at antibody affinity maturation. The technologies involved in phage display inspired Brenner and Lerner (1992) to propose that nucleic acids could be appended to organic compounds as a means of barcoding individual molecules for subsequent identification. Although first proposed in the early 1990s, about 15 years of technical advances were required before this concept could be reduced to practice, allowing DELs to be used to identify compounds that bound specifically to a target protein (see Neri and Lerner, 2018, for a review).

As practiced today, DEL is an example of an affinity selection (AS) technique. Current DELs contain millions to billions of encoded small molecules. In a typical DEL screening strategy (Figure 4.1), the target of interest is appended with an affinity tag, such as biotin, poly-histidine tags, and the like (see Chapter 1), so that it can be immobilized. The entire library, or discrete subsets, is then combined with the immobilized target. Nonspecific binding can be mitigated by including herring sperm DNA, biotin, histidine, or other blocking agents in the buffer. After a fixed incubation period, the immobilized target is subjected to a series of stringent washing steps to remove unbound and weakly bound compounds. The remaining bound compounds are eluted by changing solution conditions (pH, organic solvent, ionic strength, temperature, etc.). The DNA of each bound component is then amplified by the polymerase chain reaction (PCR). The amplified

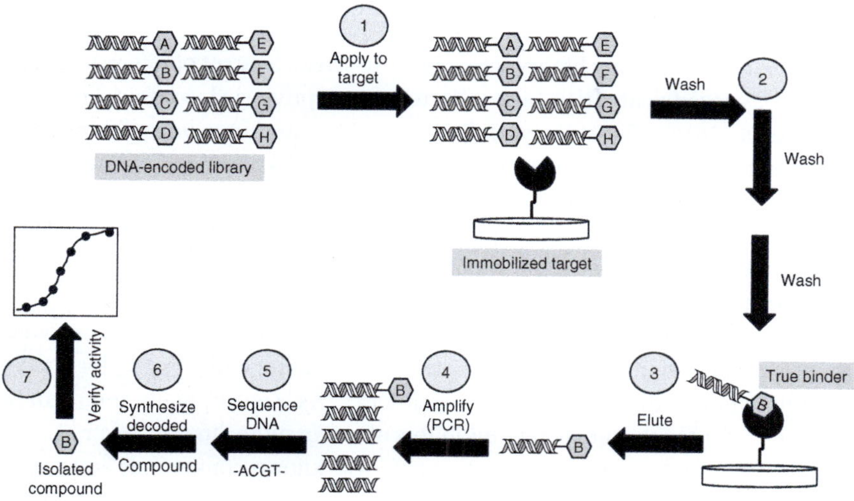

Figure 4.1 A greatly simplified schematic of the key steps in DEL screening. Each component of the library is composed of a unique double-stranded DNA sequence covalently appended to a specific organic compound (A through H in this cartoon). (1) The DNA-encoded library is mixed with the immobilized target protein. (2) After a fixed incubation time, the immobilized target is subjected to several stringent washing steps to remove unbound and weakly bound DEL components. (3) After removing unbound and weakly bound DEL components, the remaining, bound DEL components are eluted from the target. (4) The DNA constituents of the bound DEL components are amplified by PCR to generate sufficient copies for (5) DNA sequencing. The unique DNA sequence of each bound DEL component defines (decodes) the specific organic compound to which that DNA is appended. (6) The specific organic compound is then synthesized without the appended DNA, and (7) the activity of the isolated organic compound is confirmed in an assay of target binding or modulation of target activity.

DNA is then sequenced to identify the specific DNA barcode(s), and from this, the identity of the encoded small-molecule binder(s). Typically, these molecules are then synthesized without the appended DNA barcode to confirm target binding. Bioinformatic methods can then be used to suggest related chemical structures that the researcher may wish to also test as binders.

DEL screening provides a mechanism for testing ultra-large (e.g., $>10^{10}$) libraries of compounds, to effectively explore compound diversity for target binders. This can be a powerful approach to identifying chemical starting points for medicinal chemistry. The success rates of DEL screening vary considerable among users, depending on the specific targets being studied and the libraries being used. In this author's own experience, DEL screening was found to provide useful starting points for about 50% of targets, but this may not be typical of current DEL technologies. An excellent review of current DEL approaches has been provided by Gironda-Martinez et al. (2021) and includes seven examples of optimized compounds derived from DEL screening, three of which have transitioned to clinical studies.

4.5.4 Target Class-Specific Libraries

An alternative to diversity library screening is to screen a target of interest against target class-specific libraries (also referred to as privileged scaffold libraries; Welsch et al., 2010). These are typically smaller libraries (up to tens of thousands of compounds) that have been assembled based on knowledge of previously identified modulators of a specific class of targets, such as protein kinases (Kettle and Wilson, 2016), G-protein–coupled receptors (GPCRs) (Heilker et al., 2009), nuclear receptors (Wilson and Moore, 2002), methyl-lysine reader proteins (Branash et al., 2018), and the like.

As an example of this approach, since the 1990s, many industry and academic groups have had an intense focus on protein kinases as potential targets for oncology, inflammatory disease, and immune diseases. As a result, many compounds have been reported that are potent inhibitors of one or more kinases. Several organizations have pulled together collections of these known compounds and structurally related compounds into specialized libraries (see, for example, Elkins et al., 2016; Kettle and Wilson, 2016). These libraries are typically much smaller than diversity libraries, having only about 1,000–50,000 compounds making up the specialized library. However, because these libraries are highly enriched for structures that are known to inhibit kinases, the hit rate one may experience from screening a novel kinase with such a library is often much higher than for traditional diversity libraries. Additionally, because these libraries tend to be populated with known inhibitors and structural analogues from previous medicinal chemistry programs, there are often clusters of structurally related compounds among the hits, providing an early and often rich understanding of SAR.

The conceptual foundation for target class-specific libraries is based on the premise of modulator promiscuity. It is well established that the evolution of enzyme activity (and other protein functions) depends largely on the ability to generate neofunctionality, from a baseline of promiscuous substrate utilization and product formation (Copeland, 2023). That is, ancestral enzymes often displayed the ability to catalyze not only their primary reaction(s), but also collateral reactions, albeit usually at significantly reduced catalytic efficiencies. Over the course of evolution, such collateral reactions can be serially augmented through combinatorial mutagenesis, often with concomitant diminution of the primary activity of the founding enzyme – provided, of course, that the augmentation of the collateral reaction(s) is advantageous to the organism. Such neofunctionalization occurs not only in nature but today can also be performed in the laboratory to engender desirable enzyme-catalyzed reactions through directed evolution (Yuan et al., 2005; Stock and Gorochowski, 2024).

An analogous situation exists for small-molecule modulators of extant enzymes and other targets, wherein a compound designed as a selective modulator of one

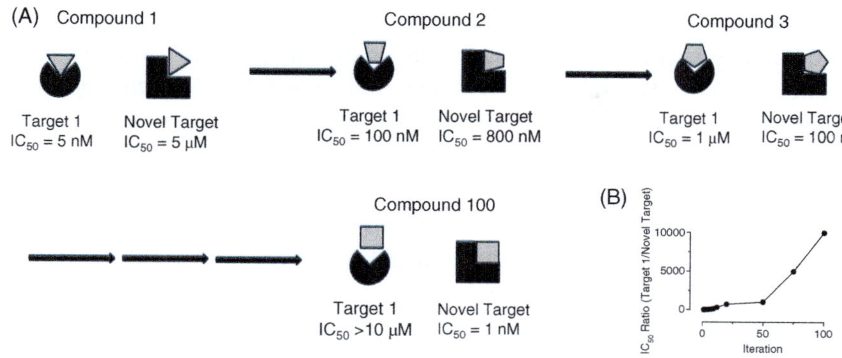

Figure 4.2 Simplified cartoon representing the process of ligand neofunctionalization. (A) The starting point for the process is a compound (compound 1) that is primarily a potent binder of Target 1 but displays promiscuity in also binding the Novel Target of interest with 1,000-fold less affinity, so that the IC_{50} ratio (Target 1/Novel Target) is 0.001. Through multiple iterations of SAR, this promiscuous activity is augmented while the affinity toward Target 1 is simultaneously diminished, ultimately resulting in a compound with high affinity for the novel target and minimal residual affinity for the original target (1). (B) A plot of IC_{50} ratio (Target 1/Novel Target) as a function of SAR iteration, illustrating the evolution of affinity toward the Novel Target and away from the original target.

target protein may have some minimal modulatory activity against a collateral, but structurally and/or mechanistically related, target protein (Figure 4.2).

This is the underlying basis for target class-specific compound library construction, although library construction is agnostic to any specific target protein and more focused on enriching for modulators of the entire target class. Using the methods of medicinal chemistry SAR, one may augment the modulatory activity toward a novel target at the expense of the original target, through systematic changes to the chemical structure of the founding compound. This "directed evolution" of modulator selectivity is greatly facilitated by structural biology methods and depends on some mechanistic similarities between the modulator-binding sites of the founding and novel target proteins. For this reason, modulator neofunctionalization is also a critical aspect of the systematic strategy we refer to as "mechanism-guided design", which is our next topic of discussion.

4.5.5 Mechanism-Guided Design

Mechanism-guided design is another approach that starts with a suboptimal ligand, such as a substrate analogue, a product analogue, or a modulator of a structurally and/or mechanistically related protein, and attempts to evolve that ligand into a potent modulator of the target of interest. Unlike target class-specific library screening, mechanism-guided design relies on detailed knowledge of the biochemical mechanism of the target and structural information about the ligand-binding pocket(s) on the target to predict new ligand structures for the target protein. Several types of design strategies fall within this general category of hit-finding. Examples of strategies subsumed within the umbrella of mechanism-guided design include transition-state inhibitor design (Radzicka and Wolfenden, 1995; Schramm, 1998, 2003), bi-substrate inhibitor design (Jencks, 1981; Broom, 1989), drug repurposing for hit-finding (Pushpakom et al. 2019), and natural ligand-based design (Copeland, 2013).

A contemporary example of mechanism-guided design is provided by DOT1L inhibitors developed by the team at Epizyme, Inc. As discussed in Chapter 2, DOT1L is a protein lysine methyltransferase (PKMT) that catalyzes methylation of histone H3 lysine 79 (H3K79) and has been implicated as a critical enzyme for mixed lineage leukemias (MLL) that are characterized by chromosomal translocations (rearrangements) involving the MLL gene. The research team at Epizyme was focused on discovering and developing inhibitors of protein methyltransferases (PMTs) as cancer therapeutics, including DOT1L inhibitors as a therapeutic mechanism for treating MLL patients. A first step toward this goal was to perform a comprehensive chemogenetic survey of human PMTs to understand the structural relatedness among members of this enzyme class, with special emphasis on the relatedness of their *S*-adenosylmethionine (SAM)-binding pockets (Richon et al., 2011). All PMTs use SAM as the methyl-donating substrate. The methyl group is transferred to a specific lysine (PKMT enzymes) or arginine (PRMT enzymes) side chain nitrogen within a protein or peptide substrate, utilizing an S_N2 reaction mechanism; this means that the two substrates must be simultaneously bound within a contiguous channel of the enzyme, for direct methyl transfer to occur. The products of this reaction are *S*-adenosylhomocysteine (SAH) and a methylated lysine or arginine within the peptide or protein (Figure 4.3). As part of their chemogenetic analysis, Richon et al. (2011) developed biochemical assays for 11 PMTs, including DOT1L, and used these assays to compare values of SAM K_M, SAH K_i, and the K_i of a generic PMT inhibitor, sinefungin. They also solved the crystal structures of human DOT1L with SAM and with SAH bound in the active site. For DOT1L, they found that both SAM and SAH bound in fully extended conformations within the enzyme active site, in stark contrast to all other human lysine methyltransferases. They also found that for DOT1L, the K_M for SAM and the K_i for SAH were quite low, at 450 and 320 nM, respectively (the K_i for SAH was first reported as ranging from 200 to 330 nM; however, subsequent measurements determined a mean value of 320 nM). Based on these collective data, the team began to design compounds that would retain the ribose–adenosine core of SAH but would replace the sulfur of the homocysteine chain and extend this chain to occupy the

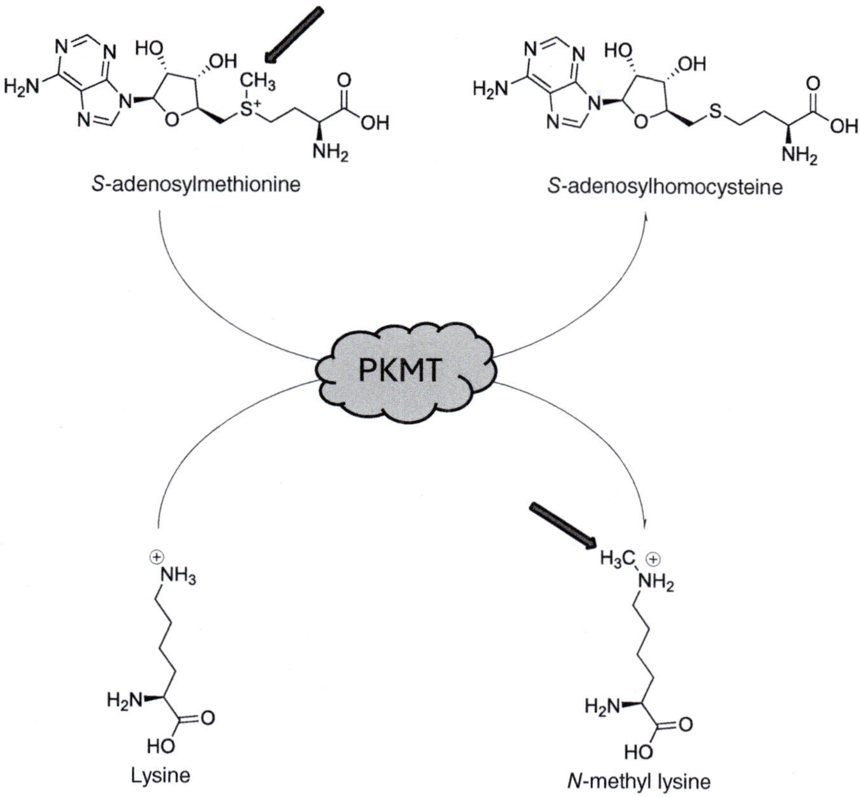

Figure 4.3 Reaction mechanism of the protein lysine methyltransferase enzymes (PKMTs), such as DOT1L. These enzymes utilize *S*-adenosylmethionine (SAM) as the methyl-donating substrate and the side chain ε-nitrogen of lysine residues within proteins as the methyl-accepting substrate. Both substrates must be simultaneously bound to the enzyme to effect direct transfer of the methyl group from SAM to the lysine side chain nitrogen. The two reaction products are *S*-adenosylhomocysteine (SAH) and *N*-methylated lysine. The bold arrows highlight the methyl group on SAM to be donated during reaction, and the resulting methyl group on the lysine ε-nitrogen as reaction product. Lysine residues in proteins may be mono-, di-, or tri-methylated by PKMTs without change to the positive charge of the ε-nitrogen. Hence, lysine residues may exist in four distinct states of methylation: non-methylated, mono-methylated, di-methylated, and tri-methylated. *Source:* Adapted from Copeland et al., 2009.

lysine-binding portion of the active site channel; thus, the team sought to create inhibitors with bi-substrate or transition state-like character.

Table 4.3 summarizes the mechanism-guided design of DOT1L inhibitors (Basavapathruni et al., 2012). Basavapathruni et al. began by replacing the sulfur and truncating the homocysteine chain to a dimethyl nitrogen (EPZ0004).

Table 4.3 Summary of the mechanism-guided design of DOT1L inhibitors as reported by Basavapathruni et al. (2012) and by Daigle et al. (2013).

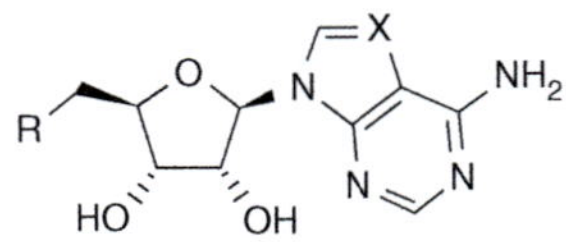

Compound	R	X	K_i (nM)
SAH	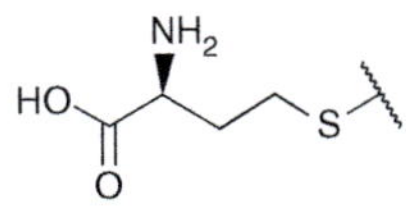	N	320
EPZ0004	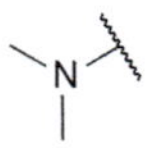	N	18,000
EPZ3647	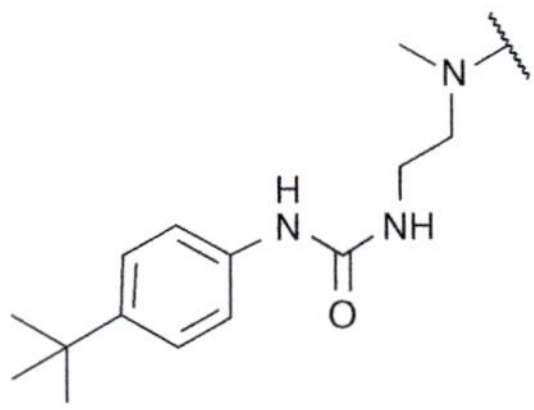	N	845
EPZ3696	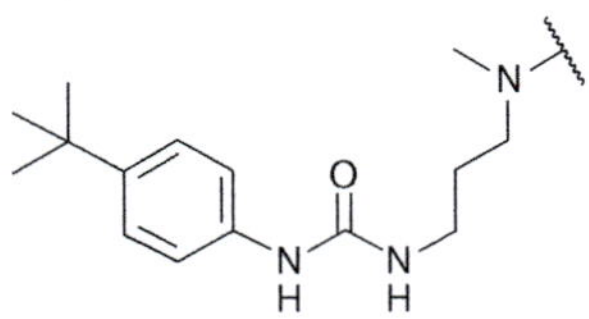	N	13
EPZ4450	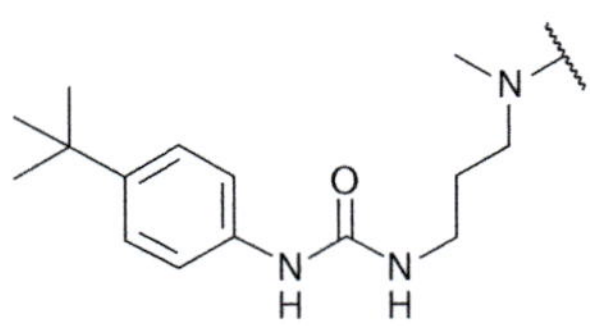	C—H	4
EPZ4777	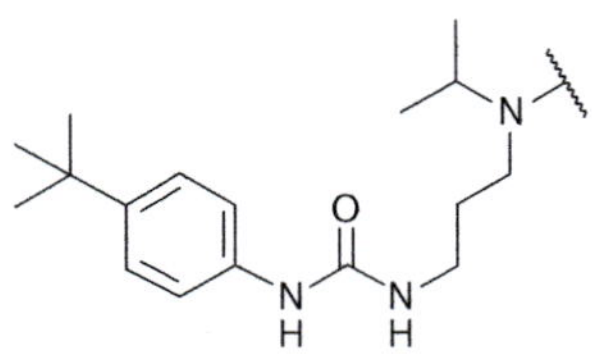	C—H	0.3
EPZ5676	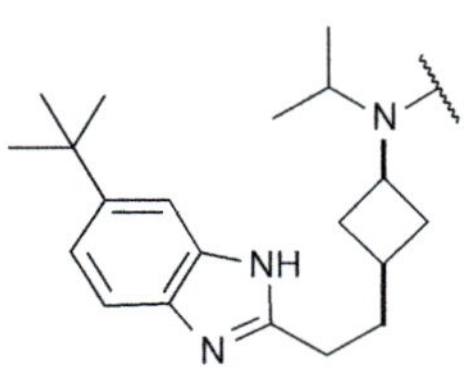	C—H	0.08

This diminished the affinity for DOT1L relative to SAH by about 100-fold (K_i = 38,000 nM). However, the compound displayed a favorable ligand efficiency, so the team continued the design efforts. Various extended replacements of one of the N-methyl groups were then tested. Compound EPZ3647 added a urea moiety, appended with a para-tert-butyl phenyl ring, to an N-ethylene chain; this brought the affinity back to within 3-fold of SAH (K_i = 845 nM). Addition of another methylene to the chain resulted in EPZ3696, which increased the affinity for DOT1L to K_i = 13 nM. In parallel, this team found that the 7-N of the nucleoside ring could be replaced by a carbon, yielding EPZ4550, which retained potent DOT1L inhibition with K_i = 4 nM. Replacement of the remaining N-methyl group with an isopropyl group led to EPZ4777 with a K_i of 0.3 nM (300 pM). Further SAR resulted in EPZ5676 (Table 4.3), also known as pinometostat (K_i = 0.08 nM), a compound that ultimately entered phase 1 clinical studies for the treatment of MLL patients (Daigle et al., 2013; Stein et al., 2018). Thus, from the starting point of EPZ4 to the final clinical compound EPZ5676, there is a 475,000-fold improvement in binding affinity (Figure 4.4). The significant improvements in binding affinity seen for EPZ4777 and EPZ5676 were largely driven by conformational changes of the DOT1L enzyme induced by compound binding; this is a topic we shall discuss further in Chapter 6.

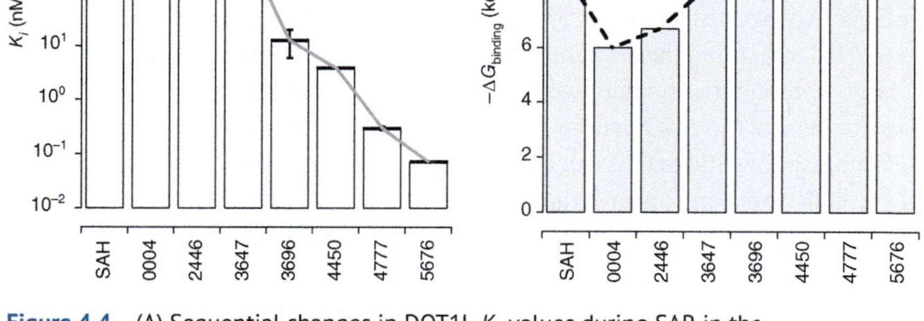

Figure 4.4 (A) Sequential changes in DOT1L K_i values during SAR in the mechanism-guided design efforts of Basavapathruni et al. (2012) and Daigle et al. (2013). (B) The corresponding negative values of the free energy of binding for the compounds shown in A.

4.5.6 Structure-Based Design and Computational Docking Methods

We discussed structure-based drug design (SBDD) and computational docking methods in Chapter 1. These methods are used today for *de novo* design of compounds, based on the three-dimensional architecture of ligand-binding pockets, as revealed by X-ray crystallography, NMR spectroscopy, and cryo-electron microscopy (Anderson, 2003; see also Chapter 1). Such efforts are augmented by computational docking methods, and most recently these methods have been further augmented by application of artificial intelligence approaches, as discussed in Chapter 1. There are many individual examples of the success of SBDD in drug discovery. In fact, the *Nature* family of journals has recently compiled a portfolio of articles from across their journals on the topic of SBDD (https://www.nature.com/subjects/structure-based-drug-design). This is an excellent resource for the reader interested in learning more about this important topic and individual success stories.

More recently, efforts have been put forth to more systematically obtain three-dimensional structures for large collections of proteins, such as for the kinome, for chromatin-modifying proteins, and the like. For example, the Structural Genomics Consortium (SGC) is a large, public–private enterprise with the stated mission of "understanding the functions of all proteins encoded by the human genome and accelerating the discovery of new medicines". A significant part of the SGC's activities is focused on obtaining three-dimensional structures of proteins and protein–ligand complexes and making these structures publicly available to researchers to stimulate chemical biology and drug discovery. Additionally, the SGC has emphasized the discovery of "probe compounds" – inhibitors and other ligands of proteins with the affinity and physicochemical characteristics to be useful research tools for exploring the pathophysiological roles of individual proteins (Arrowsmith et al., 2015). As with the structural information, the SGC is committed to making these chemical probes available to the research community at large through their chemical probes portal (Sanfelice et al., 2024). More information on the SGC can be found on their website (www.thesgc.org) and their many publications that can also be accessed through this website.

Although there are clear examples of the success of SBDD in identifying starting points for drug discovery, this author has found that the greatest strength of structural biology methods is in the optimization of validated hits and leads, during the iterative process of medicinal chemistry SAR. Indeed, the insights that are derived by iterative structural studies over the course of SAR campaigns have been critical to the success of most of the drug discovery programs with which this author has been involved. We shall discuss the use of structural biology as a springboard for medicinal chemistry in Chapters 5 and 6. However, at the stage of hit-finding, it is highly recommended that one begin investing in structural biology to have baseline structures of the target protein in the ligand-free state, with natural ligands bound (e.g., substrates, products, agonists, antagonists, allosteric modulators, and the like), and, where available, with tool compounds bound. Structural biology often involves significant trial and error before optimal crystallization, NMR, or cryo-EM conditions are identified. By starting these studies early – during hit-finding – one can be best prepared to leverage structural information as soon as validated hits have been identified.

4.6 Assay Formats

In the context of molecular targeted drug discovery, the most common methods of hit-finding involve cell-free assays using recombinant target protein constructs (see above). Generally, these assays fall into two categories: binding assays and assays of target function. Here we shall briefly describe both approaches. Additional information on binding and functional assay formats can be found in Kenakin (2004), Copeland (2013, 2023), and references therein.

4.6.1 Binding Assays

Kenakin (2004) and Copeland (2023) have reviewed several experimental approaches to measuring binding interactions between a protein and small-molecule ligand. However, to be useful as a primary screen for HTS purposes, the assay must be formattable in microtiter plates and executable with a throughput that allows large libraries to be screened in a reasonable amount of time. Hence, binding assays for HTS typically involve formats in which the protein and compound are mixed, allowed to reach equilibrium, and then binary protein–ligand complex concentration is somehow quantified. Two general categories of binding assays have found widespread use in HTS: (1) assays based on physical separation of bound and free compound and (2) competitive displacement assays using a radiometric or spectroscopic label on the primary ligand.

Assays that rely on physical separation of bound and free compound include the DEL library method described in Section 4.5.3, AS mass spectrometry (ASMS), microscale thermophoresis (MST), and surface plasmon resonance (SPR). In ASMS, the library of compounds is divided into convenient size pools, with each pool containing hundreds to thousands of compounds. Each pool is mixed with the target protein in the wells of a microtiter plate and incubated at ambient temperature for 30–45 minutes. Samples from each well are then loaded onto a micro high-performance liquid chromatography (μHPLC) size exclusion chromatography (SEC) column, equilibrated at 4 °C (to retard dissociation of the protein–compound complex during chromatography), and chromatography is performed with phosphate-buffered saline as the isocratic mobile phase. The void volume, containing protein and protein–compound binary complexes, is collected and applied to a reverse phase C18 column equilibrated at 60 °C. Chromatography is performed with a gradient of mobile phase from low to high acetonitrile/methanol. This denatures the target protein and liberates the bound compounds. The eluents from the reverse phase column are diverted to a mass spectrometer, and the identity of each compound is determined from time-of-flight mass spectrometry (Prudent et al., 2020; Copeland, 2023).

SPR is a method by which either the target protein or a known ligand thereof is immobilized on a chip in direct contact with a gold surface. Irradiation of the chip with plane-polarized light allows one to detect changes in the refractive index due to binary complex formation (see Copeland, 2023 for further technical details on SPR). In a typical SPR assay, the target protein is covalently immobilized to the chip surface. Buffer is flowed over the chip, and the refractive index changes are monitored as a function of time. After having established a stable baseline reading with buffer, the mobile phase is replaced with buffer containing a known concentration of a potential ligand. The approach to equilibrium (association) can thus be monitored as a function of time. Once equilibrium is established, the mobile phase is again replaced with ligand-free buffer to effect dissociation of the bound compound from the immobilized target. Once ligand dissociation is complete, the chip is typically washed with modified buffer to remove any residual bound or adventitiously adhered compound (regeneration), before the next set of experiments are initiated (Figure 4.5). In this manner, multiple compounds can be tested as potential ligands for the target protein, or a single compound can be titrated to extract the on-rate, off-rate, and/or K_d value (see also Chapter 1). Thus, SPR can be used in both kinetic and equilibrium modes to extract information on compound

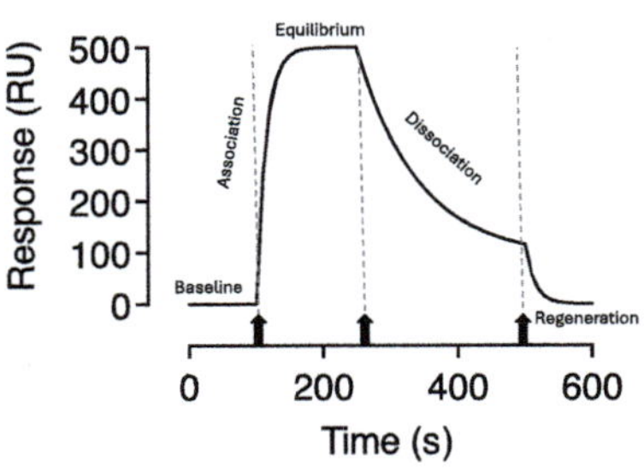

Figure 4.5 Typical kinetic response plot for an SPR experiment illustrating the baseline, association, equilibrium, dissociation, and regeneration phases of the kinetic trace. The up-pointing arrows along the *x*-axis indicate time points at which the mobile phase is changed. *Source:* Copeland (2023) / with permission of John Wiley & Sons.

binding to a target protein. SPR is most used in drug discovery for detailed studies of selected compounds during hit-to-lead and lead-optimization SAR campaigns. However, the SPR technology has been adapted to running in microtiter plate formats, making it amenable for compound library screening as well.

MST measures the motion of molecules induced by a thermal gradient. Typically, the target protein is labeled with a fluorescent tag to follow protein movement along the thermal gradient. The thermal gradient is induced with an infrared laser focused through a microscope objective, and a second laser is used to excite fluorescence. The binding of small-molecule ligands to the protein causes a change in the rate of protein movement along the gradient, and this is the basis of the measurement of compound binding. The application of microfluidic chambers allows MST to be performed in high throughput. For example, recently Zimmermann et al. (2022) demonstrated that a fluorescent adenosine analogue, 5-FAM-triazolyl-adenosyl-DAB (FTAD), can bind to DNA- and protein-methyltransferases within the SAM-binding pocket. They used this fluorescent ligand with unlabeled DNA methyltransferase 2 to measure MST movement of the ligand, instead of the protein, along the thermal gradient. They then screened a library of compounds to identify ligands that would competitively displace FTAD from the enzyme.

4.6.1.1 Competitive Displacement Assays

The example of the MST assay format of Zimmermann et al. (2022) is a good segue to our second type of binding assay format, competitive displacement of a labeled, known ligand from the target protein (Copeland, 2023). For most target proteins one can identify a natural ligand or tool compound that is already known from the literature. Such ligands can be synthesized with radioisotopes incorporated, such as ^{3}H, ^{14}C, ^{125}I, and the like. The incorporation of a radioisotope provides a means of following ligand binding to a protein after separation from the unbound ligand population by nitrocellulose filter binding, SEC, and other separation methods (Copeland, 1994, 2023). Alternatively, scintillation proximity assays (SPAs) can be used to measure radiolabeled ligand binding to a target protein without the need for a physical separation step. Radioactive decay releases high-energy particles (e.g., β-particles and Auger electrons) that can excite light emission in certain organic molecules (e.g., diphenylanthracene) and lanthanide metals (e.g., europium). The emitted light is referred to as scintillation, and the molecules that emit light when excited by radioactive decay particles are therefore referred to as scintillant. In SPA, the protein target is modified with an affinity tag, such as biotin (see Chapter 1) so that it can be bound to scintillant-containing beads (typically of polyvinyl toluene or polystyrene) covalently coated with the complementary affinity partner (e.g., streptavidin in the case of biotin). The protein-associated beads are then mixed with radioligand in solution. When the labeled ligand binds to the bead-associated protein, the radioisotope comes into close enough proximity to the scintillant within the bead to excite light emission. In this way, a unique signal (emission) is associated only with protein-bound radioligand molecules.

In addition to radioisotopes, ligands may be appended with fluorescent or luminescent moieties to introduce a spectroscopic label into the ligand. Binding of such labeled ligands to the target protein can then be detected by methods such as direct fluorescence (Copeland, 1994, 2023), fluorescence polarization (Copeland, 1994, 2023), or either Förster resonance energy transfer (FRET; Copeland, 2023) or homogeneous time-resolved fluorescence (HTRF; Degorce et al., 2009), when the protein is also labeled with a fluorescent donor species. For example, many proteins bind the product of ATP hydrolysis, ADP (e.g., kinases, kinesins, helicases, and the like). The fluorescent molecule N-methyl anthraniloyl (Mant) can be appended to one of the ribose oxygens of ADP to form Mant-ADP, a commercially available molecule that emits intense fluorescence at a wavelength of 448 nm when irradiated with ultraviolet light of wavelength 355 nm. When Mant-ADP binds to proteins, the change from the polar, aqueous buffer to the more nonpolar binding pocket on the target protein significantly increases the fluorescence intensity of the Mant group. This change in fluorescence intensity can then be used to follow protein binding of the ligand. Detailed protocols for the development of protein-binding assays for HTS can be found in several literature sources. For example, a detailed review of receptor-binding assays for HTS and drug discovery has been provided by Auld et al. (2018).

Typically, assay buffer and conditions are experimentally optimized for individual assay formats (Auld et al., 2018, but see below for additional considerations). Then, the K_d is determined for complex formation between the target protein and the labeled ligand (also referred to as the analyte ligand). The concentration of protein and labeled ligand are then fixed to ensure adequate signal and signal reproducibility, and such that the ligand is present at or near its K_d value. With these conditions established and fixed, an important control experiment is to titrate the system with an unlabeled version of the same ligand, to ensure that the unlabeled ligand can competitively displace the labeled ligand, and that the K_d value of the labeled ligand is not significantly perturbed from that of the unlabeled ligand; this control is especially important in the case of fluorescent or luminescent labels, where the appended species may cause nonspecific binding to the protein. Often, an estimate of the K_d for the unlabeled ligand will be available from the literature. As described in Appendix 1, if we identify the labeled ligand as L, and its K_d as K_L (which is known from the previous experiment), and identify the competing (unlabeled) ligand as X and its K_d as K_X, we can determine the value of K_X from

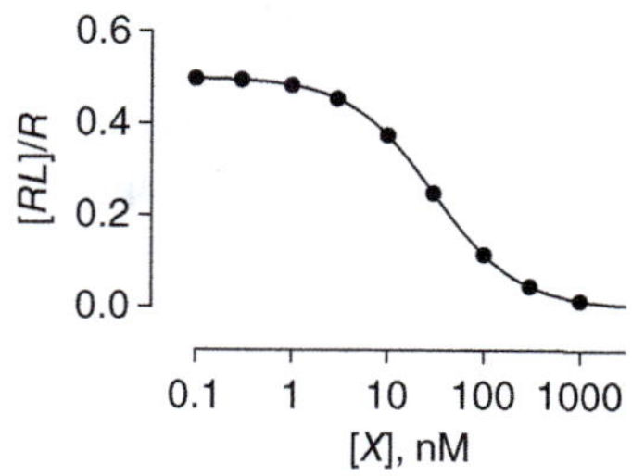

Figure 4.6 The diminution in protein-labeled ligand concentration [*RL*] relative to total protein concentration [*R*] as a function of the concentration of a competing, unlabeled ligand [*X*], according to Equation 4.4. See current text and Appendix 1 for further details.

the diminution in RL complex concentration (where *R* is the target protein) as a function of *X* concentration as follows (Equation (4.4) and Figure 4.6).

$$\frac{[RL]}{[R]} = \frac{1}{1 + \frac{K_L}{[L]}\left(1 + \frac{[X]}{K_X}\right)} \tag{4.4}$$

It should be noted that the analyte ligand used in developing binding assays need not be a small-molecule ligand. Radio- and fluorescence-labeling strategies can equally be applied to peptides, proteins, nucleic acids, and other biological binding partners of the target protein.

Having thus validated the competitive displacement assay, one can now use this to screen compound libraries to identify novel binders of the target protein. In library screening mode, the protein and analyte ligand (e.g., the labeled ligand) concentrations are fixed, and the library components are tested at a fixed concentration (again, typically 10 or 30 μM). The value of the signal corresponding to [*RL*]/[*R*] in control wells, where no library compound is added, is defined as 100% signal. The corresponding signal from wells to which library compounds are added is then compared to the control signal to calculate a percent inhibition of [*RL*] formation, as will be described in more detail below.

There are two caveats to bear in mind for assays based on competitive displacement of a ligand. First, any screening based on binding assays will identify compounds that bind to the target protein. Binding, per se, does not necessarily equate with modulation of protein function, which is our aim (except, perhaps in the case of PROTACs and glues as discussed in Chapter 2). Hence, screening hits from such assays must be subsequently tested in assays of target protein function, as we shall discuss below. Second, generally, competitive displacement assays will only identify compounds that bind in a way that displaces the analyte ligand, by either competing for the same binding pocket on the protein as the analyte ligand or by allosterically affecting analyte ligand binding. Any compounds that bind the protein and modulate function by an alternative mechanism (e.g., noncompetitive inhibitors as discussed in Chapter 2) would fail to be identified as hits in such an assay.

4.6.2 Assays of Target Function

The goal of most targeted drug discovery programs is to identify compounds that will modulate the biochemical activity of the target protein in such a way as to lead to disease-altering phenotypic outcomes. Hence, hit-finding assays that are based on target function are the preferred method of screening in most situations. As discussed in Chapter 2 and elsewhere (Copeland, 2013, 2023), a large majority of FDA-approved small-molecule therapeutics elicit their pharmacological activity by inhibition of specific enzyme targets. Hence, we will illustrate assays of target function here with the example of developing an assay of enzyme activity for screening of compound libraries for inhibitors.

As discussed in Chapter 2, enzymes catalyze specific biochemical reactions by binding substrate(s) and chemically converting them into product(s) molecules. Enzymes generally catalyze chemical changes to the structure of the substrate in converting it to product. Some enzymes hydrolyze chemical bonds (e.g., proteases, aminopeptidases, nucleases, and the like) or form new covalent bonds between substrates (e.g., ligases). Other enzymes transfer chemical groups from one substrate to another (e.g., kinases, methyltransferases, acetyltransferases), and still others perform redox chemistry (e.g., oxidoreductases). The key point for our current discussion is that the substrate(s) and product(s) of enzymatic reactions are chemically distinct. In many cases, the chemical distinction between substrate(s) and product(s) can be exploited in the design of assays to follow the enzyme-catalyzed transition from substrate to product as a function of time. In some cases, the substrate or product of the enzymatic reaction can be detected directly, based on some unique spectroscopic or other physicochemical feature (direct assay: Copeland, 2023). In other cases, the substrate or product molecule can react chemically with another reagent to form a spectroscopically detectable species (indirect assay: Copeland, 2023). Finally, in yet other cases, the product of the enzyme can be used as the substrate for a second enzymatic reaction that leads to a detectable product species in what is referred to as a coupled reaction (Copeland, 2023). As an example of a coupled reaction, consider the protein kinases, which represent a large class of enzymes, many of which are interesting targets for drug discovery. Protein kinases catalyze the transfer of a phosphate group from ATP to a specific amino acid side chain of a protein substrate, yielding ADP and phospho-protein as the products of their enzymatic reactions. A common way to assay kinases, and other ATP-hydrolyzing enzymes, is the ADP-glo assay marketed by Promega Corporation (www.promega.com). In this system, the kinase reaction is allowed to run for a defined length of time, during which ATP is converted to ADP by the enzyme. Then, a proprietary reagent is added to the assay well that degrades any remaining ATP in the solution. The ADP generated by the kinase is then converted to ATP by another proprietary

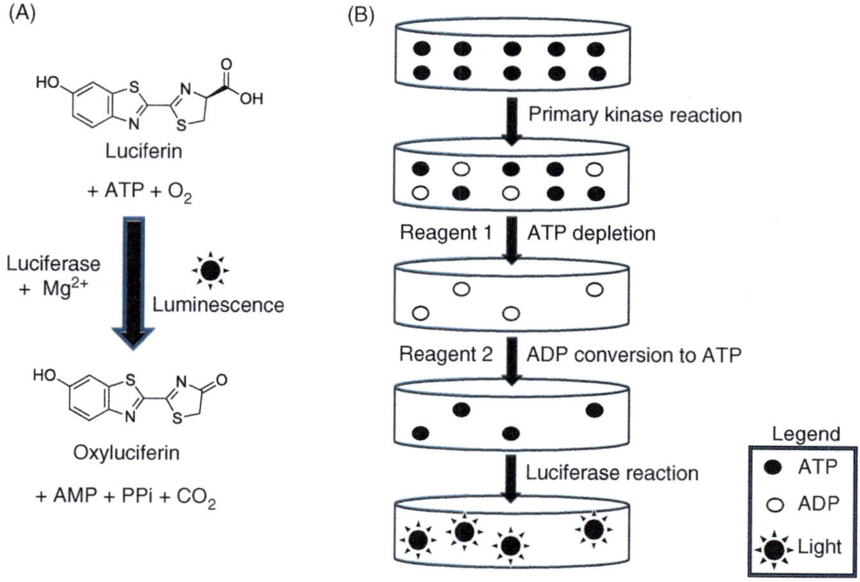

Figure 4.7 (A) Oxidation of luciferin to oxyluciferin catalyzed by firefly luciferase. (B) Schematic diagram of the steps involved in a coupled kinase assay using the ADP-Glo method from Promega Corporation. Reagents 1 and 2 are proprietary, and their composition has not been disclosed. Panel **B** is redrawn based on data on the Promega.com website.

reagent, and the ATP thus generated is used as a substrate, together with luciferin, for a genetically optimized version of the firefly enzyme luciferase. Luciferase converts oxygen, ATP, and luciferin to AMP, pyrophosphate, carbon dioxide, and oxyluciferin, with the production of bioluminescence. The light emitted during this process provides a quantitative measure of the ADP produced during the primary kinase reaction (Figure 4.7).

Regardless of the detection method, enzyme activity assays must conform to some general characteristics to be amenable to HTS. First, it is critical to run the assay under steady-state conditions (Copeland, 2013), where the concentration of product generated is linear with time (a plot of product concentration as a function of enzyme reaction time is referred to as a progress curve, as illustrated in Figure 4.8A). These conditions typically occur during the early phase of enzymatic reaction up to the point where about 10–20% of the initial concentration of substrate has been consumed. Next, the researcher should determine the K_M value for each substrate by experimental titration of substrate concentration and its effect on reaction velocity. For compound screening, it is best to set the concentration of substrate(s) at their K_M value, or as close to K_M as is practically possible, based on signal strength, substrate solubility, and other pragmatic aspects of assay

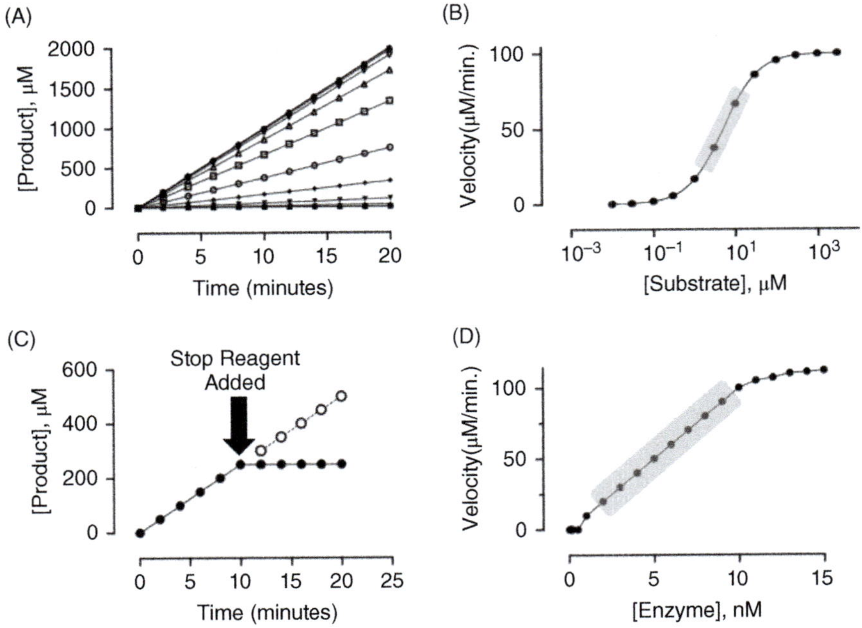

Figure 4.8 Verification of necessary characteristics of enzyme assays for HTS. (A) Linearity of reaction progress curves at multiple substrate concentrations. (B) Velocity as a function of substrate concentration to determine the K_M value. The shadowed region of the curve represents the range of acceptable substrate concentrations for use in HTS assays. Ideally, the substrate concentration for screening would be fixed at the K_M value. (C) Verification of the ability of a stop reagent to halt completely the continued generation of product. The black circles represent the assay before and after addition of the stop reagent at the time point indicated by the broad, down-facing arrow (10 minutes after initiation of reaction). The open circles represent the continued production of product in the absence of stop reagent addition. (D) The effect of enzyme concentration on velocity, illustrating the experimentally determined linear range (shaded region of the curve). Note that although velocity should be a linear function of enzyme concentration, in practice this linear phase is usually limited, as illustrated here.

design (Figure 4.8B). By poising the substrate concentration at K_M, one ensures that inhibitors of multiple modalities (competitive, noncompetitive, uncompetitive; Chapter 2) are most likely to score as hits in the assay, as seen in Chapter 2.

For large library screening efforts, one is unlikely to follow the kinetic progress curve for every assay well. Instead, most researchers use an end point (also referred to as discontinuous) assay format for screening. This refers to reading signal at a single, fixed time point within the linear range of the enzyme progress curve (Copeland, 2003, 2013). For simplicity in processing multiple plates during screening, one wishes to stop (or quench) the enzyme reaction, at the selected end point, so that no further product formation occurs over the time required to collect and read multiple plates. This requires a reagent that instantly halts enzymatic activity by denaturing the enzyme, sequestering a key substrate or cofactor, or by completely inhibiting the enzyme. The stop (quench) reagent must not interfere with the detection method used to read the plates, and must not decompose the product formed, so that the signal amplitude is constant over time after the stop reagent is added (Figure 4.8C). Examples of stop reagents include ethylenediaminetetraacetic acid (EDTA) for metal-dependent enzymes, strong acids, organic solvents (e.g., ethanol or acetone), high salt (to disrupt protein–protein or protein–nucleic acid interactions), and inhibitors of the specific enzyme or enzyme class (Copeland, 2013, 2023). A final consideration is the enzyme concentration to be used in the assay. Theoretically, the reaction velocity should be linearly dependent on enzyme concentration (Copeland, 2023). However, in practice, one finds that this linear relationship holds only over a finite concentration range of enzyme (Figure 4.8D). Often deviations from linearity are observed at very low enzyme concentrations due to the limits of detection (Ingle, 1974) for the method being used, protein denaturation at low concentration, requirement for dimerization or higher-order oligomerization for optimal activity, and other factors. Likewise, at very high enzyme concentrations, deviations from linearity may occur due to enzyme aggregation or nonproductive oligomerization, limits of detection for the method, and departure from steady-state conditions. Hence, the linear range of enzyme concentrations should be determined by experimental titration for each enzyme and assay format. The concentration of enzyme to be used in screening should then be fixed at a low concentration, consistent with signal quality, and well within the linear range. Having defined the enzyme and substrate concentrations and solution conditions (see below), one should redetermine the linearity of the progress curve, under these specific conditions, and reaffirm signal stability after addition of the stopping reagent.

4.6.3 Solution Conditions for Assays

Solution conditions can make dramatic differences in the activity of target proteins, as reviewed previously (Copeland, 2013, 2023). Buffer composition, pH, ionic strength, viscosity, detergent, and carrier protein (inert proteins that are added to assays to minimize target protein adsorption to vessel walls and other surfaces) concentration can all affect assay performance. As with other considerations (see above), the default position should be to attempt to mimic pathophysiological conditions as much as possible. For example, if the target protein is known to function within the cytosol of a mammalian cell, one may wish to begin assay optimization at solution conditions that mimic those of the cell cytosol (Table 4.4). However, one often finds that some variation from physiological pH, ionic strength, and/or other solution conditions may be necessary

Table 4.4 Some typical values of physiological conditions for mammalian cell cytosol.

Parameter	Value
pH	7.3–7.5
Ionic strength	150 mM
Cation composition	
Potassium	139 mM
Sodium	12 mM
Magnesium	0.8 mM
Calcium	<0.0002 mM
Water content	70%

Source: Copeland (2013) / with permission of John Wiley & Sons.

to optimize assay performance. For example, in Table 4.4, we see that cytosolic ionic strength is about 150 mM. However, for target proteins that bind other proteins or nucleic acids through electrostatic interactions (e.g., salt bridges), the ionic strength may need to be lowered to facilitate interactions between the target and its macromolecular ligand(s). Similarly, cytosolic pH is between 7.3 and 7.5, but for aspartyl proteases, optimal enzymatic activity is observed at pH below 6.0.

Compounds to be screened as binders or modulators of the target protein are typically stored as high-concentration stock solutions in apolar solvents, most commonly DMSO. Hence, when compound is added to the assay solution, a defined volume of apolar solvent is also introduced to the system. To correct for this perturbation, an equal volume of neat DMSO (i.e., without any compound) is added to all control wells. Most proteins will tolerate somewhere between 0.5% and 5% (v:v) DMSO without significant loss of activity. However, every protein is unique in this regard, and the tolerance range for DMSO must be determined empirically, as part of assay solution optimization studies for each target protein. Additional information on determining DMSO tolerance and other assay solution conditions can be found in Copeland (2013) and references therein.

4.7 Measures of Assay Performance

Whether screening compounds based on binding or target function, the ability to identify bona fide hits depends critically on the assay conforming to well-established criteria of performance. These performance criteria generally focus on the ability to distinguish true signal from random noise in the assay, to quantitatively relate signal changes to changes in concentration of a specific analyte, and to define a statistically valid cutoff with which to declare hits based on the change in signal, relative to controls, that is elicited by the presence of the compound. In this section, we shall review some of the performance criteria that are commonly applied to screening assays.

4.7.1 Calibration Curves

For competitive displacement binding assays and assays of target function, a first step in establishing assay performance is to define the linear dynamic range of analyte concentration over which the signal being monitored changes linearly with analyte concentration (Copeland, 2013). During the actual execution of assays, it is critical to ensure that the signal being measured remains within this linear dynamic range, so that one can be confident that the signal amplitude is proportional to the concentration of the analyte species being generated. The analyte to be assessed may be the binary target–ligand complex from which ligand is displaced by competing compounds (Section 4.6.1.1), the substrate or direct product of an enzyme assay, or an indirect product of a coupled assay system (Section 4.6.2). As described above, the signal being detected may be fluorescence, fluorescence polarization, luminescence, radioactivity, and the like.

Ideally, one can purchase, synthesize, or otherwise generate the analyte species in pure form and use this to prepare a stock solution of high and precisely known concentration. The calibration curve is then generated by making a series of dilutions of the stock solution into assay buffer, ideally containing the quenched target protein (i.e., denatured, or in the presence of saturating amounts of unlabeled ligand for displacement assays, or in the presence of the stop reagent for enzymatic assays; see above), and measuring the signal generated by each dilution in replicate measurements (typically 3–5 replicates). In addition to this series of dilutions of the analyte, it is important to also measure the signal generated from the assay solution alone, without any analyte present. This is referred to as a blank sample and is important for establishing any background signal in the absence of analyte. The replicate measurements are used to determine the mean (average) signal value and the standard deviation for each analyte concentration (including zero, the blank). Figure 4.9 illustrates a typical calibration curve for assay signal readout as a function of analyte concentration. Several parameters of assay quality can be defined from such a calibration curve. First, the y-intercept of the calibration curve provides a measure of any static background signal that may be associated with the assay solution that must be corrected for by subtracting this value from the total signal observed at each nonzero concentration of analyte. The standard deviation of the blank signal also gives some estimate of the noise level in

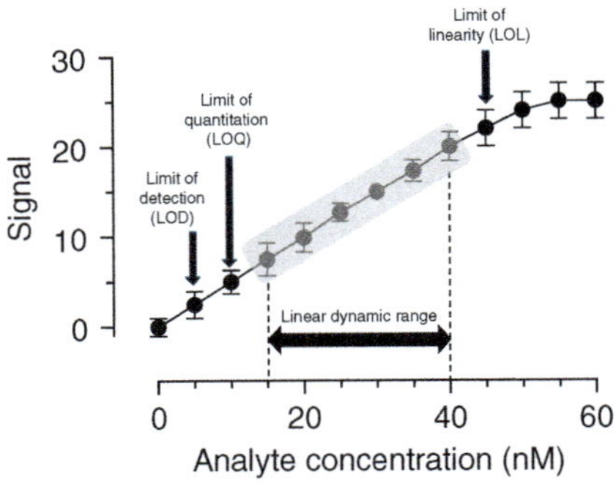

Figure 4.9 A typical calibration curve for an assay system. The closed circles represent the mean signal value at each analyte concentration, and the vertical lines represent the standard deviation of each signal value. Key quality measures (LOD, LOQ, LOL, as defined in the text) are identified by down-pointing arrows. The linear dynamic range is shaded and represents the analyte concentration range from the LOQ to the LOL.

the assay. The limit of detection (LOD) of the assay is defined as the lowest analyte concentration that yields a signal amplitude that can be statistically distinguished from the blank (background) signal; this is typically defined as a signal that is 2- or 3-fold greater than the standard deviation of the blank sample signal (Ingle, 1974). The limit of quantitation (LOQ) is defined as the lowest analyte concentration yielding a signal with a predefined level of precision. This is typically defined as the lowest analyte concentration for which the coefficient of variation of signal is <20% (Ingle, 1974), and this may or may not be coincident with the LOD. As analyte concentration is further titrated, one often observes deviations from linearity at high analyte levels, as seen in Figure 4.9. This deviation from linearity can have many sources, such as saturation of the detection device, the solubility limit of the analyte in assay buffer, and other physical properties (see Copeland, 1994, 2013 for further details). Regardless of the cause, the highest concentration of analyte that conforms to a linear relationship between analyte concentration and signal, defines the limit of linearity (LOL) of the assay system. The range of analyte concentration between the LOQ and the LOL values define the linear dynamic range of the assay (shaded area in Figure 4.9).

4.7.2 Time-Dependent Background

In addition to any static background signal seen in the construction of a calibration curve, one can also encounter an increasing background over time during the execution of binding or functional assays, due to target protein-unrelated reactions in the assay solution. Copeland (2013) discusses this in detail in the context of enzyme activity assays. Let us consider a concrete example for which the target protein is a protease – an enzyme that hydrolyzes specific peptide bonds in substrate proteins and peptides. Let us say that we are using a peptide substrate where the peptide bond to be hydrolyzed by the enzyme occurs in the middle of the peptide sequence. We may append the peptide with a fluorescent group on the N-terminus and a fluorescence-quenching group on the C-terminus. In this way, the fluorescence of the peptide remains low until the N- and C-terminal portions of the peptide are separated by enzymatic hydrolysis of the central peptide bond. Hence, over time, the catalytic action of the protease results in increasing fluorescence signal as substrate peptide is cleaved into two separate peptides, one containing the fluorescent group and the other containing the fluorescence-quenching group (i.e., the quenching of fluorescence only occurs when the substrate peptide is intact and the fluorescent and quenching groups are covalently connected by the contiguous amino acid sequence of the substrate peptide). However, in addition to protease-catalyzed peptide hydrolysis, a very low rate of water-mediated peptide hydrolysis can also occur, and this would add an increasing background signal to the assay over time. Hence, to correct for this, the total signal (S_t) generated by the active protease should be measured at a specific time point within the linear progress curve of the enzyme (see above), and the background signal (S_b) using pre-stopped enzyme conditions (see above) should be measured at the same time point. The specific signal (S_s) from enzyme catalysis at that time point is then calculated by subtracting S_b from S_t (i.e., $S_s = S_t - S_b$).

4.7.3 The Z′ Statistic

As defined above, the value of S_t and S_b will each display some variation in replicate measurements, thus one can calculate a mean value (μ_t and μ_b, respectively) and a standard deviation (σ_t and σ_b, respectively) for both values (Figure 4.10).

From these mean and standard deviation values, Zhang et al. (1999) derived a simple statistic, referred to as the Z′ score, as a measure of assay quality. The Z′ score is defined as follows.

$$Z' = 1 - \frac{(3\sigma_t + 3\sigma_b)}{|\mu_t - \mu_b|} \tag{4.5}$$

Note that in this equation the denominator represents the absolute value of the difference between the mean values for total and background signal. The Z′ value can range from 0 to 1, with 1 representing a perfect assay in which the standard deviations for total and background signals are both zero. It is generally agreed that to qualify for HTS purposes, an assay must demonstrate a Z′ value of >0.5. Additional statistical practices for HTS have been reviewed by Malo et al. (2006).

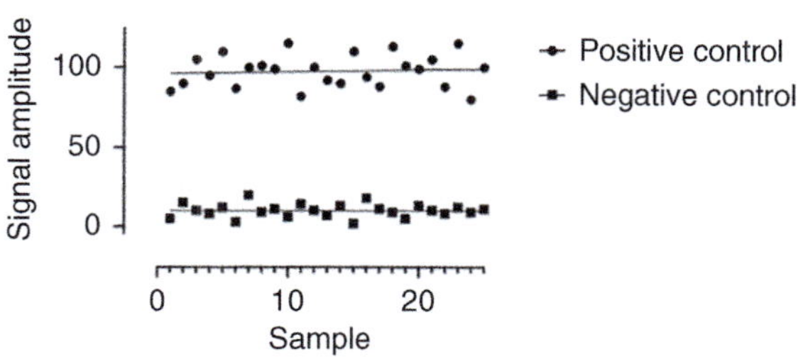

Figure 4.10 Hypothetical example illustrating the variation in total positive control (S_t) and negative control (S_b) signal amplitude. The horizontal lines represent the mean values of S_t and S_b (μ_t and μ_b, respectively).

4.7.4 Definition of Hits

Having qualified an assay for HTS, one next performs the screen and measures the ability of library components to either bind to or modulate the activity of the target protein. As stated above, screening is done at a single concentration of each library component (typically, 10 or 30 μM), and signal is measured at a fixed time point after mixing the assay components and compound. The percent bound, percent inhibition, or percent activation are then calculated based on the mean signals for the positive and negative controls and the signal obtained in the presence of the fixed concentration of compound. A reasonable expectation is that most library compounds will not impact the target protein, so that the percent modulation (i.e., binding, inhibition, or activation) for most library compounds will conform to a Gaussian distribution centered around a mean value of zero modulation, with a full width at half maximum of 2.355σ (where σ is again the standard deviation). In practice, the mean value of the distribution is typically close to but not necessarily equal to zero (i.e., typically between −5% and + 5%). With this profile, hits are defined either by an arbitrary cutoff or by statistical analysis. The researcher may decide to arbitrarily declare as hits any compound that displays >50% modulation. More commonly, hits are declared as compounds displaying a percent modulation greater than 2σ (95% confidence limit) or 3σ (99.73% confidence limit) away from the mean value (Motulsky, 1995). For example, let us say that we run a hypothetical screen for inhibitors of a target enzyme and find that the mean value for library compounds is 3% inhibition with a standard deviation of 10% (Figure 4.11). At the

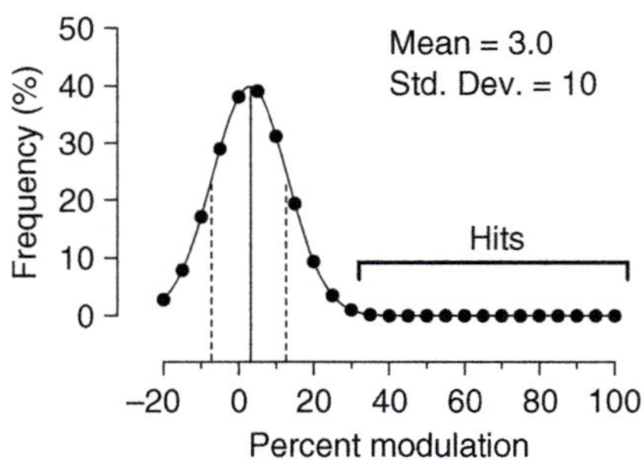

Figure 4.11 Hypothetical data illustrating the typical Gaussian distribution of percent modulation values among diversity library compounds in an HTS. In this example, the mean value of the distribution is 3% modulation of target function, and the standard deviation is 10%. The mean value is represented by a solid vertical line intersecting the peak of the distribution. The dashed vertical lines represent one standard deviation (1σ) above and below the mean value. Hits, at the 99.73% confidence limit, are defined as any compound demonstrating a percent modulation $>3\sigma$ above the mean (i.e., $3 + (3\times10) = 33\%$ modulation, in our example).

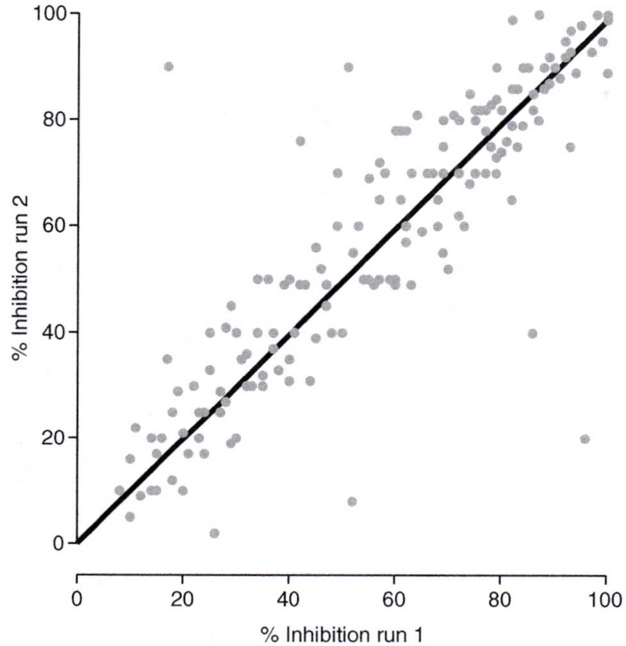

Figure 4.12 Hypothetical example of a correlation plot between two runs of an HTS assay looking for inhibitors of an enzyme target. The solid line through the data represents the line of identity between the two assays. See text for further details. *Source:* Copeland (2013) / with permission of John Wiley & Sons.

99.73% confidence limit, we would declare as hits any compounds that displayed >33% inhibition of enzyme activity (mean value plus 3σ).

Typically, the entire screening library is run in the same assay a second time to test for reproducibility. The percent modulation for each compound in run 2 is plotted against that obtained in run 1 (Figure 4.12). In an ideal case, this plot would yield a diagonal line with a slope of 1 (the line of identity between the two runs; also, sometimes referred to as the line of equality). In practice, one sees deviations from this ideal behavior, both because of random error and because some compounds do not reproducibly modulate the target protein. Generally, only compounds that score statistically (see above) as hits in both runs of the assay are declared as reproducible hits and are taken forward for further consideration.

4.7.5 Post-Screening Prioritization of Reproducible Hits

Even with a modest hit rate of 0.01–0.1%, screening a large diversity library can result in 50 to several hundred (or more) initial hits. This is usually too many compounds for the research teams to evaluate in detail. Hence, some prioritization of screening hits must be applied.

For those hits that reproduce in replicate runs of the primary assay, the researcher can next reconfigure the plate map to collect concentration–response data for each remaining compound. This allows researchers to rank compounds in terms of potency based on EC_{50} values rather than merely percent modulation at a single compound concentration. At the same time, the concentration–response data also allows the researcher to identify compounds that deviated from the expected simple 1:1 stoichiometry of binding and modulation; that is, one molecule of modulatory ligand binding to one molecule of target protein. In our presentation of the Hill–Langmuir equation in Chapter 1 (Equations 1.4 and 1.5), we made the common and simplifying assumption of a 1:1 stoichiometry between ligand and receptor leading to effect. A more complete version of this equation includes a slope term, known as the Hill coefficient (h), that accounts for the steepness of the concentration–response curve and is related to the number of ligand molecules binding to a single molecule of target protein and any cooperativity among these ligand-binding events (Copeland, 2023). Accounting for the Hill coefficient, Equation 1.5 becomes

$$E = \frac{E_{max}}{1 + \left(\frac{EC_{50}}{[L]}\right)^h} \tag{4.6}$$

In fitting concentration–response data with the values of E_{max}, EC_{50}, and h as fitting parameters, one expects that most well-behaved modulators will display a unity (1.0) value of h. Given random error in experimental data, values of h within the range of 0.8–1.2 are generally assumed to be consistent with a simple 1:1 stoichiometry. There are bona fide biochemical situations in which the value of h can be greater or less than unity (Copeland, 2013, 2023), but these are not often encountered among initial screening hits. Hence, whenever a screening compound exhibits a value of $h > 1.5$, or $h < 0.8$, the compounds should be flagged for greater scrutiny. Thus, reproducible screening hits may be rank-ordered by EC_{50} values, and the researcher may decide to prioritize well-behaved compounds that exhibit Hill coefficients close to unity.

Additional post-screening hit triage can be performed based on the purported chemical structures of the screening hits. Hit compound presumed structures are evaluated computationally to ensure that they are consistent with medicinal chemistry tractability standards of molecular weight, cLog(P), number of rotatable bonds, and the like (Pearce et al., 2006; Harpel, 2009). Structures are also scrutinized for undesirable structural moieties, such as redox-active species (Lor et al., 2007), alkylating and acylating agents, metal-chelating functionalities, and other problematic functionalities (Figure 4.13), and nonspecific hits, known as PAINS and promiscuous inhibitors (see below) or "bad actors" (Copeland, 2013;

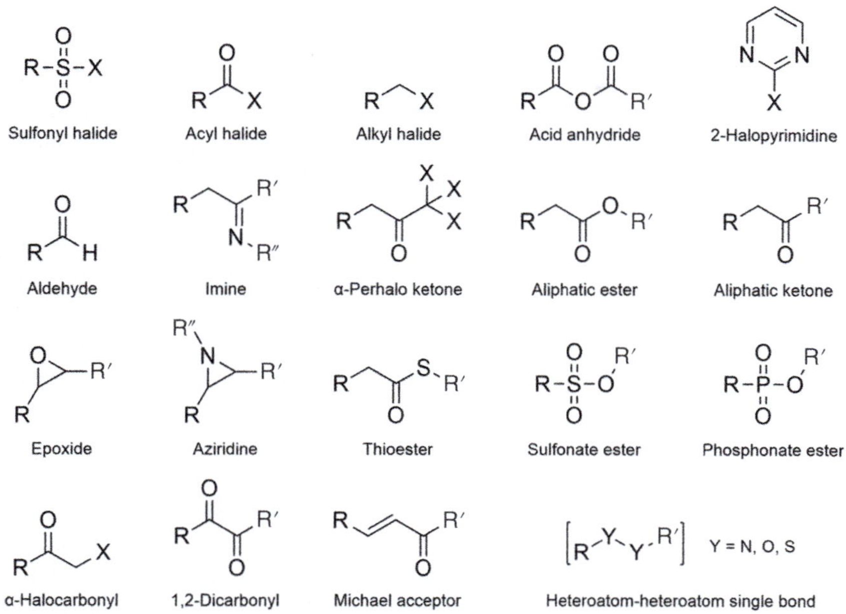

Figure 4.13 Examples of reactive functional groups to be avoided in screening hits. *Source:* Copeland (2013) / with permission of John Wiley & Sons.

Harpel, 2009). Compounds that are deemed intractable for any of these reasons are deprioritized as hits.

As HTS became a more common approach to drug discovery, researchers (McGovern et al., 2002; Ryan et al., 2003; Seidler et al., 2003; Feng et al., 2005; Shoichet, 2006; Feng et al., 2007; Coan et al., 2009) noted that certain types of compounds commonly appeared as hits against a variety of unrelated target proteins. In all cases, these compounds elicited their effects on target function through pharmacologically intractable mechanisms. For example, many of these compounds caused nonspecific aggregation of proteins, or of themselves to form micelle-like species. In other cases, the exact basis of interference with target function remains unclear. Nevertheless, these "promiscuous inhibitors" need to be identified and deprioritized as screening hits. Several simple tests for promiscuous inhibition have been reported (see a summary in Copeland, 2013, and in the other references cited earlier in this paragraph), and these should be used to weed out promiscuous inhibitors from the collection of reproducible screening hits.

At this point, the hits that have passed these various quality control gates are again reviewed by medicinal chemists to look for clusters of compounds that may share common structural features that may give some early hints of SAR. In favorable cases, this surveying for structurally related clusters may also identify the key structural element required for target modulatory activity; this structural element is referred to as the pharmacophore, and a group of compounds that share this structural element is referred to as a pharmacophore series (see below). The chemist may also survey the full screening library for structurally related compounds that did not score as hits in the primary assay. Some of these may prove to be false negatives (i.e., compounds that are target modulators but for various reasons failed to score in the assay). Other structurally related compounds may not be target modulators, and this knowledge also provides information on the structural requirements for target engagement. At this point, the chemist may also search for structurally related compounds from chemical supply companies (a process referred to in the industry as "SAR by catalogue") and purchase these as an early step in hit expansion (see below).

4.8 Orthogonal Assays and Mode of Modulation

All screening assays, no matter how well designed, will be susceptible to artifacts that can cause a compound to appear to be a hit against the target, when in fact it is not a true hit (a false positive). Depending on the nature of the signal detection system and the nature of the artifact, these false positives can reproducibly appear as hits in the primary HTS assay. Hence, it is critically important to verify reproducible hits from the primary screens in distinct assay systems that use alternative detection methods and potentially different assay formats. In many cases, an assay of target function may already be known that is less susceptible to artifacts but is too slow or laborious for HTS. Having winnowed the number of hits to a workable number, however, such lower-throughput assays may now be practical as orthogonal assays. For example, one may have used a high-throughput fluorescence assay for HTS. Compounds that contain multiple aromatic rings, or other extended π–electron systems, can reabsorb the fluorescence photons before they can reach the detector, thus giving the appearance of inhibition of target function. An alternative assay for the same target may be based on LC/MS separation and quantitation of species and would thus be impervious to this artifact. Another alternative assay format may be to follow binding or function by radiometric means. The key point is to verify the target modulatory activity of screening hits in one or more orthogonal assays that utilize distinct assay methodologies. Compound potency should be assessed in the primary assay and the orthogonal assay(s) by titrating compound to quantify the concentration–response relationship and determine the K_d or EC_{50} values. Similar potency values (typically K_d and/or EC_{50} values within 3–5-fold agreement between assays) are further reassuring of hit-validation (but see below).

Some researchers prefer to use a binding assay as the orthogonal assay format for primary screens based on modulation of target function and vice versa. This is generally a good strategy because compounds that bind to a target may not necessarily result in target modulation. On the other hand, if the primary HTS assay were run with an assay of target function, one would expect bona fide modulators to also score as hits in orthogonal binding assays, such as SPR.

An important caveat to this last statement is that agreement between a function and binding assay will depend on the presence of all the same assay components at similar concentrations between the two assays. As we saw in Chapter 2, compounds that are competitive with a natural ligand of the target will show reduced apparent binding affinity as the concentration of the natural ligand is increased. In counterpoint, compounds that are uncompetitive with a natural ligand will bind with greater apparent affinity as the concentration of natural ligand is increased and may in fact show no binding in the absence of the natural ligand. Hence, it is also important to determine the modality of modulation of hit compounds prior to selecting conditions to be used in orthogonal assays and prior to attempting any structural studies (i.e., x-ray crystallography) with the compound (Chapter 2 and references therein). In addition to natural ligand concentration, the concentration of other assay components can also influence the apparent potency of screening hits. Compound affinity can be affected by ionic strength, pH, detergents, the concentration of carrier proteins, cation identity, and the like. Thus, primary and orthogonal assays should be run under similar conditions to the extent possible. When differences in solution conditions are necessary between two assays, these should be carefully noted. In such cases, any discrepancy between the apparent potency of compounds between the two assays must be carefully investigated to avoid deprioritizing a screening hit inappropriately.

A question that often arises at this stage is whether there are advantages to a particular mode of target modulation with respect to cellular and eventual clinical utility. The advantages and disadvantages of different modalities of enzyme inhibition (e.g., competitive, noncompetitive, or uncompetitive with respect to enzyme substrate) have been reviewed in detail elsewhere (Copeland, 2013; Buker et al., 2019). For example, a disadvantage of competitive modulators is that their cellular potency may be diminished by high concentrations of the natural ligand of the target. If the target is an enzyme, and the drug acts as a competitive inhibitor, the intracellular concentration of competing substrate may increase due to inhibition of enzymatic conversion of substrate to product. In such a case, the eventual buildup of substrate concentration will diminish the ability of the inhibitor to effectively compete for binding to the enzyme target. Nevertheless, the view of this author is that one should not eliminate compounds from further consideration based on modulation modality alone. Rather, I recommend that modulators of different modalities be brought forward into cellular assays of target function and phenotypic consequences. Then, discrimination of compound series can be made based on the quantitative phenotypic translation of target occupancy, as will be discussed further in Chapter 5. The key point here is that modulators of all modalities (including competitive inhibitors) have demonstrated utility as human therapeutics.

4.9 Confirmation of Chemical Purity and Identity

Library components for HTS are commonly stored both as dry powder and as concentrated solutions in DMSO at low temperature – conditions intended to maximize the stability of the compounds. Nevertheless, some decomposition of compounds can occur during storage. In rare cases, compounds can be misplaced on assay plates so that the identity of the compound in a particular well may be incorrect. One step toward verification of chemical identity is to have a fresh DMSO stock prepared from dry powder and to rerun the primary assay using this fresh sample. However, a better and more common way to verify compound purity and chemical identity is to purchase the putative compound from a chemical supplier that provides NMR and/or LC/MS verification of structure and purity. If the compound (or close structural analogues) is not available for purchase, then the chemist can have the compound synthesized within their own laboratory or through a contract research organization (CRO). Because of the time, effort, and expense of resynthesis and analytical testing, these activities are often deferred until after the pool of screening hits has been curated through the various filters described above, so that only a small number of highly validated hits are thus confirmed. The repurchased or resynthesized compound is then used to verify the binding or modulatory activity seen with the original screening hit.

4.10 Hit Expansion

With the collection of screening hits now winnowed to a workable number of highly validated hits, medicinal chemists will begin to expand on their early SAR studies by designing additional, structurally related compounds. The goal of these studies is to prepare a small number (typically 10–30 compounds per validated hit) of compounds with systematic changes in chemical structure from the original hit, to establish that positive and negative SAR can be realized within the pharmacophore series. If minor structural changes from the original hit all ablate binding or functional modulation, the compound may be deemed to display a narrow SAR and may therefore be deprioritized. On the other hand, if systematic changes in chemical structure lead to corresponding changes in binding or modulatory potency, then the medicinal chemists may have a high degree of confidence that further structural manipulations are likely to lead to improved potency and other pharmacological properties.

For validated hits that display favorable SAR, structural biologists may attempt to solve the structure of the target protein with a member of the pharmacophore series bound. For crystallization trials, compound and target protein are usually dissolved in aqueous buffer at high concentrations. Hence, favorable aqueous solubility of the compound greatly increases the probability of success in these experiments. If successful, this early foray into structural biology can yield great insights into how to modify the chemical structure of compounds to improve potency and other properties, provide greater confidence in the tractability of the chemical series, and provide information on the recognition elements within the binding pocket that are critical to compound engagement. Note again, that knowing the modulation modality (i.e., competitive, noncompetitive, uncompetitive, etc.) of the compound is critical for attempting structural studies of the target-compound complex. For instance, this author has seen multiple examples where co-crystallization studies have failed because the compound was uncompetitive with respect to one of the natural substrates of the target, and that substrate was not included in the samples for crystallization studies.

Finally, for validated hit series that are of high priority, the medicinal chemists, together with a patent attorney, may conduct a thorough search of the patent literature to ensure that the chemical series to be prosecuted further is not encumbered by existing patents that might limit the researcher's freedom to operate in that chemical space for the target of interest.

During hit expansion, some, but not all, researchers also explore some early aspects of the compound metabolic stability and cell permeability (see Table 4.2). Metabolic stability is initially assessed *in vitro* by following the kinetic stability of the compound in the presence of whole human blood and human hepatocyte microsomes. Passive and transporter-mediated cell permeability is initially assessed using the Caco-2 system. These assays will be discussed further in Chapter 5.

4.11 Hit-Validation Funnel

As one can gather from the above discussions, the process of hit-finding and hit-validation requires multiple steps of testing and assessment. Different research groups perform these various activities in different sequences to continuously sift and refine the number, validation level, and overall quality of compounds for further evaluation. In some groups, chemical purity and identity are confirmed prior to performing orthogonal assay confirmation of hits. In other groups, the order is inverted, and in still other groups these activities may occur in parallel.

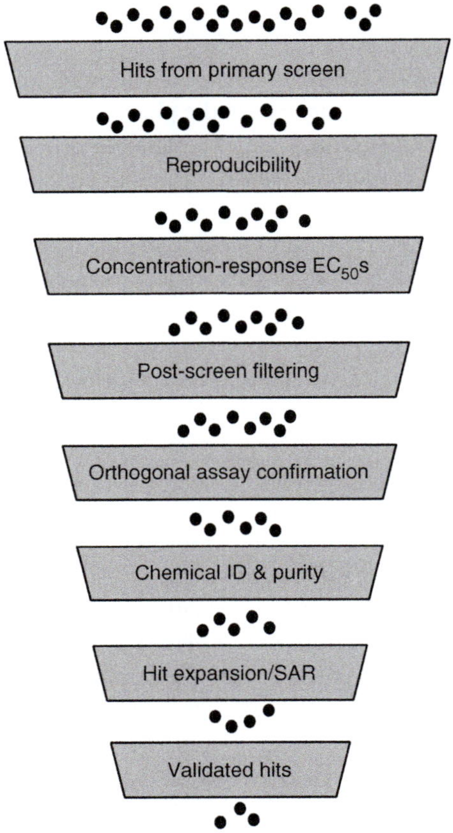

Figure 4.14 Schematic illustration of a typical hit-validation funnel. The small, black circles are intended to represent individual compounds. The number of these compounds remaining as hits is continuously reduced as compounds are sifted downward through the various gating filters, illustrated as gray trapezoid shapes. See text for further details.

The choice of order of operations depends on the number of hits to be processed, resources available to the team, and the personal preferences of the research team.

Figure 4.14 illustrates an example sequence of hit-validation studies that is commonly referred to as a "hit validation funnel". This figure is intended merely to illustrate the concept of such an operational funnel and not to advocate for a specific sequence of operations. What is most important is that the research team agree upon the sequence of operations that they will use for hit-validation studies, preferably prior to beginning the screen. At the same time, research teams need to be opportunistic and not hold rigorously to a specific order of operations if a particularly interesting hit emerges or the data otherwise compel a deviation from the fixed sequence.

4.12 Summary

In this chapter, we have described the approaches commonly used today to identify chemical starting points for molecular targeted drug discovery. These chemical starting points are referred to as hits. As we have seen, there are several strategies commonly employed for hit-finding, as well as multiple sources of compounds from which to identify hits. Of these, the most common approach to hit-finding, especially for novel targets, is diversity library HTS. Regardless of the hit-finding strategy and sources of chemical matter, hits identified from the primary screen must be subjected to a rigorous battery of tests to validate the compounds as true, reproducible hits that conform to a chemically and pharmacologically tractable mechanism of target modulation, that are free of undesirable chemical functionalities, and that are likely to be optimizable through medicinal chemistry SAR. We saw in this chapter that it is common for medicinal chemists to engage in some early SAR during hit expansion efforts. In the next two chapters, we shall see how SAR campaigns continue from validated hits toward the next goal of identifying compounds with sufficient potency, selectivity, and cell permeability to demonstrate intracellular target modulation. Compounds that achieve this goal are commonly referred to as leads. Lead compounds are further optimized during the lead-optimization stage of drug discovery to improve further the potency, selectivity, phenotypic response, plasma stability, oral bioavailability, pharmacokinetics, and a host of additional pharmacological and pharmaceutical properties, with the ultimate goal of producing a compound worthy of clinical investigation. Although, for pedagogical purposes, we divide compound optimization steps into the discrete stages of hit-finding, hit-to-lead, and lead optimization, in practice these stages really represent an intellectual continuum of hypothesis generation, testing, and refinement.

References

Anderson, A. (2003) The process of structure-based drug design. *Chem. Biol.* **10**, 787–797.

Arrowsmith, C.H., Audia, J.E., Austin, C., Baell, J. et al. (2015) The promise and perils of chemical probes. *Nat. Chem. Biol.* **11**, 536–541.

Auld, D.S., Farmen, M.W., Kahl, S.D., Kriauciunas, A. et al. (2018) Receptor binding assays for HTS and drug discovery. In *Assay Guidance Manual [Internet]* (S. Markossian, A. Grossman, M. Arkin et al., Eds.). Bethesda, MD. https://www.ncbi.nlm.nih.gov/books/NBK91992

Baell, J. and Walters, M.A. (2014) Chemistry: chemical con artists foil drug discovery. *Nature* **513**, 481–483.

Bajusz, D., Racz, A. and Heberger, K. (2015) Why is Tanimoto index an appropriate choice for fingerprint-based similarity calculations? *J. Chem.* **7**, 20.

Basavapathruni, A., Jin, L., Daigle, S.R., Majer, C.R.A., Therkelsen, C.A., Wigle, T.J., Kuntz, K.W., Chesworth, R., Pollock, R.M., Scott, M.P., Moyer, M.P., Richon, V.M., Copeland, R.A. and Olhava, E.J. (2012) Conformational adaption drives potent, selective and durable inhibition of the human protein methyltransferase DOT1L. *Chem. Biol. Drug Des.* **80**, 971–980.

Branash, K.D., James, L. and Frye, S.V. (2018) Target class drug discovery. *Nat. Chem. Biol.* **13**, 1053–1056.

Brenner, S. and Lerner, R.A. (1992) Encoded combinatorial chemistry. *PNAS* **89**, 5381–5383.

Broom, A.D. (1989) Rational design of enzyme inhibitors: multisubstrate analogue inhibitors. *J. Med. Chem.* **32**, 2–7.

Buker, S.M., Boriack-Sjodin, P.A. and Copeland, R.A. (2019) Enzyme-inhibitor interactions and a simple, rapid method for determining inhibition modality. *SLAS Discov.* **24**, 515–522.

Capecchi, A., Probst, D. and Reymond, J.-L. (2020) One molecular fingerprint to rule them all: drugs, biomolecules, and the metabolome. *J. Chem.* **12**, 43.

Coan, K.E.D., Maltby, D.A., Burlingame, A.L. and Shoichet, B.K. (2009) Promiscuous aggregate-based inhibitors promote enzyme unfolding. *J. Med. Chem.* **52**, 2067–2075.

Congreve, M., Carr, R., Murray, C. and Jhoti, H. (2023) A "rule of three" for fragment-based lead discovery. *Drug Discovery Today* **8**, 876–877.

Copeland, R.A. (1994) *Methods for Protein Analysis: A Practical Guide to Laboratory Protocols*. Chapman & Hall, New York.

Copeland, R.A. (2003) Mechanistic considerations in high-throughput screening. *Anal. Biochem.* **320**, 1–12.

Copeland, R.A. (2013) *Evaluation of Enzyme Inhibitors in Drug Discovery: A Guide for Medicinal Chemists and Pharmacologists*, 2nd Edition, Wiley, Hoboken, NJ.

Copeland, R.A. (2023) *Enzymes: A Practical Introduction to Structure, Mechanism and Data Analysis*. 3rd Edition. Wiley, Hoboken, NJ.

Copeland, R.A., Solomon, M.E. and Richon, V.M. (2009) Protein methyltransferases as a target class for drug discovery. *Nature Rev. Drug Discov.* **8**, 724–732.

Daigle, S.R., Olhava, E.J., Therkelsen, C.A., Basavapathruni, A. et al. (2013) Potent inhibition of DOT1L as treatment of MLL-fusion leukemia. *Blood* **122**, 1017–1025.

Degorce, F., Card, A., Soh, S., Trinquet, E. et al. (2009) HTRF: a technology tailored for drug discovery – a review of theoretical aspects and recent applications. *Curr. Chem. Genom.* **3**, 22–32.

Eder, J., Sedrani, R. and Wiesmann, C. (2014) The discovery of first-in-class drugs: origins and evolution. *Nature Rev. Drug Discov.* **13**, 577–587.

Ehrlich, P. and Hata, S. (1910) *Die Experimented Chemotherapie der Spirillosen*. Jilius Springer, Berlin.

Elkins, J.M., Fedele, V., Szklarz, M., Azeez, K.R.A. et al. (2016) Comprehensive characterization of the published kinase inhibitor set. *Nature Biotech.* **34**, 95–103.

Feng, B.Y., Shelat, A., Doman, T.N., Guy, R.K. and Shoichet, B.K. (2005) High-throughput assays for promiscuous inhibitors. *Nat. Chem. Biol.* **1**, 146–148.

Feng, B.Y., Simeonov, A., Jadhav, A., Babaoglu, K., Inglese, J., Shoichet, B.K. and Austin, C.P. (2007) A high-throughput screen for aggregation-based inhibition in a large compound library. *J. Med. Chem.* **50**, 2385–2390.

Gironda-Martinez, A., Donckele, E.J., Samain, F. and Neri, D. (2021) DNA-encoded chemical libraries: a comprehensive review with successful stories and future challenges. *ACS Pharmacol. Transl. Sci.* **4**, 1265–1279.

Hajduk, P.J. and Greer, J. (2007) A decade of fragment-based drug design. *Nature Rev. Drug Discov.* **6**, 211–219.

Harpel, M.R. (2009) Wheat from chaff: general and mechanistic triage of screening hits for enzyme targets. In *Handbook of Drug Screening*, 2nd ed., R. Seethala and L. Zhang, Eds., Informa Healthcare, New York, pp. 269–297.

Heilker, R., Wolff, M., Tautermann, C.S. and Bieler, M. (2009) G-protein-coupled receptor-focused drug discovery using a target class platform approach. *Drug Discovery Today* **14**, 231–240.

Ingle, J.D. Jr., (1974) Sensitivity and limit of detection in quantitative spectrophotometric methods. *J. Chem. Educ.* **51**, 100–105.

Jencks, W.P. (1981) On the attribution and additivity of binding energies. *PNAS* **78**, 4046–4050.

Keller, T.H., Pichota, A. and Yin, Z. (2006) A practical view of druggability. *Curr. Opin. Chem. Biol.* **10**, 357–361.

Kenakin, T. (2004) *A Pharmacology Primer: Theory, Application, and Methods*. Elsevier, San Diego, CA.

Kettle, J.G. and Wilson, D.M. (2016) Standing on the shoulders of giants: a retrospective analysis of kinase drug discovery at AstraZeneca. *Drug Discovery Today* **21**, 1596–1608.

Koutsoukas, A., Paricharak, S., Galloway, W.R.J.D., Spring, D.R., Ijzerman, A.P., Glen, R.C., Marcus, D. and Bender, A. (2014) How diverse are diversity assessment methods? A comparative analysis and benchmarking of molecular descriptor space. *JCIM* **54**, 230–242.

Lipinski, C., Lombardo, F., Dominy, B.W. and Feeney, P.J. (2001) Experimental and computational approaches to estimate solubility and permeability in drug discovery and development settings. *Adv. Drug Delivery Rev.* **46**, 2–26.

Lor, L.A., Schneck, J., McNulty, D.E., Diaz, E., Brandt, M., Thrall, S.H. and Schwartz, B. (2007) A simple assay for detection of small-molecule redox activity. *J. Biomol. Screening* **12**, 881–890.

Malo, N., Hanley, J.A., Cerquozzi, S., Pelletier, J. and Nadon, R. (2006) Statistical practice in high-throughput screening data analysis. *Nat. Biotechnol.* **24**, 167–175.

McGovern, S.L., Castelli, E., Grigoriefe, N. and Shoichet, B.K. (2002) A common mechanism underlying promiscuous inhibitors from virtual and high-throughput screening. *J. Med. Chem.* **45**, 1712–1722.

Motulsky, H. (1995) *Intuitive Biostatistics*. Oxford University Press, New York.

Neri, D. and Lerner, R.A. (2018) DNA-encoded chemical libraries: a selection system based on endowing organic compounds with amplifiable information. *Annu. Rev. Biochem.* **87**, 479–502.

Pearce, B.C., Sofia, M.J., Good, A.C., Drexler, D.M. and Stock, D.A. (2006) An empirical process for the design of high-throughput screening deck filters. *JCIM* **46**, 1060–1068.

Prudent, R., Annis, D.A., Dandliker, P.J., Ortholand, J.-Y. and Roche, D. (2020) Exploring new targets and chemical space with affinity selection-mass spectrometry. *Nat. Rev. Chem.* **5**, 62–71.

Pushpakom, S., Iorio, F., Evers, P.A., Escott, K.J. et al. (2019) Drug repurposing: progress, challenges and recommendations. *Nature Rev. Drug Discov.* **18**, 41–58.

Radzicka, A. and Wolfenden, R. (1995) Transition state and multisubstrate analog inhibitors. Methods Enzymol. **249**, 284–312.

Rees, D.C., Congreve, M., Murray, C.W. and Carr, R. (2004) Fragment-based lead discovery. *Nature Rev. Drug Discov.* **3**, 660–672.

Richon, V.M., Johnston, D., Sneeringer, C.J., Jin, L. et al. (2011) Chemogenetic analysis of human protein methyltransferases. *Chem. Biol. Drug Design* **78**, 199–210.

Rishton, G.M. (1997) Reactive compounds and in vitro false positives in HTS. *Drug Discovery Today* **2**, 382–384.

Ryan, A.J., Gray, N.M., Lowe, P.N. and Chung, C.-W. (2003) Effect of detergent on "promiscuous" inhibitors. *J. Med. Chem.* **46**, 3448–3451.

Sanfelice, D., Antolin, A.A., Crisp, A., Chen, Y., Antolin, A.A., Bellenie, B., Brennan, P.E., Edwards, A., Müller, S., al-Lazikani, B. and Workman, P. (2024) The chemical probes portal – 2024: update on this public resource to support best-practice selection and use of small molecules in biomedical research. *Nucleic Acids Res.* **53**, D1663–D1669.

Schramm, V.L. (1998) Enzymatic transition states and transition state analog design. *Annu. Rev. Biochem.* **67**, 693–720.

Schramm, V.L. (2003) Enzymatic transition state poise and transition state analogues. *Acc. Chem. Res.* **36**, 588–596.

Schuffenhauer, A., Schneider, N., Hintermann, S., Auld, D. et al. (2020) Evolution of Novartis small molecule screening deck design. *J. Med. Chem.* **63**, 14425–14447.

Seidler, J., McGovern, S.L., Doman, T.N. and Shoichet, B.K. (2003) Identification and prediction of promiscuous aggregating inhibitors among known drugs. *J. Med. Chem.* **46**, 4477–4486.

Seiffert, D., Bradley, J.D., Rominger, C.M., Rominger, D.H. et al. (2000) Presenilin-1 and -2 are molecular targets for γ-secretase inhibitors. *J. Biolumin. Chemilumin.* **275**, 34086–34091.

Shoichet, B.K. (2006) Interpreting steep dose-response curves in early inhibitor discovery. *J. Med. Chem.* **49**, 7274–7277.

Southey, M.W.Y. and Brunavs, M. (2023) Introduction to small molecule drug discovery and preclinical development. *Front. Drug Discovery* **3**, 1314077.

Stein, E.M., Garcia-Manero, G., Rizzieri, D.A., Tibes, R. et al. (2018) The DOT1L inhibitor pinometostat reduces H3K79 methylation and has modest clinical activity in adult acute leukemia. *Blood* **131**, 2661–2669.

Stock, M. and Gorochowski, T.E. (2024) Open-endedness in synthetic biology: a route to continual innovation for biological design. *Sci. Adv.* **10**, eadi3621.

Veber, D.F., Johnson, S.R., Cheng, H.Y., Smith, B.R., Ward, K.W. and Kopple, K.D. (2002) Molecular properties that influence the oral bioavailability of drug candidates. *J. Med. Chem.* **45**, 2615–2623.

Welsch, M.E., Synder, S.A. and Stockwell, B.R. (2010) Privileged scaffolds for library design and drug discovery. *Curr. Opin. Chem. Biol.* **14**, 347–361.

Wilson, T.M. and Moore, J.T. (2002) Genomics versus orphan nuclear receptors – a half-time report. *Mol. Endocrinol.* **16**, 1135–1144.

Yuan, L., Kurek, I., English, J. and Kennan, R. (2005) Laboratory-directed protein evolution. *Microbiol. Mol. Biol. Rev.* **69**, 373–392.

Zhang, J.H., Chung, T.D.Y. and Oldenburg, K.R. (1999) A simple statistical parameter for use in evaluation and validation of high throughput screening assays. *J. Biomol. Screening* **4**, 67–73.

Zheng, W., Thorne, N. and McKew, J.C. (2013) Phenotypic screens as a renewed approach to drug discovery. *Drug Discovery Today* **18**, 1067–1073.

Zimmermann, R.A., Schwickert, M., Meidner, J.L., Nidoieva, Z., Helm, M. and Schirmeister, T. (2022) An optimized microscale thermophoresis method for high throughput screening of DNA methyltransferase 2 ligands. *ACS Pharmacol. Transl. Sci.* **5**, 1079–1085.

5

From Hits to Leads

Key Learning Points

- The integrated activities that drive compound optimization define a continuum from validated hits to optimized clinical candidate molecules. An important milestone along this continuum is the optimization of validated hits to leads.
- For pedagogic purposes, we define a lead as compound(s) displaying target potency, target selectivity, cell permeability, solubility, and other properties sufficient to demonstrate modulation of intracellular target function, leading to cell-selective phenotypic outcome.
- The hit-to-lead stage of drug discovery is thus an exercise in multiparametric optimization, simultaneously seeking to optimize compounds with respect to target potency, target selectivity, solubility, cell permeability, free-fraction clearance, and several other parameters related to effective modulation of intracellular target function.
- Concomitantly, researchers begin to optimize parameters related to *in vivo* compound adsorption, distribution, metabolism, and elimination (ADME), and chemical tractability during the hit-to-lead stage.
- Relating back to the receptor theory (Chapter 1), a lead series of compounds should display correlated target potency, intracellular pharmacodynamic potency, and cellular phenotypic potency, in what has been termed the elements of translational chemical biology.

With hit molecules having been identified and validated, the research team now turns its attention to the optimization of these hits to create molecules that combine the potency, selectivity, chemical, and pharmacological characteristics to be worthy of clinical testing in human subjects. A key milestone along the path

Molecular Targeted Drug Discovery: A Guide to How Modern Medicines are Created,
First Edition. Robert A. Copeland.

to this ultimate goal is the demonstration that the chemical structures of validated hits can be modified to yield compounds that alter target function within cells (pharmacodynamics [PD]), and that this, in turn, leads to the desired phenotypic outcome in specific disease-associated cells. Compounds that demonstrate such cellular activity are termed leads (or lead molecules) for the purpose of our present discussion. In this chapter, we shall discuss the multiple characteristics of compounds that must be simultaneously optimized to facilitate compound action within cells. We shall also see that during the hit-to-lead stage of drug discovery, researchers also begin preliminary optimization and characterization of the adsorption, distribution, metabolism, and elimination (ADME) properties of the compounds, as well as their chemical tractability. As with every stage of drug discovery, success in these efforts requires the integration of multiple scientific disciplines. Hence, with a validated target and validated hits against that target in hand, research organizations typically form a more formal project matrix team at this stage of discovery.

5.1 The Project Matrix Team

Drug discovery is one of the most complex and interdisciplinary endeavors that one can envisage. Despite the common portrait painted by the film industry, academic public relations, and others, the discovery and development of a novel medicine is never the result of a single researcher in an epiphanic flash of brilliance. Rather, it is a team effort involving systematic integration of a multitude of scientific disciplines. Some of the key disciplines represented in a typical project matrix team are illustrated in Figure 5.1. This figure is not meant

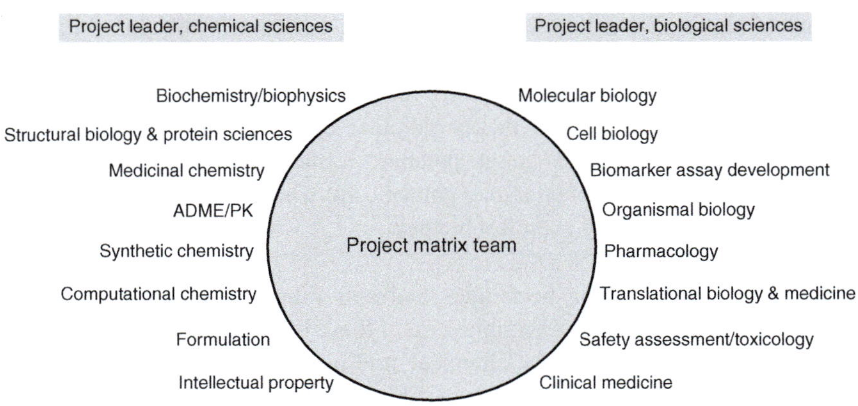

Figure 5.1 Some of the scientific disciplines typically integrated into a drug discovery project matrix team.

to be comprehensive but rather to give some sense of the breadth of disciplines required for drug discovery.

The degree of engagement of each discipline may wax and wane at different stages of discovery. For example, during the hit-to-lead stage, formulation, organismal biology, safety assessment, and clinical medicine will have relatively minor roles that will become amplified as the team progress on to lead-optimization and investigational new drug (IND)-enabling studies (see subsequent chapters and appendices). Nevertheless, it is important to engage these disciplines early in the project lifetime in planning for success. Also, as illustrated in Figure 5.1, typically two researchers are enlisted to colead the project matrix team – one with expertise in the chemical sciences and the other with expertise in the biological sciences. The coleaders are primarily tasked with setting the overarching goals of the team (discussed more in Chapter 6), integrating the efforts and results of the various team members toward the specific goals, and reporting team progress to the scientific and organizational leadership of the enterprise (e.g., the company, institute, or other organization; the board of directors; the scientific advisory board; the clinical advisory board; and others). Another critical leadership need for any project matrix team is to negotiate with organizational leadership to set aggressive timelines and budgets for each stage of drug discovery, to oversee these nonscientific aspects of team performance, and to report any deviations to organizational leadership in a timely manner. In some organizations, these duties are also assigned to the scientific project coleaders and in others they are the purview of project management professionals. Regardless of the details of team structure, the question arises as to what may be reasonable expectations for the time and cost of each stage of drug discovery. Little has been published on this topic, but Paul et al. (2010) have published average values for large pharmaceutical companies. The mean costs and cycle times for each stage are summarized in Table 5.1. Interestingly, in 2021, Dr. Bruce Booth of Atlas Venture and this author performed an informal survey of biotechnology companies and found that the mean costs and cycle times for each stage were similar in magnitude to what was reported for large pharmaceutical companies by Paul et al. (2010). Of course, each individual project will move at its own cadence and costs, but the values provided by Paul et al. (2010) are likely to be reasonable estimates of averages for the industry at large (appropriately adjusted for inflation).

The concept of a project matrix team is introduced here because it serves as a valuable construct for efficient and effective operational governance of systematic drug discovery, and because this construct is often unfamiliar to newcomers to drug discovery. Having introduced this organizational concept, we now return our attention to the multiple properties of small molecules that must be simultaneously optimized to yield cell-active lead compounds.

Table 5.1 Mean direct costs and cycle times for different stages of drug discovery. Note that costs quoted in this table reflect direct costs only and do not account for internal employee salaries, benefits, overhead, and other indirect costs.

Stage	Mean cost (in millions of US dollars)	Mean cycle time (years)
Target-to-hit	1.0	1.0
Hit-to-lead	2.5	1.5
Lead-to-candidate	10.0	2.0
IND-enabling	5.0	1.0
Totals	18.5	5.5

Source: Adapted from Paul et al. (2010).

As discussed in earlier chapters, at each stage of drug discovery, it is important to prospectively define specific (and where practical, quantitative) metrics by which success can be objectively judged. Hence, one of the first tasks to be taken up by the project matrix team is the establishment of a set of criteria that should be met to declare a compound or series as leads. As with hit-validation criteria (Chapter 4), the set of criteria for lead declaration is somewhat subjective and varies from group to group. They should also be viewed as guidelines for lead declaration and not as immutable and rigid requirements. Nevertheless, it is important that the project matrix team define these criteria prospectively, so that the team has a common view of the target for which they are aiming. Table 5.2 provides an example set of lead declaration criteria for a project focused on reversible, small-molecule modulators of a molecular target. These are divided into three major categories of criteria: *in vitro* properties, physicochemical properties, and ADME properties.

5.2 Target Potency and Selectivity

For each chemical-structure series (referred to as a chemotype), originating from a validated hit, medicinal chemists will initiate systematic structure–activity relationship (SAR) studies with the goal of driving potency to a level sufficient for cellular studies. The target potency required to realize cellular target modulation and phenotypic outcome will vary, depending on the mechanism of modulation, compound modality, and other features of the molecules. Generally, however, one would wish to drive the target-modulator K_d values to <500 nM and preferably <100 nM, before attempting any cellular assays. This may seem like a stringent threshold, based on some of the published literature. However, experience has

Table 5.2 Example of typical lead declaration criteria for a reversible, small-molecule modulator of a molecular target.

Property	Criterion
***in vitro* properties**	
Target potency (EC_{50})	<500 nM, preferably <100 nM
Target selectivity	>10-fold
Concentration–response behavior	Well-behaved with minimum % modulation = 0 ± 10%, maximum % modulation = 100 ± 10%, Hill coefficient = 1.0 ± 0.2
Reversibility	Reversible as demonstrated by jump dilution, SPR, or other methods
Modulation modality (e.g., competitive, noncompetitive, uncompetitive)	Experimentally determined to ensure proper evaluation of potency and other properties
Cellular PD potency (EC_{50})	<2 μM
Cellular phenotypic potency (EC_{50})	<2 μM
General cytotoxicity (IC_{50})	>10 μM or >100-fold phenotypic EC_{50}
Co-crystal structure bound to target protein	Highly desirable, but not required
Demonstrable SAR	Required, including inactive compounds within the pharmacophore series
Physicochemical properties	
Molecular weight	<500 Da
Lipophilicity:	
Log (P) or Log ($D_{7.4}$)[a]	<5 or 1–3
Solubility (aqueous)	>10 μM
Adherence to Lipinski rule of 5	Generally desired
Clear intellectual property position	Required
ADME properties	
Plasma protein binding (PPB)	<99%
hERG inhibition (IC_{50})	>10 μM
Whole blood or plasma stability ($t_{1/2}$)[b]	>30 min.
Liver microsome stability ($t_{1/2}$)[b]	>15 min.
Caco-2 permeability (cm/s)	$>2 \times 10^{-6}$

a) Log (D) is the apparent partition coefficient (between octanol and water) for ionizable compounds, measured at a specific pH for the aqueous solution (here pH = 7.4).

b) $t_{1/2}$, or half-life, refers to the amount of time required to reduce the concentration of starting compound by 50%.

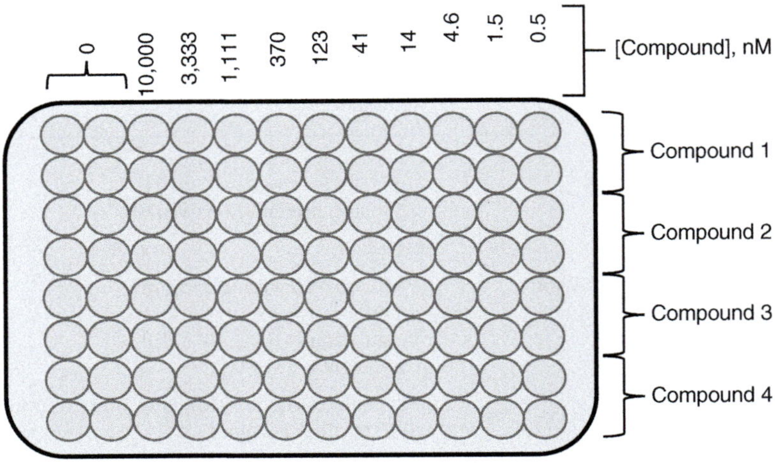

Figure 5.2 Schematic diagram of compound titration, based on a 10-point, 3-fold dilution scheme (Copeland 2013) with a top concentration of 10 μM (10,000 nM), in 96-well microtiter plates. Each plate contains 8 rows of 12 wells per row. In each row, the two leftmost wells are reserved for controls (zero compound). The remainder of the wells in a row titrate the compound in 3-fold descending concentration from 10 μM, going from left to right. By this scheme, four compounds can be titrated over the same range of compound concentration, each in duplicate. *Source:* Copeland, 2013 / with permission of John Wiley & Sons.

shown that it is often difficult to distinguish target-specific cellular effects from nonspecific effects at compound concentrations greater than about 10–50 μM. In performing cellular assays, it is seldom sufficient to test compounds at a single concentration, despite this also being a common practice in the literature. Instead, one wishes to demonstrate a clear concentration–response relationship in cellular assays that conforms to the Hill–Langmuir isothermal behavior discussed in previous chapters. Considering the form of the Hill–Langmuir isotherm, this requires titration of compound concentration from about 100-fold below the K_d value to 100-fold above the K_d value. This is readily accomplished by using a 3-fold dilution scheme (Copeland 2013) starting at the highest concentration to be tested, as illustrated in Figure 5.2 with an example titration from a top compound concentration of 10 μM.

It can become difficult to perform a complete concentration–response titration at top concentrations much above 10–50 μM, as nonspecific effects on cells are often seen at higher concentration (as stated above) and because results can also be confounded by approaching or exceeding the solubility limit of the compound at these higher concentrations. We shall discuss cellular assays in greater detail below. For now, we may consider 100–500 nM as a rough or ballpark estimate of the target affinity threshold for progressing to cellular assays. This threshold should be viewed as a generic guideline and not as a rigid rule. The researcher should feel free to experiment with what range of compound concentrations may be appropriate for cellular studies on a case-by-case basis.

5.2.1 Comparative Measures of Target Potency

During SAR studies, two situations arise in which one will wish to make comparisons of target–compound affinity among compounds: among compounds within a common chemotype (intra-series comparisons) and among compounds of different chemotypes (inter-series comparisons). Intra-series comparisons are the initial basis of SAR for any chemotype series. In this case, the modulation modality (i.e., competitive, noncompetitive, uncompetitive; see Chapter 2) should be determined for the founder compound (i.e., the validated hit) of the series and should be confirmed from time to time during SAR for sentinel compounds (i.e., compounds of particular interest in terms of SAR, structural insights [see below], and/or progression to other assays). Knowing the modulation modality and fixing the natural ligand concentration at a specific value relative to its K_d (e.g., $[S]/K_M$ for an enzyme substrate), one can safely compare relative target affinity among compounds of the same chemotype series using the EC_{50} values from the binding or functional assay. These values can be readily converted to K_d values by use of the appropriate Cheng–Prusoff equation for the modulation modality of the series (see Chapter 2).

The situation is a bit more complicated for making inter-series comparisons. This often arises when one wishes to simultaneously perform SAR on more than one chemotype series, for example when one has multiple validated hits from hit-finding efforts. For practical convenience, the binding or functional assay used for performing SAR studies will be fixed in terms of solution conditions, including the concentration of natural ligand relative to its K_d value. However, if the two chemotype series being compared conform to different modulation modalities (e.g., one is competitive and the other is uncompetitive), direct comparison of EC_{50} values may no longer be valid. In this case, one should calculate the K_d value for each compound, from the appropriate Cheng–Prusoff equation (Chapter 2), and make SAR comparisons of target affinity based on this metric. Let us work through an example to illustrate the concern raised here.

Suppose that we wish to compare the potency for two compounds, **1** and **2**, from two different chemotype series, that are both inhibitors of a target enzyme. Let us say that our primary assay of enzyme function is run under conditions of $[S]/K_M = 0.1$, because of solubility limitations of the substrate. Suppose the series from which compound **1** is derived is competitive with substrate, but that from which compound **2** is derived is uncompetitive with substrate. Suppose further that both compounds are equipotent in binding to their respective enzyme species

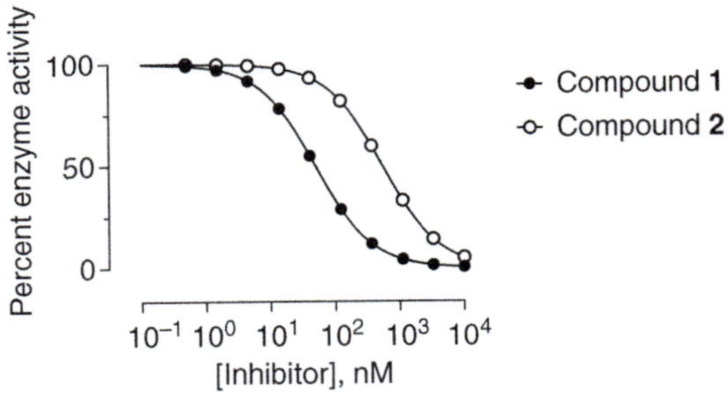

Figure 5.3 Concentration–response plots for two hypothetical inhibitors of a target enzyme that are equipotent with K_d values of 50 nM, measured in an enzyme functional assay at $[S]/K_M = 0.1$. Compound **1** is a competitive inhibitor and displays an EC_{50} value of 55 nM. Compound **2** is an uncompetitive inhibitor and displays an EC_{50} of 550 nM. This example illustrates the issue with making comparisons of target affinity based on EC_{50} values alone for compounds of different chemotypes. See text for further details.

(free enzyme in the case of the competitive inhibitor and enzyme–substrate complex in the case of the uncompetitive inhibitor; Chapter 2), with K_d values (K_i and αK_i, respectively) of 50 nM. Based on the Cheng–Prusoff equations for competitive and uncompetitive inhibition (Chapter 2), we can calculate the EC_{50} values that we would expect to observe for each compound under the fixed assay conditions being used (Figure 5.3). For the competitive inhibitor, we would expect the EC_{50} to be 55 nM, but for the uncompetitive inhibitor, the EC_{50} would be expected to be 550 nM. Hence, if we compared these two compounds based on EC_{50} alone, without accounting for their different inhibition modalities, we would erroneously assume that compound **2** is 10-fold less potent than compound **1**.

Similarly, when comparing the potency of a single compound toward more than one protein (see below), there can be differences between the assays of the two proteins with respect to solution conditions and the like. Hence, again making comparisons based on EC_{50} values can be misleading, and one is usually best served by instead making such comparisons based on K_d values (Copeland 2013, 2023).

Another motive for making inter-series comparisons is when comparing the target affinity of a chemotype series under study with that of a standard compound of a different chemotype. For example, there may be a literature or patent report of a modulator of the target protein from a competitor laboratory that is a tool compound, part of the competitor's own preclinical research efforts, already in clinic studies, or even an approved therapeutic. Whenever practical, the researcher should purchase or synthesize the competitor's compound and test this in the same assay system used for internal chemotype SAR studies. In this manner, the researcher can ensure that comparisons are made under the same solution conditions (e.g., pH, ionic strength, $[S]/K_M$, and the like), confirm the modulation modality of the competitor's compound, and again make comparisons based on K_d values.

Unfortunately, there can be situations in which the chemical structure of a competitor's compound is unclear, and only a literature value of EC_{50} is available. This can happen, for example, when a competitor has advanced a compound to clinical studies without revealing its exact chemical structure but has reported EC_{50} values for members of the chemotype series in a patent application. In such circumstances, the group filing the patent may purposely attempt to obscure the identity of the clinical compound among the compounds reported in the patent application. Now, one has no choice but to make comparisons based on the reported EC_{50} value in the patent application. However, appropriate caution should be exercised in making such comparisons, as one may not know the exact solution conditions or assay format used to generate the competitor's EC_{50} value. The appropriate caveats for such comparisons should be made clear when reporting such comparisons in the literature, at scientific conferences, or at internal meetings with organizational leadership.

A question that often arises in comparing EC_{50} or K_d values is, "What is a meaningful change in value from one compound to the next, or from one protein to the next". A rule of thumb commonly used in drug discovery is that values within 3-fold of one another are usually within the noise of the assay and are therefore not considered different. A 10-fold change in value is usually considered to be a meaningful difference, but again this is an arbitrary threshold based merely on common experience. To determine if two values (say for sample 1 and sample 2) are statistically different, the mean values (EC_{50} or K_d) and their associated standard errors can be used to calculate the t statistic (Equation 5.1) as a test for statistical significance.

$$t = \frac{|K_1 - K_2|}{\sqrt{\mathrm{SE}_1^2 + \mathrm{SE}_2^2}} \tag{5.1}$$

where K_1 and K_2 are the K_d values for samples 1 and 2, respectively, and SE_1 and SE_2 are the standard errors of samples 1 and 2, respectively. The standard error is calculated as the standard deviation divided by the square root of n, the sample size. Further details of the t-test can be found in any text on statistics and data analysis (e.g., Motulsky and Christopoulos, 2004).

5.2.2 Selectivity Profiling

In addition to primary target potency, the selectivity of a compound for its intended target is another critical aspect of pharmacological tractability. The collateral modulation of nontarget proteins is commonly referred to as "off-target activity" and is often the molecular underpinning of off-target toxicity in animals. Hence, it is critical to understand the degree to which a compound will modulate nontarget proteins, relative to its potency for the target of interest. This is quantified by the selectivity ratio, sometimes also referred to as the fold selectivity.

The selectivity ratio is commonly calculated as the ratio of the EC_{50} value for a nontarget protein over that for the target protein. However, as just described, EC_{50} values are significantly dependent on assay conditions, natural ligand concentration, and modality of target modulation. Hence, these values are inherently unreliable as true measures of compound affinity for any protein. What is most germane to safety concerns is to understand the relative degree of modulation (or drug occupancy) for the target of interest and any potential toxicity-causing collateral protein(s) at a fixed concentration for *in vivo* dosing. In favorable cases, one may know the modality of interaction, the K_d for both proteins (target and collateral binder), and the physiological concentrations of natural substrates or agonists against which the drug may need to compete *in vivo*. With this information in hand, one could calculate the expected fractional occupancies of each protein at a given systemic exposure (i.e., concentration) of drug. Hence, the selectivity under physiological conditions could thus be determined. Unfortunately, it is rare that one has this depth of information available, especially in the early stage of hit-to-lead. Therefore, during preclinical compound assessment, the selectivity ratio is most appropriately defined by the ratio of K_d values of the compound for the collateral protein as follows.

$$\text{Selectivity} = \frac{K_d^{\text{collateral}}}{K_d^{\text{target}}} \tag{5.2}$$

Thus, from Equation 5.2 if a compound displayed a K_d of 10 nM against a target protein and 1 μM (1000 nM) against a collateral protein, one would say that the compound displayed 100-fold selectivity for the target. Of course, this only represents the selectivity with respect to a single collateral protein. In practice, a selected panel of collateral proteins is typically screened, and the panel size and constituents vary with the target and the stage of drug discovery.

At the early stage of drug discovery, it would be prohibitively expensive and time-consuming to do exhaustive selectivity analysis against a broad panel of proteins. More typically, one would assess selectivity against a relatively small panel of proteins that are structurally and/or functionally similar to the target of interest. If, for example, the target protein was a receptor tyrosine kinase, in the hit-to-lead stage one might decide to assess selectivity with a panel of 5 to 10 other receptor tyrosine kinases that share the greatest sequence homology or substrate protein preference with the target enzyme. The reasoning here is that if one can achieve selectivity against these closely related proteins, then the risk of off-target activity against more structurally unrelated proteins is much lower. Of course, Nature has a way of surprising drug discovery scientists, and this working principle does not always prove to be correct. Nevertheless, this is a reasonable and practical approach to early assessment of compound selectivity. Later, usually during lead optimization (Chapter 6), one would expand the repertoire of proteins against which selectivity is assessed. This is typically done in two stages. First, the panel of structurally and/or functionally related proteins is expanded to cover a broader representation of the target family of proteins. Again, using the example of a receptor tyrosine kinase as the target, one might expand the selectivity panel to include additional protein kinases, especially kinases for which inhibition is associated with clinical toxicity, regardless of their amino acid sequence relatedness to the target of interest. This may be done in-house or contracted out to contract research organizations (CROs) that specialize in broad screening of specific protein classes. In the second stage of selectivity assessment, one would expand the selectivity panel even further and beyond the protein class (e.g., kinases) to which the target belongs. This broader screening for off-target binding or modulation is usually reserved for later in the lead-optimization stage of drug discovery, and we shall therefore defer further discussion of this until Chapter 6.

5.2.3 Iterative Structural Biology

Throughout compound optimization – from validated hit through clinical candidate selection – iterative structural biology is a powerful springboard for medicinal chemistry. Having an atomic-resolution structure of the target protein with compound bound provides a bounty of information that can be used to drive chemical structure optimization along three vectors: target potency, target selectivity, and ADME optimization.

The structure of the target–compound complex provides details of the recognition elements on the protein and the corresponding elements on the compound with which these engage. This provides a starting point for SAR hypothesis generation relating to compound structural changes that may strengthen existing interactions and/or augment binding affinity by engaging additional, proximal recognition elements. Amino acid side chains can engage the compound through hydrogen bonding, van der Waals forces, π–π interaction, and π–cation interactions. Protein backbone amides can also engage in hydrogen bond interactions with the compound, and bound water molecules can act as hydrogen bonding bridges, creating indirect interactions between protein and compound recognition elements. All these types of interactions can be strengthened and new interactions created by specific changes to compound structure to drive improvements in target affinity. In Chapter 4 (Table 4.3), we summarized the SAR of *S*-adenosylhomocysteine (SAH) mimetic inhibitors of the protein lysine methyltransferase, DOT1L (Basavapathruni et al. 2012). The ability to generate iterative, co-crystal structures of DOT1L–ligand complexes

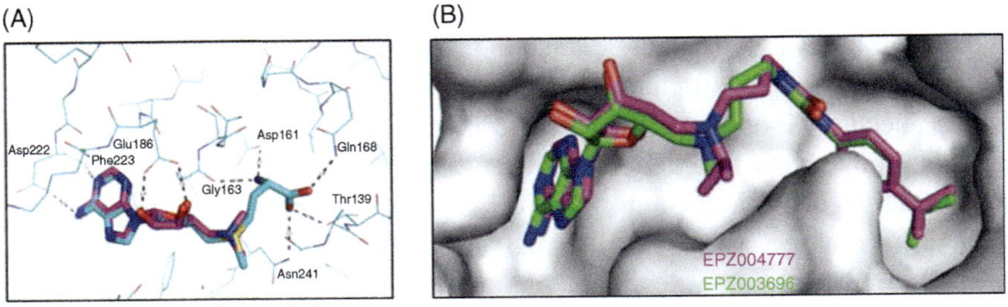

Figure 5.4 Co-crystal structures of the enzyme DOT1L with various ligands bound in the active site. (A) DOT1L active site structure with SAH (carbon atoms in light blue) and EPZ4 (carbon atoms in maroon) bound, showing some of the major interactions between the enzyme and ligands. (B) DOT1L active site structure with EPZ3696 (green) and EPZ4777 (magenta) illustrating the neomorphic side pocket induced by binding of EPZ4777 and other extended compounds. *Source:* Basavapathruni et al. (2012) / with permission of John Wiley & Sons.

was critical to driving SAR for this program and revealed an unexpected basis for optimization of target potency. Early in the program, the co-crystal structures of DOT1L complexes with the substrate and product of the enzymatic reaction, SAM and SAH, were solved and showed that both ligands bound in the enzyme active site in an extended conformation, in stark contrast to that of all other known protein lysine methyltransferases (e.g., SAM in Figure 5.4A, carbon atoms in light blue). As summarized in Table 4.3, Basavapathruni et al. (2012) used this information to design SAH mimetic inhibitors. Initially, compounds, such as EPZ4 and EPZ3696, engaged the same set of recognition elements as SAH (EPZ4 in Figure 5.4A, carbon atoms in maroon; EPZ3696 in Figure 5.4B, green structure). Subsequently, this team sought to extend the homocysteine mimetic side chain to engage additional recognition elements within the methyl-accepting lysine–binding pocket; EPZ4777 is an example of such an extended compound and indeed this compound lowered the K_i value, relative to EPZ3696, from 13 to 0.3 nM (Table 4.3). When the co-crystal structure of the DOT1L–EPZ4777 complex was solved, the team was surprised to find that, contrary to the design strategy, the extended compound did not reach into the methyl-accepting lysine–binding pocket. Instead, the binding of extended compounds like EPZ4777 induced a conformational change of the enzyme, creating a neomorphic pocket (i.e., a pocket that was not evident in the structures of DOT1L bound to more truncated ligands) into which the compound bound, engaging completely novel recognition elements (Figure 5.4B, magenta structure). This unexpected discovery dramatically shifted the emphasis of medicinal chemistry, ultimately resulting in pinometostat (EPZ5676), which entered phase 1 clinical studies (Stein et al. 2018).

Selectivity against closely related proteins can be a challenge, especially in the hit-to-lead stage where the SAR knowledge base is still rather limited. A powerful tool to aid in achieving selectivity is the application of structural biology. By way of example, let us consider inhibitors selective for cyclooxygenase 2 (COX2) over the structurally similar cyclooxygenase 1 (COX1). Nonsteroidal anti-inflammatory drugs (NSAIDs), such as aspirin, ibuprofen, indomethacin, and the like, inhibit both COX isozymes. The inhibition of COX2 by these drugs leads to their analgesic, antipyretic, and other anti-inflammatory efficacy. However, their additional inhibition of COX1 is thought to lead to the gastric and renal damage associated with high-dose, chronic use of these drugs. Hence, the discovery of the second isozyme, COX2, led several research groups to launch programs aimed at identifying selective inhibitors of COX2. Human COX1 and COX2 share a high degree of sequence similarity, especially in the active site region of the enzymes (Smith et al. 1996; Blobaum and Marnett 2007). Nevertheless, COX2-selective inhibitors were quickly identified from compound screening (see, for example, Copeland et al. 1994), and several of these led to clinical stage compounds (Blobaum and Marnett 2007). The selectivity for COX2 displayed by these compounds depends on a slow, induced fit mechanism of inhibition – a mechanism which we shall discuss further in Chapter 6. The structural basis for selectivity is made clear from the crystal structures of the two isozymes and their co-crystal structures with inhibitors bound (Luong et al. 1996; Blobaum and Marnett 2007). The COX2-selective inhibitors bind within a side pocket off the main substrate-binding pocket of the enzyme; this side pocket is much smaller and less accessible to compounds in the COX1 isozyme. Three amino acid differences between COX1 and COX2 facilitate the selective binding of compounds to the side pocket of COX2 (Figure 5.5). Ile523 of COX1 is replaced by Val in COX2, providing greater access to the side pocket due to the reduced steric bulk of Val compared to Ile. Ile434 of COX1 is also replaced by a Val in COX2, and this replacement opens a path for the nearby Phe 518 to swing out of the way of the mouth of this side pocket. Finally, His513 of COX1 is replaced by Arg in COX2, and the side chain of Arg513 provides a locus for productive interactions with the polar sulfonamide or methylsulfonyl groups that are common features of COX2-selective inhibitors (Blobaum and Marnett 2007; Copeland 2021). The structural insight into the mechanism of COX2 selectivity for inhibitors was retrospective to the identification of such compounds. However, this example is illustrative of how structural biology can be used prospectively to develop SAR hypotheses for achieving selectivity for a target of interest over structurally and/or mechanistically related collateral proteins.

Finally, iterative structural biology, starting early in validated hit optimization, can often reveal trajectories leading from the compound-binding pocket to the solvent-exposed surface of the protein. This can provide medicinal chemists with important information for appending groups onto compounds not for the

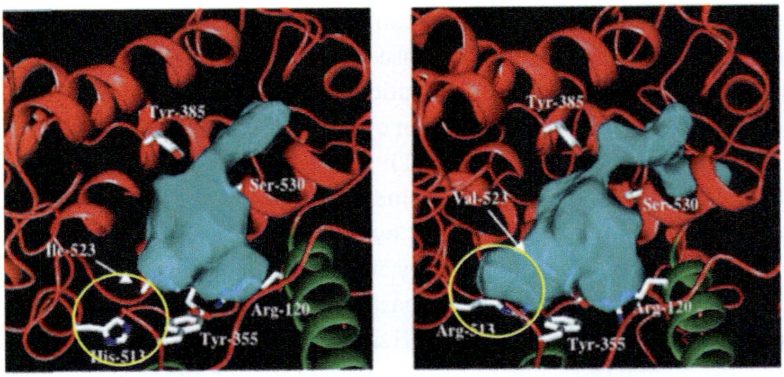

COX-1 COX-2

Figure 5.5 Crystal structures of COX1 and COX2 highlighting the solvent accessible surface areas (in blue) of the catalytic domains (in red) of the two isozymes. The side pocket into which selective COX2 inhibitors bind is highlighted by the yellow circles in the two images. Note the changes in COX1 Ile523 to Val in COX2 near the mouth of the side pocket and COX1 His513 to Arg in COX2 at the bottom of this pocket. COX1 Ile434 is also changed to Val in COX2, but this change is not visible in this view. *Source:* Adapted from Blobaum and Marnett 2007.

purpose of improving target affinity, but rather for improving other physicochemical properties of the compound, such as aqueous solubility, Log(*P*), oral bioavailability, or greater pharmacokinetic stability. Improvements in all these properties are important aspects of the multiparametric optimization of hits toward cell-active and eventually *in vivo* active compounds. Additionally, a trajectory from the binding pocket to the solvent-exposed surface can likewise provide a basis for appending a linker group to the pharmacophore for the purpose of developing a heterobifunctional target degrader (i.e., a PROTAC), as discussed in Chapter 2.

5.2.4 Solubility and Lipophilicity

The solubility of a compound is critical for determining the appropriate range of concentrations that can be used in cell-free and cellular assays in the early stages of hit optimization. Solubility is commonly measured by the shake flask method, in which an excess amount of solid compound is added to a specific amount of buffer, and this saturated solution is agitated (i.e., shaken) for 24 hours to establish equilibrium. The solution is then separated from the residual undissolved compound, and the concentration of dissolved compound is determined by spectroscopic, mass spectral, potentiometric, light scattering, or other analytical methods (Aderniji and Adejare, 2008; Sou and Bergstrom 2018). It is important to recognize that solubility varies with temperature and pH. Hence, solubility measurements

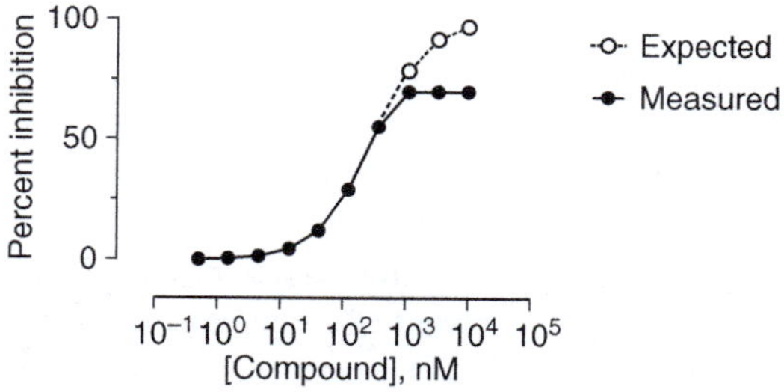

Figure 5.6 Example of a concentration–response plot for an enzyme inhibitor with a solubility limit of 1 μM. The open circles represent the expected behavior of an inhibitor in an enzyme assay, approaching 100% inhibition as the inhibitor is titrated to higher concentrations. The closed circles represent the experimentally measured apparent inhibition, which reaches an abrupt and premature percent inhibition plateau at concentrations at or above 1 μM (i.e., 10^3 nM).

for *in vitro* studies should be performed at pH 7.4 (or the pH to be used for *in vitro* assays) and a temperature of 25 or 37 °C.

Concentration–response studies using *in vitro* biochemical and biophysical assays should be framed by the solubility limit of the compound to be tested. It is often apparent when one has exceeded the solubility limit of a compound in such assays, as the response will reach an apparent premature plateau at concentrations exceeding the solubility limit. Consider the example of an enzyme inhibition assay for a compound with a solubility limit of 1 μM under the conditions of the enzyme assay (Figure 5.6). Our reasonable expectation is that the limits of percent inhibition will be zero as the compound concentration approaches zero, and 100% as the compound concentration approaches complete saturation (open circles in Figure 5.6). However, once the solubility limit of the compound is reached, no further compound concentration will be available in solution to inhibit the enzyme, and the system will reach an abrupt and premature plateau at less than 100% inhibition (in the present case, 70% inhibition). In more extreme cases, the insoluble compound in the assay system will cause light scattering that will affect the detection readout, leading to an over- or underestimation of analyte concentration.

Exceeding the solubility limit of a compound is also problematic in cellular assays. In phenotypic assays, for example, insoluble compound can exert a nonspecific impact on cell growth, causing the appearance of compound-induced cell killing. In other cases, limited compound solubility may result in a bell-shaped PD or phenotypic effect curve in cellular assays. In these cases, the readout may initially increase with increasing compound concentration, then drop dramatically as the solubility limit of the compound is approached. Thus, one must be cautious in ensuring that *in vitro* assays are performed over a concentration range of compound that does not exceed the solubility limit.

Compound solubility in aqueous media is clearly not only necessary for preclinical testing but also for compound dissolution in the gastrointestinal tract, bloodstream, and other aqueous bodily fluids. Aqueous solubility, however, must be balanced with compound lipophilicity that will allow compounds to distribute into tissue and to permeate cell membranes. As mentioned briefly in previous chapters, drug lipophilicity is commonly measured as the logarithm of the partition coefficient between aqueous and nonpolar solvent (P). The most common nonpolar solvent used for this purpose is 1-octanol. The partition coefficient is thus defined as follows.

$$P = \frac{[\text{Drug}]_{\text{octanol}}}{[\text{Drug}]_{\text{aqueous}}} \tag{5.3}$$

The use of P and of Log (P) is appropriate for unionizable compounds. However, if a compound contains ionizable groups, only the unionized version will be taken up by the nonpolar solvent. To account for ionization, researchers have defined an alternative parameter known as the distribution coefficient, D, defined at a specific pH of the aqueous solution.

$$D_{\text{pH}} = \frac{[\text{Drug}_{\text{unionized}}]_{\text{octanol}}}{[\text{Drug}_{\text{ionized}}]_{\text{aqueous}} + [\text{Drug}_{\text{unionized}}]_{\text{aqueous}}} \tag{5.4}$$

Knowing the pK_a of the ionizable group(s) on the compound, one could fix the pH of the aqueous phase to a value that would ensure predominately unionized compound, in which case Log (D) = Log (P). However, most researchers instead fix the pH at 7.4 to reflect the pH of whole blood, as this is the most relevant lipophilicity value for *in vivo* studies. Hence, it is common to see the lipophilicity of ionizable compounds defined by the Log ($D_{7.4}$).

Rather than experimentally determining the Log (P) or Log ($D_{7.4}$) of every compound, researchers rely on methods for calculating these values based on the chemical structure of the compound and training sets of compounds for predicting how different chemical groups will influence partitioning (see Aderniji and Adejare, (2008), for a review of various calculation algorithms). The calculated values of Log (P) and Log ($D_{7.4}$) are denoted as cLog (P) and cLog ($D_{7.4}$), respectively, the lower case "c" indicating that these are calculated values. Experimental determination of Log (P) or Log ($D_{7.4}$) is reserved for sentinel compounds during SAR campaigns.

5.3 Cellular Assays

Despite our earlier-stated guideline to reserve cellular testing for compounds displaying target–compound K_d values of <500 nM, in practice one will want to spot-check compounds for cellular permeability, PD effects, and phenotypic effects throughout the course of pharmacophore series SAR. In some favorable cases, one may be pleasantly surprised to find high cell permeability in compounds of more modest target affinity. Before testing compounds for PD and phenotypic effects, researchers generally assess cell permeability and active efflux in a generic cell line assay. The Caco-2 cell line, derived from an epithelial colon carcinoma, is the most used cell line for this purpose.

Compounds can enter the cell through three main mechanisms: passive diffusion, facilitated transport, and active transport (Roberts 2021). Passive diffusion occurs with compounds of sufficient lipophilicity to permeate the phospholipid cell membrane directly. Facilitated transport refers to the situation in which compounds bind to an integral membrane protein (a carrier protein) on the extracellular surface and are translocated across the cell membrane and into the cytosol of the cell. Carrier proteins generally belong to a group of proteins composed of 12 membrane-spanning helices. These carrier proteins function to transport otherwise impermeant metabolites, such as glucose and urea, into cells. Finally, active transport refers to the influx (or efflux) of small molecules, against their concentration gradient, by membrane-bound transporter proteins (also referred to as pumps) that utilize the energy of ATP hydrolysis, pH gradients, or ion concentration gradients to drive the movement of small molecules into or out of the cell. There are three superfamilies of drug transporter proteins: the ABC transporters, the SLC transporters, and the organic anion transporting polypeptide (OATP) transporters. Of these, the ABC transporter superfamily is the most mechanistically well understood and is exemplified by the common drug efflux transporter P-glycoprotein (PGP, also known as multidrug resistance protein 1 or MDR1).

5.3.1 Caco-2 Permeability and Active Efflux

Drugs that are administered orally (per os, PO) must be absorbed through the luminal surface enterocytes of the intestines to reach the mesenteric vein and ultimately systemic circulation. Intestinal permeability is thus a major determinant of the oral bioavailability of a drug. The Caco-2 cell line is used *in vitro* to assess the potential for intestinal permeability of compounds, determine the potential for active efflux of compounds, and provide a general measure of cell membrane permeability.

The Caco-2 cell line has the unique feature of spontaneously differentiating into monolayers of polarized cells when cultured. The cells form a tight monolayer on surfaces within the culture vessel, such as a porous membrane, with the apical (intestinal lumen) and basolateral (blood) surfaces of all cells oriented in the same direction (Figure 5.7).

Compound is added to the apical side chamber, and permeability is determined by measuring the amount of compound that permeates to the basolateral side

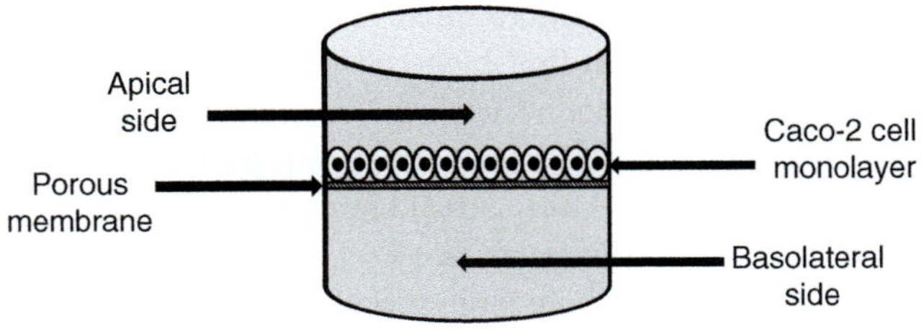

Figure 5.7 Schematic diagram of the experimental setup for measuring drug permeability/transport through Caco-2 cells. See text for further details.

chamber as a function of time. The apparent permeability (P_{app}) is calculated as follows.

$$P_{app} = \frac{V_A \times \frac{\partial Q}{\partial t}}{A \times C_0} \tag{5.5}$$

where V_A is the volume (mL = cm^3) of solution in the drug-accepting chamber (the basolateral chamber in the above example), $\partial Q/\partial t$ is the rate of transport, calculated as the molar concentration of compound appearing in the drug-accepting chamber divided by the interval of time (typical units = μM/s), A is the surface area (cm^2) of the monolayer, and C_0 is the molar concentration (e.g., μM) of compound initially added to the drug-donating chamber (the apical chamber in the above example). The resulting units of P_{app} are cm/s.

The apical (A) to basolateral (B) side experiment just described provides an estimate of directional transport from the intestinal lumen to the mesenteric vein. To understand the transport mechanism fully, one would also do the reverse experiment in which compound is added to the basolateral chamber and transport to the apical chamber is measured over time. If transport is due to passive diffusion of compound, the permeability should be the same when measured in either direction. If active transport through membrane-bound transporter proteins occurs, permeability will be greater in the A-to-B direction than in the B-to-A direction. In counterpoint, compounds may also efflux from the cell if they are recognized as substrates for the brush border membrane efflux pump, PGP. If significant efflux occurs for the test compound, the B-to-A permeability can be greater than that in the A-to-B direction. Further details of the Caco-2 assay can be found in Copeland (2013) and references therein.

5.3.2 Pharmacodynamic Assays

The term PD is often defined as "the effect that a drug has on the patient". However, for preclinical studies, a more appropriate definition would be the biochemical or biological effect of target engagement by a drug in cells, tissues, or organs of interest. Recall that in our discussion of the receptor theory of drug action (Chapter 1), we articulated a fundamental tenet of molecular targeted drug discovery: the binding of a drug to a specific receptor (i.e., target), and the attendant modulation of target function, is the predicate of the pharmacological action of the drug. Whereas target–drug binding ultimately leads to a phenotypic outcome that is disease modifying (i.e., pharmacological action), an intervening step in this cause-and-effect relationship is the modulation of biochemical pathways (PD) due to target engagement. Thus, we can define the causal relationship between drug binding to a specific target, ensuing PD effects, and the resultant phenotypic outcome in disease by the following simple Scheme 5.1.

Scheme 5.1 The causal relationship among specific target engagement (binding), modulation of cellular pharmacodynamics and phenotypic outcome that is the *sine qua non* of the receptor theory of molecular targeted therapeutics.

Although this scheme is generally correct in disease-associated cells, the causal relationship between PD and phenotypic outcome may not always hold for normal (i.e., non-disease) cells, as we shall explain momentarily.

Because target function in cells is very often part of a biochemical pathway, PD assays may be based on a proximal effect of target function or a more distal or downstream effect. As an example, suppose the goal of a research project is to discover and optimize inhibitors of the protein kinase BRAF. In response to extracellular epidermal growth factor (EGF) binding to its cell-surface receptor (EGFR), BRAF becomes activated and in turn phosphorylates mitogen-activated protein kinase kinase (MAPKK, otherwise known as MEK), and the thus activated MEK phosphorylates mitogen-activated protein kinase (MAPK, otherwise known as ERK). MEK-catalyzed phosphorylation of ERK activates the latter kinase, which then phosphorylates other cytosolic proteins, but is also translocated to the cell nucleus, where it can directly and/or indirectly phosphorylate transcription factors, such as myelocytomatosis (MYC) (a direct substrate of ERK) and cAMP response element-binding protein (CREB) (an indirect substrate). The phosphorylated transcription factors then induce transcription of a program of proteins that promote cell proliferation. [*Nota bene*: The above is a highly simplified description of the mitogen-activated signal transduction cascade that is useful for our pedagogic purposes. This signaling cascade is in fact much more complex, with multiple, intersecting pathways and multiple isoforms of rapidly accelerated fibrosarcoma (RAF), MEK, and ERK.] The most proximal PD effect that would result from inhibition of BRAF would be a reduction in the level of intracellular phospho-MEK, as this is the direct product of BRAF catalysis. A more downstream PD effect would be a reduction in the level of intracellular phospho-ERK, and even further downstream PD effects would be reductions of phospho-MYC or phospho-CREB, or diminished transcription (measured as mRNA level) of transcriptional targets of MYC and/or CREB. In general, the more proximal a PD effect is to the target of interest, the more reliably it will report on modulation of target function without (or with less of) the confounding issue of biological redundancy that can impact further downstream signals in a target-independent manner. In other words, the further downstream from the target a PD readout is, the more likely it may be impacted by biochemical mechanisms unrelated to the target of interest (e.g., in the example here, perhaps other kinase pathways could also contribute to augmented gene transcription). Despite this concern, in some cases, technical issues may make it more practical to measure distal PD effects because of better-quality reagents (e.g., antibodies), the stability of the PD mark, or other factors. Whenever practical, however, one should prioritize PD readouts that are as proximal to the target as possible.

Recall from Chapter 3 that in many diseases the drug target itself is not genetically altered but displays a selective essentiality in disease cells due to a synthetic lethal relationship with another protein. In other situations, disease-selective essentiality may be due to a unique alteration of the target (e.g., mutation) in disease cells, but the wild-type target function is dispensable in normal cells. In either of these cases, drug binding to the target will elicit the same proximal PD effects in both disease and normal cells, but the translation of this PD effect into phenotypic outcome may only occur in the disease cells (i.e., the basis of selective essentiality). Hence, while the scheme shown above remains intact for the disease cells, the causal relationship between PD effect and phenotypic outcome is not realized for normal cells. Table 5.3 provides some examples of this situation for cancer targets, where the PD effect is observed in both cancer and noncancer cells, but the antiproliferative phenotype of target inhibition by drugs is only seen in the cancer cells.

The reader may at first view the disconnect between PD and phenotypic readouts, exemplified in Table 5.3, as an issue. However, in practice, this can be beneficial in using a PD readout to detect intracellular target engagement for *in vivo* animal studies and human clinical trials in a noninvasive or minimally invasive manner. In many cases, the diseased tissue may be inaccessible to the researcher or physician without invasive methods, such as major surgery. Hence, PD readouts from the actual disease tissue may not be practical on a routine or longitudinal basis. If, however, drug treatment induces the same changes in PD readout in healthy tissue, this provides a basis for measuring PD in animals and/or patients without resorting to invasive measures. In this case, PD may be read out by sampling blood cells, cheek swabs, hair follicles, skin cells, and the like, using

Table 5.3 Some examples of cancer targets for which a proximal PD effect is seen in cancer and noncancer cells, but the phenotypic outcome (antiproliferation) is only seen in specific cancer cells.

		Impact of target modulation by compound			
		PD outcome		Phenotypic outcome	
Target	Proximal PD effect	Cancer	Noncancer	Cancer	Noncancer
DOT1L	H3K79me2 reduction	Yes	Yes	Yes	No
EZH2	H3K27me3 reduction	Yes	Yes	Yes	No
DHX9	circBIRP1 increase[a)]	Yes	Yes	Yes	No
METTL3	m^6A reduction[b)]	Yes	Yes	Yes	No

a) circBRIP1 is a circular RNA transcript.
b) $m^6A = N^6$-methyl adenosine within mRNA transcripts.

categorical or semiquantitative readouts, such as immunohistochemistry (IHC), or more quantitative readouts such as enzyme-linked immunosorbent assays (ELISAs), qPCR, and the like. The use of PD assays as biomarkers of biochemical response to drug in animal models and clinical studies will be presented further in Chapter 6.

In addition to measuring PD per se, intracellular assays of drug binding to target protein have more recently been developed and utilized in drug discovery and development (Schurmann et al. 2016; Stefaniak and Huber 2020). For example, the technique of bioluminescence resonance energy transfer (BRET) has been applied to intracellular drug–target interactions. Here, cells are transfected with the target of interest fused to a bioluminescence acceptor such as luciferase, and a cell-permeable small-molecule ligand is covalently appended with a fluorescent donor moiety (referred to as the tracer). When the tracer binds to the target within cells, one observes energy transfer from the tracer to the acceptor and resultant bioluminescence. Compounds can then be tested in this cellular system for their ability to competitively displace the tracer molecule. Similarly, a cell-permeable ligand for the target of interest can be tagged with a fluorescent group, and the fluorescence polarization (i.e., anisotropy) can be measured intracellularly. Again, the ability of other compounds to disrupt binding of the fluorescent ligand can then be determined from the changes in fluorescence polarization they cause. The ability of ligand binding to stabilize proteins to thermal denaturation (cellular thermal shift assay or CETSA) and proteolytic digestion (drug affinity responsive target stability or DARTS) have also been used as methods to detect intracellular compound binding to target proteins. The reviews by Schurmann et al. (2016) and Stefanik and Huber (2020) provide additional details on all these methods and references to the primary literature.

The methods used to measure intracellular target engagement are useful as alternatives when assays of bona fide PD are impractical, or as confirmatory assays in parallel with PD assays. However, because these assays do not report directly on the functional consequences of intracellular target engagement, and because they often require unnatural engineering of the cell system, they should be used cautiously. As we shall discuss below, whether using assays of intracellular target engagement or PD assays, it is critical to establish a quantitative relationship between compound potency in these assays, that seen in assays of cell-free target binding or function, and with that seen in cellular assays of phenotypic outcome.

5.3.3 Phenotypic Assays

If the biological hypothesis upon which target validation was based is correct, then small-molecule modulators of target function should elicit the same phenotypic outcome as target knockout or knockdown (Chapter 3). Phenotypic assays for validated hit optimization vary depending on the therapeutic area of interest and the specific disease to be addressed. Desired phenotypic outcomes of drug treatment can consist, for example, of changes in cell morphology; changes in cytokine, chemokine, or protein secretion; changes in cytotaxis; or changes in cell proliferation.

For infectious diseases and cancer, the most desired phenotypic outcome of drug treatment is selective cell killing. Typically, researchers quantify this by measuring cell proliferation in culture and the effect of compound concentration on proliferation. Cell proliferation proceeds through cell division that results in a doubling of the number of cells after each division (Figure 5.8).

Under a fixed set of environmental conditions, and provided that sufficient nutrients and other required factors are available in nonlimiting quantities, cells will proliferate by consecutive doubling (i.e., cell division) over a period referred to as log-phase growth (for reasons that will become clear below). During log-phase growth, the number of cells present in culture increases with time according to the following equation.

$$N_t = N_0 \times 2^{\left(\frac{t}{t_D}\right)} \tag{5.6}$$

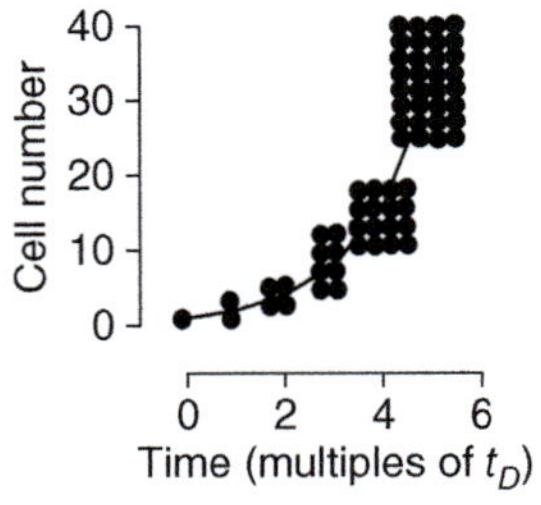

Figure 5.8 Cartoon of cell proliferation as a function of time (in multiples of the doubling time, t_D). At time = 0, a single cell (illustrated as a black circle) is placed into culture. Over the course of 5 doubling time periods, the population has increased to 32 cells.

where N_t is the number of cells at a time point (t) after initiation of the observation period, N_0 is the number of cells at time zero (i.e., the initiation of the observation period), and t_D is the doubling time (the time interval required for cell doubling) for the specific cell line under study. Equation (5.6) can be converted into an exponential equation in base e, using the equality 0.693 = ln(2).

$$N_t = N_0 e^{\left(\frac{0.693t}{t_D}\right)} \tag{5.7}$$

We can define the rate constant for cellular proliferation, k_p, as $0.693/t_D$, so that Equation 5.7 can be expressed as

$$N_t = N_0 e^{(k_p t)} \tag{5.8}$$

Equation 5.8 can be linearized by taking the natural logarithm of each side.

$$\ln(N_t) = \ln(N_0) + k_p t \tag{5.9}$$

Thus, from Equations 5.8 through 5.9, we see that if we seed a fixed number of cells into culture and measure the cell number as a function of time, we will observe exponential growth throughout the log phase of cell growth (where nutrients and other growth factors are in nonlimiting supply). If we titrate the concentration of a compound that inhibits cell growth, we will observe a set of exponential growth curves of decreasing amplitude as the concentration of compound increases (Figure 5.9A). We also see from Equation 5.9 that if we plot the cell number on a logarithmic scale, the exponential growth curves will appear linear, making it easy to derive the value of k_p from the slope of such plots (Figure 5.9B).

To quantify the potency of a compound in inhibiting cell growth, researchers typically measure the cell number at a single, specific time point (N_t^i, where t is often the last time point to be measured) within the log phase, and calculate the percent residual cell growth (%RCG) by normalizing this value to the corresponding value of the zero-compound control (N_t^0, e.g., treatment with dimethyl sulfoxide (DMSO) only, instead of compound dissolved in DMSO), correcting for any background signal, and multiplying by 100. Alternatively, some researchers prefer to report data as the percent inhibition of cell growth caused by compound (%Inhibition = 100−%RCG).

$$\%\text{RCG} = 100 \times \left(\frac{N_t^i}{N_t^0}\right) \tag{5.10}$$

The %RCG or %Inhibition is then plotted as a function of compound concentration to determine the EC_{50} for inhibition of cell proliferation (Figure 5.9C).

Although it is common practice to measure %RCG at a single, fixed time point, several authors argue that this can be misleading in that it can overestimate true compound potency and does not account for the complete time dependence of cell growth inhibition (Knutson et al. 2012; Hafner et al. 2016; Harris et al. 2016; Jarrett

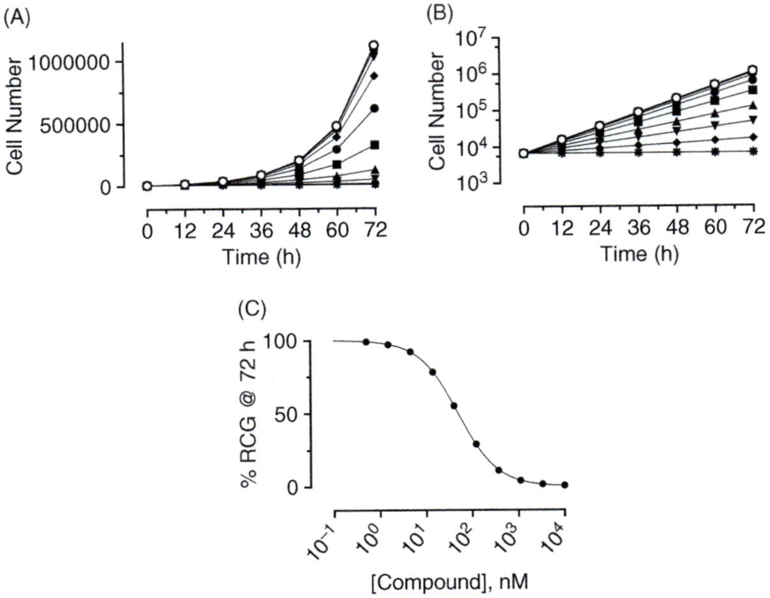

Figure 5.9 Measuring the effect of cell growth inhibitory compounds on cell proliferation. (A) Cell growth curves in the presence of vehicle (DMSO; open circles) and varying concentrations of a growth inhibitory compound (various closed symbols). (B) As in A but plotted on a semilog scale (logarithmic scale for cell number). (C) Concentration–response plot of %RCG as a function of compound concentration. See text for further details.

et al. 2018). For example, some inhibitors of cell proliferation, such as compounds targeting certain epigenetic targets, display a significant lag phase before their antiproliferative activity begins to be realized in cell culture assays. In such cases, it is recommended that one limit the observation period to start at the earliest time point at which compound effect is observed and end at the last measurement taken. Rather than measuring %RCG at a single time point, these authors instead suggest that the k_p value, or the related growth rate value (Hafner et al. 2016), be determined at each concentration of compound, and that this be used for the construction of concentration–response plots.

A compound applied to growing cells can affect proliferation by either of two distinct ways: (1) by inhibiting further cell division (cytostasis) or (2) cell killing (cytotoxicity). If a compound is cytostatic, increasing concentrations of compound will reduce the value of k_p over the observation period until no further cell division occurs, at which point, k_p becomes zero. In contrast, in the presence of a cytotoxic compound, the value of k_p will be influenced by the absolute values of two rate constants: the rate constant for residual cell proliferation in the presence of compound

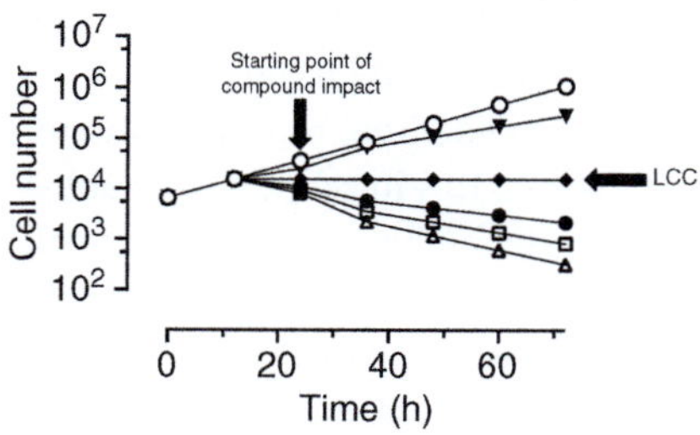

Figure 5.10 Example of the effects of titration of a cytotoxic compound on the cell proliferation after a lag period of 24 hours. Open circles represent the DMSO control data, and closed symbols represent different concentrations of the cytotoxic compound. The slope values (k_p) in such a case are measured from the starting point of compound impact to the end of the growth curve. The concentration of compound that results in a slope of zero is designated as the lowest cytotoxic concentration (LCC) of the compound. See Knutson et al. (2012) for further details on the calculation of LCC. *Source:* Adapted from Knutson et al. (2012).

(k_g) and the rate of cell killing by the compound (k_d).

$$k_p = |k_g| - |k_d| \tag{5.11}$$

At low concentrations of a cytotoxic compound, the value of k_g will be greater than that of k_d, so that the value of k_p will be positive (i.e., > 0). As compound concentration is increased, the rate of cell killing will exceed that of residual cell growth; hence, the value of k_p will now be negative (< 0). Knutson et al. (2012) have defined the lowest cytotoxic concentration (LCC) for cytotoxic compounds as the concentration at which the value of $k_p = 0$ and suggested that this value is a more informative metric of compound potency for cell killing (Figure 5.10).

5.3.4 Intervening Biological Responses

As described above, medicinal chemistry progress in understanding and driving SAR will be largely based on assays of proximal PD and ultimate phenotypic outcome. However, in most cases, there will be other biological consequences of drug action that occur between the intracellular target engagement and ultimate phenotypic outcome. We have already seen an example of this in the context of signal transduction cascades, wherein inhibition of an upstream kinase will not only result in diminished phosphorylation of the direct substrate of that kinase (a proximal PD effect) but will also result in diminished phosphorylation of proteins downstream of that substrate kinase, eventually leading to an altered program of gene transcription. We also saw in Chapter 4 that inhibition of the RNA helicase DHX9 selectively induces R-loop structures in MSI-H cancer cells that result in increased replication stress upon which selective MSI-H cancer cell killing is predicated.

Other examples of intervening biological consequences include cell cycle arrest and induction of apoptosis as precursors of drug-induced cell killing.

These intervening biological consequences can provide important information on the biological mechanism of drug action in the cell. It is important to establish that these intervening biological consequences are shared between siRNA knock-down and drug treatment (in the case of antagonists or inhibitory compounds), or gene amplification and drug treatment (in the case of agonists or activators). The signals with which these biological consequences are measured must be quantitative and titratable with compound concentration. However, in some cases, such as induction of apoptosis, the researcher may find that such signals display very high Hill coefficients, so that the concentration–response behavior is very abrupt, appearing more like a step function (i.e., a threshold effect) than the smooth, S-shaped curve that we have come to expect.

In some cases, assays for intervening biology may be available in convenient formats for high-throughput screening (HTS), so that these signals may be tested for every compound that the chemists may synthesize. However, since SAR is largely driven by proximal PD and phenotypic readouts, it is also often the case that readout of intervening biology is only conducted on sentinel compounds during SAR development, especially in cases where these assays are more complex, time-consuming, or otherwise burdensome.

5.3.5 Characteristics of Useful Cellular Assays

In addition to the general standards for any assay format (see Chapter 4), cell-based assays for drug discovery purposes must be quantitative and, in general, formattable for microtiter plate-based assays. Qualitative or categorical assays, based on low-throughput methods such as Western blotting, and like assays, are fine for early target-validation studies and/or as gold standards for development and validation of higher-throughput assays. However, for medicinal chemistry SAR efforts, one needs assays of the robustness, reproducibility, and throughput to prosecute tens to hundreds of compounds per month in replicate concentration–response studies. To conduct concentration–response studies, PD and phenotypic cellular assays must be quantitative, and whenever possible, based on calibration curves of analyte to be measured, as was discussed for cell-free assays (see above). Also, for PD assays, whenever possible it is desirable to be able to convert analyte signal into molar concentration of the PD species under study, as a means of accurate molecular accounting of the effects of compound treatment; this ensures that the assays make good biological, chemical, and physical sense. Let us consider an example to illustrate this last point.

Suppose that the PD marker we select for a certain target is induction of specific species, such as target-mediated formation of a metabolite, or modification of a protein or nucleic acid. Let us say we have set up an ELISA, based on a well-behaved antibody selective for the species induced by our target. We may generate a genuine, pure sample of the species (e.g., by chemical synthesis, or recombinant protein expression) with which to generate a calibration curve for antibody-mediated capture and quantification of the analyte species. Suppose that in untreated cell samples (i.e., treated only with DMSO), the analyte is not present, hence the baseline value of the signal should be zero. If we perform a concentration–response study of PD signal with a compound that should induce formation of the analyte species, the resulting data should appear as shown in Figure 5.11A for the raw signal from the assay. This looks like a well-behaved concentration–response plot. However, when we use our calibration curve to convert signal to molarity of analyte, we observe the concentration–response plot shown in Figure 5.11B. At first glance, this again looks like a well-behaved concentration–response plot. Closer inspection, however, reveals that the low compound concentration plateau is nonzero, suggesting that at compound concentration approaching zero, there is a nonzero amount of analyte present. Clearly, something here does not make biological nor physical sense. This is a common observation in cellular assays of PD. The nonzero plateau often results from some background signal that has not been adequately corrected for in the data analysis. Less frequently, this can also indicate that there is more than one compound-mediated, titratable event that is impacting PD, such as compound binding to a nontarget (i.e., off-target) protein that can also contribute to PD (Copeland 2023). In this case, the concentration–response plot is biphasic, but we have not titrated compound to low enough concentrations to reveal fully the high-affinity (low concentration of compound) binding event. Regardless

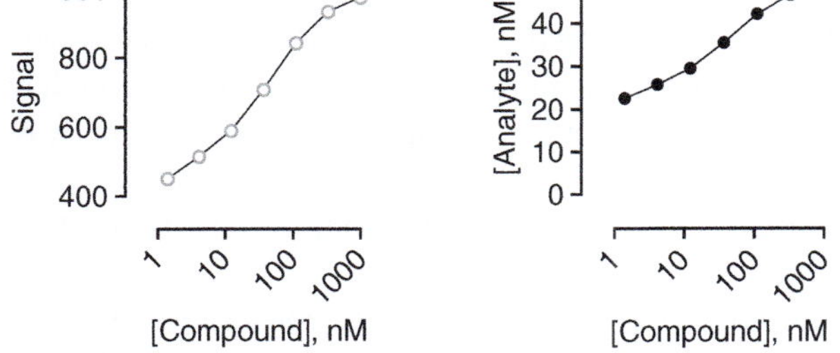

Figure 5.11 Comparison of concentration–response data for a cell phenotypic assay based on (A) uncorrected signal and (B) molar concentration of analyte, based on a proper calibration curve. At a brief glance, both plots appear to display a well-behaved concentration–response relationship. However, by converting the signal to molar units and plotting the *y*-axis full scale, it becomes evident that the lower compound concentrations are not approaching the true minimum limit of zero.

of the underlying cause, data of the type illustrated in Figure 5.11 call for further investigation to understand the cause of the unexpected result so that the concentration–response data can be properly analyzed. This caveat holds not only for cellular assays, but for any assay to be used for concentration–response data (Copeland 2013). However, this issue does tend to show up more commonly for cell-based assays.

5.3.6 Positive and Negative Cell Types

In our discussion of target selection (Chapter 3), we emphasized the need for a selective essential profile, wherein disease cells are dependent on target function for survival, proliferation, or some other required phenotypic outcome, but target function is dispensable in nondisease cells. This differential dependence on target function should be recapitulated by compounds that modulate target function. To assess this, one needs to titrate compound in cells that are expected to be dependent on target function (positive cell types), as well as in cells expected to be refractory to target modulation (negative cell types), to compare the phenotypic outcomes of compound titration in both cell types. Although both positive and negative cell types are expected to display PD readouts of intracellular target engagement by compound, there should be a meaningful, quantitative difference in phenotypic outcome of compound titration (measured as the EC_{50}) between the two cell types. To make this comparison rigorous, multiple cell lines representing positive and negative types should be assessed. We have already encountered examples of this type of analysis for inhibitors of the enzymes WRN and EZH2 in Chapter 3 (see Figure 3.7). Below we present yet another example of such data for a selective inhibitor of mutant BRAF in which valine 600 has been mutated to glutamic acid ($BRAF^{V600E}$). This mutation is commonly associated with melanoma and also occurs in subsets of colorectal, thyroid, and lung cancers. Yang et al. (2010) reported data for the $BRAF^{V600E}$-selective compound RG7204 (also known as PLX4032). Across cell lines representing multiple cancer types, they found that the antiproliferative activity of the compound, as quantified by the EC_{50} value, was much greater (i.e., lower EC_{50}) in cell lines bearing the V600E mutation, relative to cells expressing wild-type BRAF (Figure 5.12) or cells bearing mutations in BRAF other than V600E (not shown).

Demonstration of the expected differential phenotypic response of positive and negative cell lines is an important component of lead compound(s) validation. Reflecting on the data presented in Figures 3.7 and 5.12, one observes that there can be a range of EC_{50} values for both positive and negative cell lines. In fact, we also observe that there can be specific cell lines that deviate from expectations, such that some positive cell lines may be refractory to compound and/or some negative cell lines may be responsive to compound. Such exceptions

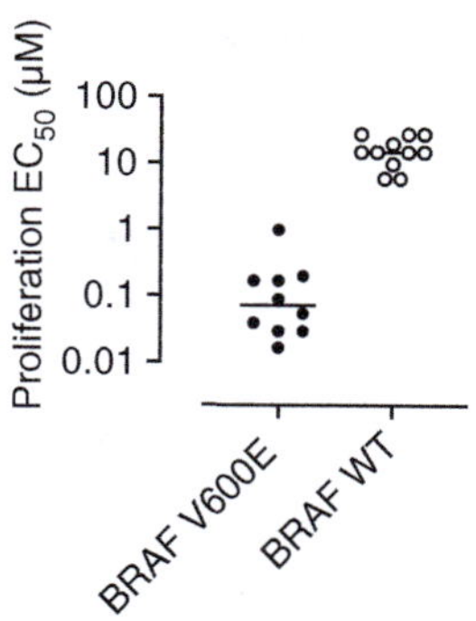

Figure 5.12 Comparison between the EC_{50} values of the $BRAF^{V600E}$-selective inhibitor RG7204, for inhibiting cell proliferation in cancer cells that bear the BRAF V600E mutation or wild-type BRAF. The horizontal lines represent the median value for each population of cells. *Source:* Adapted from Yang et al. (2010).

are commonly seen in experimental studies because of additional – often unknown – determinants of cell response to compounds. Hence, for lead compound(s) validation, it is not necessary that every cell responds as expected in such studies, but rather that the overarching trend in cell responsiveness to compound is consistent with the selective essentiality hypothesis. Those cells that deviate from expectations may be an opportunity to understand additional biological determinants of responsiveness through further studies; such studies may provide important insights into which patients may most likely benefit from a specific therapeutic strategy.

5.4 ADME Properties

5.4.1 Plasma Stability

The plasma component of blood contains a variety of proteins, including proteases, esterases, and other hydrolytic enzymes, that can potentially degrade drug molecules within systemic circulation. During SAR studies, the susceptibility of compounds to plasma-mediated degradation is measured by incubating the compound in human (and other preclinical species, such as mouse, rat, dog, and monkey) plasma at 37 °C and measuring any loss of the compound. After a specific incubation time, a sample of plasma is removed and an organic solvent is added to this to precipitate proteins. The sample is then centrifuged, and the supernatant is then analyzed by liquid chromatography with mass spectroscopy detection (LC/MS/MS) to determine the concentration of original (parent) compound remaining. Two approaches are typically used for plasma stability studies: (1) measurement of remaining parent compound after a fixed incubation time (typically 30 or 60 minutes) and (2) measurement of remaining parent compound after multiple incubation times (typically 0, 15, 30, 60, and 120 minutes). The single time-point assay is often used early in SAR studies to get a rough estimate

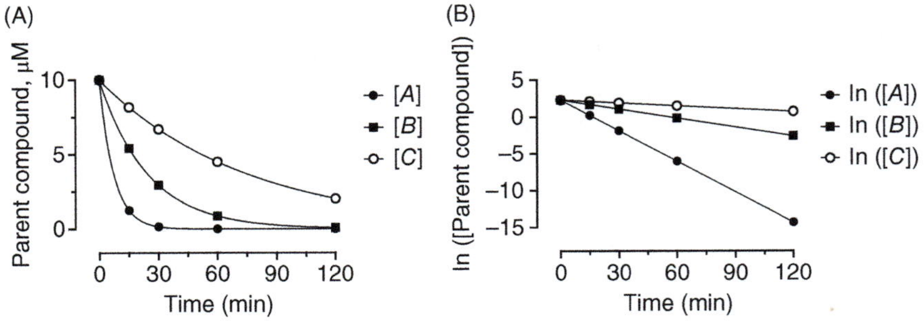

Figure 5.13 Human plasma stability data for three hypothetical compounds A, B, and C. (A) Kinetics of compound decay in plasma plotted on a linear y-axis scale and fit to Equation 5.12 by nonlinear curve fitting. (B) As in A but using the natural logarithms of compound concentration at each time point. Data for ln ([Parent Compound]) as a function of time are fit to a linear equation: $\ln([C]_t) = \ln([C]_0) - kt$.

of the relative stability of different compounds, measured as the percent parent compound remaining at the fixed time point. The kinetic method (multiple time points) allows one to determine the half-life of compound in plasma and is generally considered to be a more accurate assessment of plasma stability. Figure 5.13A illustrates the type of kinetic data obtained from a human plasma stability study for three compounds within a common pharmacophore series.

For each compound, the concentration of parent compound remaining as a function of plasma incubation time can be modeled as a simple exponential decay.

$$[C]_t = [C]_0 e^{-kt} \tag{5.12}$$

where $[C]_t$ is the concentration of parent compound at time t, $[C]_0$ is the starting concentration of parent compound (i.e., at $t = 0$), and k is the rate constant for compound decay. The half-life ($t_{1/2}$) is defined as the amount of time required to reduce the concentration of parent compound to half of its starting value (i.e., $0.5[C]_0$). It can be shown that the half-life of decay is inversely related to the rate constant (k) as follows (Copeland 2013).

$$t_{1/2} = \frac{0.6931}{k} \tag{5.13}$$

Data as in Figure 5.13A can be directly fit to Equation (5.12) by nonlinear curve fitting, and this is the preferred way to determine k and $t_{1/2}$ today. However, before the ubiquitous availability of computer software for curve fitting, scientists sought to linearize data such as that in Figure 5.13A. If one plots the natural logarithm of $[C]_t$ as a function of time, one obtains a linear plot with slope equal to the negative value of the rate constant ($-k$) and y-intercept equal to ln ($[C]_0$), as illustrated in Figure 5.13B. By either graphic method, the value of k can be derived from curve fitting, and from this, the half-life may be calculated.

Table 5.4 Summary of human plasma stability data for hypothetical compounds A, B, and C, based on the data illustrated in Figure 5.13.

Compound	Rate constant, k (min^{-1})	Half-life (min)
A	0.1386	5
B	0.0408	17
C	0.0133	52

For the three hypothetical compounds illustrated in Figure 5.13, we obtain the comparative data summarized in Table 5.4. From this table, we see that the medicinal chemistry team is making progress in optimizing plasma stability through SAR, in going from compound A to B to C.

5.4.2 Plasma Protein Binding

When drug molecules enter the bloodstream, they partition between the free (unbound) form and drug bound to proteins present (mainly) in the plasma. It is generally assumed that only free compound in the blood can be distributed to the tissues of the body, so that plasma protein binding (PPB) creates a drug sink that diminishes, or slows, the distribution of drugs to the site(s) of their pharmacological action. It is therefore generally desirable to mitigate excessive PPB as part of compound optimization through SAR. The reader should note, however, that despite the near-universal acceptance of the above assumption, many examples of successful therapeutics that are highly bound by plasma proteins are known (e.g., warfarin – >99% bound, vinblastine – 99% bound, and paclitaxel – 95–98% bound: see Copeland 2013).

In earlier chapters and Appendix 1, we derived the equations for equilibrium binding between a ligand (L) and a receptor (R), with the simplifying assumption that the ligand is present in great excess over the concentration of the receptor. Under these conditions, the diminution of free ligand concentration ($[L]_f$) due to receptor binding is insignificant and can be ignored. Hence, $[L]_f$ is well approximated by the total ligand concentration, $[L]_T$, and we saw that the midpoint (i.e., EC_{50}) of a Hill–Langmuir binding isotherm provides a good estimate of the K_d for ligand binding (Chapter 2). However, Strauss and Goldstein (1943) showed that when this simplifying assumption does not apply, a more general relationship between the EC_{50} and K_d is given by Equation (5.14).

$$EC_{50} = K_d + \frac{1}{2}[R]_T \tag{5.14}$$

where $[R]_T$ is the total concentration of receptor (i.e., $[R]_f + [RL]$). Thus, when $[R]_T$ is much smaller than K_d, the $\frac{1}{2}[R]_T$ term in Equation 5.14 becomes insignificant and can often be ignored. However, as $[R]_T$ approaches and exceeds the value of K_d, this term contributes more to Equation 5.14. Thus, when $[R]_T$ is much greater than K_d, the K_d term becomes insignificant, and equilibrium formation of the $[RL]$ complex is driven almost entirely by $[R]_T$. This becomes a critical issue whenever the protein concentration is very high, or when the K_d value approaches the concentration of target protein in cells or in *in vitro* assays. In fact, under either of these conditions, the equilibrium binding of drug is no longer well modeled by the Hill–Langmuir equation and must instead be modeled by a more complex quadratic equation (Copeland 2013).

$$[RL] = \frac{([R]_T + [L]_T + K_d) - \sqrt{([R]_T + [L]_T + K_d)^2 - 4[R]_T[L]_T}}{2} \tag{5.15}$$

The above discussion becomes relevant when considering drug molecules in circulation binding to plasma proteins, such as human serum albumin (HSA) and alpha-1 acid glycoprotein (AAG), both of which are found at high micromolar concentration in human plasma. In healthy adult humans, the concentration of HSA in whole blood ranges from 291 to 417 μM, and that of AAG ranges from 7 to 30 μM. Thus, even for a compound of low intrinsic affinity for PPB, a large portion of the drug in systemic circulation may be protein bound due to the very high concentrations of such proteins.

A variety of biochemical/biophysical methods have been used to study drug binding to plasma proteins, as reviewed by Cohen (2004). Three methods are most used for this purpose and are familiar to regulatory agencies, such as the Food and Drug Administration (FDA): equilibrium dialysis, ultrafiltration, and analytical ultracentrifugation. In all these methods, a known concentration of compound is added to a sample of plasma. The sample is incubated until equilibrium is established, then the free and protein-bound populations of compound are separated by a semipermeable barrier (dialysis and filtration) or centrifugal force (centrifugation). A sensitive analytical method (e.g., mass spectroscopy) is then applied to quantify the concentration of compound in the bound fraction, the free fraction, or both. Additional information on measurement of PPB can be found in the review by Cohen (2004).

For SAR purposes, researchers often measure PPB in plasma from a single species, most typically human. However, for compounds that are progressing to *in vivo* studies, it is important to measure PPB in all species in which the compound may be tested (e.g., mouse, rat, dog, monkey), as PPB values can vary considerably among species. It is also important to recognize that PPB not only occurs *in vivo* but can also impact *in vitro* assays of drug PD and phenotypic effect in cells. This is because most mammalian cells are cultured in serum-containing media (e.g., fetal bovine serum) that includes plasma proteins. Because the concentration of plasma proteins in such media is diluted (e.g., 10%), relative to whole plasma, the impact of PPB on drug activity is lessened. Nevertheless, with compounds displaying high PPB, this becomes a confounding factor in the assessment of compound cellular potency and SAR. Hence, cellular EC_{50} values often need to be corrected for the effect of PPB (due to the presence of plasma proteins in the culture medium) to accurately assess compound SAR.

5.4.3 Microsomal Stability: An Initial Indication of Hepatic Metabolism

Biotransformation of xenobiotics (including drug molecules) can occur in systemic circulation and in any tissue of the body. However, much of drug metabolism is mediated by intracellular enzymes within liver cells (hepatocytes), with the overarching goal of increasing the aqueous solubility of these lipophilic molecules so that they may be excreted in the bile and thereby eliminated in feces. Hepatic metabolism of drugs is a major contributor to overall drug clearance from the body and therefore an important aspect of drug optimization.

The reactions of hepatic, metabolic enzymes are categorized as two broad groups of phase I and phase II reactions. Phase I reactions increase the water solubility of drug molecules through oxidation, reduction, and hydrolytic transformations. Among the phase I enzymes is a family of heme-dependent redox enzymes known as the cytochrome P450s or CYPs. These enzymes play major roles in drug metabolism, as we shall discuss in more detail in Chapter 6. Phase II reactions involve conjugation of endogenous moieties to specific chemical functionalities that either exist on the parent molecule or have been appended to the parent molecule by phase I reactions. Examples of phase II reactions include glucuronidation of hydroxyl or carboxylate groups; amino acid conjugation of carboxylate groups; sulfation, acetylation, and methylation of hydroxyl and amine groups; and glutathione conjugation of sulfhydryl groups.

When purified hepatocytes are disrupted by, for example, sonication, they form small membrane vesicles known as microsomes that can be isolated by differential centrifugation. These microsomes retain many (but not all) of the hepatic metabolic enzymes. Microsomes are also very stable and can be stored frozen for long periods, making them a convenient source of hepatic drug-metabolizing enzymes.

In the early stages of SAR, microsomes are commonly used to test the stability of compounds toward hepatic metabolism. The microsomes are supplemented with NADPH and/or a NADPH-regenerating system to supply this required cofactor for CYP activity. Compounds are incubated with microsomes at 37 °C and typically sampled at 0, 15, 30, 45, 60, and 120 minutes. The amount of remaining parent compound at each time point is assessed by analytical methods such as LC/MS/MS (see above), and this is plotted as a function of time, as in Figure 5.13, to determine the half-life of microsomal metabolism. Detailed protocols for *in vitro* studies of drug metabolism with microsomes, intact hepatocytes, and related systems can be found in Ackley et al. (2004) and Wu and McKown (2004). Microsomal stability provides a general assessment of hepatic metabolism of compounds that is sufficient for early SAR efforts toward compound optimization. As we shall see in Chapter 6, subsequent compound optimization requires a more nuanced understanding of the specific enzymes involved in drug metabolism and whether compounds may inhibit or augment the activity of metabolic enzymes in a manner that can impact the pharmacokinetics of the drug itself, or of other drugs that might also be taken by the patient (i.e., drug–drug interactions; see Chapter 6).

5.4.4 Assessment of hERG Channel Inhibition

Torsades de points (TdP) is a potentially fatal, polymorphic ventricular tachycardia that can be induced by certain drug molecules. TdP is characterized by a prolongation of the time interval between the Q trough and T peak features of an electrocardiogram (QT prolongation; Figure 5.14). Mutations in the human Ether-a-go-go-Related Gene (hERG) potassium ion channel are associated with several congenital syndromes that share the common feature of prolonged QT interval and attendant TdP. Also, retrospective studies have found that essentially all drugs that induce TdP are inhibitors of the ion channel activity of hERG. Together, these data imply a causal relationship between hERG inhibition, QT prolongation, and resultant TdP. Indeed, several drugs have been withdrawn from the market due to their potential for causing TdP, leading to sudden death in patients (Copeland 2013).

The FDA and other regulatory agencies view QT prolongation by INDs as a serious liability, and the agency has issued guidelines for its assessment during preclinical studies. Hence, early in SAR studies, researchers begin to assess the potential of compounds to act as inhibitors of hERG. If a compound is found to be an inhibitor of this ion channel, an early assessment of the likelihood of dialing out hERG inhibition through SAR must be made.

To test compounds for hERG binding and inhibition of ion channel function, the gene encoding the channel is constitutively expressed in Chinese hamster ovary (CHO) cells, or other mammalian cells (e.g., human embryonic kidney [HEK] cells), to generate a functioning voltage-gated potassium ion channel within the cell-surface membrane. These cells are used to assess compound interactions with hERG in two common formats. Binding of compounds to the hERG channel is assessed using a competitive displacement assay with fluorescently labeled dofetilide (a known inhibitor of hERG). This assay is formattable in 384-well microtiter plates for HTS. However, the correlation between

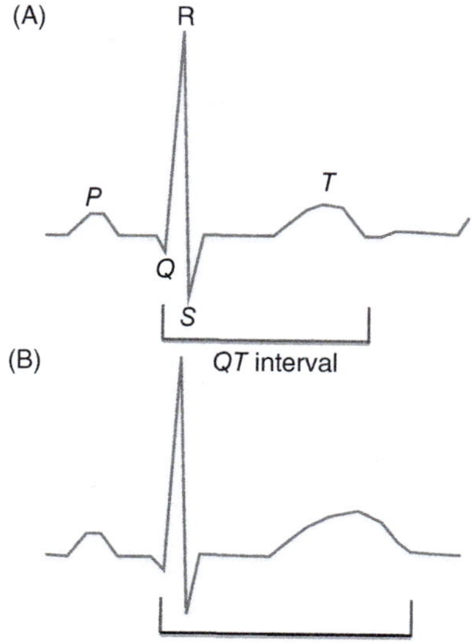

Figure 5.14 (A) Simplified illustration of a normal human electrocardiogram, illustrating the major peaks and troughs that are commonly observed. The time interval between the *Q* trough and the *T* peak (the *QT* interval) is highlighted. (B) As in A for a system in which the *QT* interval is prolonged. *Source:* Copeland, (2013) / with permission of John Wiley & Sons.

binding and functional inhibition of hERG has been inconsistent, depending on the pharmacophore series under study. Hence, the preferred approach is to test compounds in a functional assay of hERG ion channel activity. For this purpose, the "gold standard" assay is an electrophysiological method known as the patch-clamp assay (Figure 5.14). This method measures directly the current due to hERG-mediated potassium ion efflux from a single cell when a voltage step is applied (Figure 5.15; Copeland 2013).

This is a highly specialized and low-throughput assay, that is best performed by experts in electrophysiology. Several CROs provide fee-for-service testing of compounds in this assay. At the early stages of SAR, one typically runs the patch-clamp assay at only two concentrations of test compound: 0 and 10 μM. The percent inhibition at 10 μM is then used to assess compounds and make comparisons within and between pharmacophore series. For more advanced and sentinel compounds, full concentration–response data are collected to determine the IC_{50} for hERG inhibition. Because of the low throughput of traditional patch-clamp methods, efforts have been made to develop more moderate-throughput adaptation of this assay. This has resulted in a multiplexed version of patch-clamp known as PatchXpress. Data obtained with the PatchXpress method and with the traditional, single-cell patch-clamp assay appear to correlate well (Dubin et al. 2005). The moderate-throughput PatchXpress assay is therefore gaining increased use in compound testing. However, for key compound, especially for a development candidate (Chapter 6), researchers typically verify the IC_{50} for hERG inhibition with the single-cell patch-clamp assay.

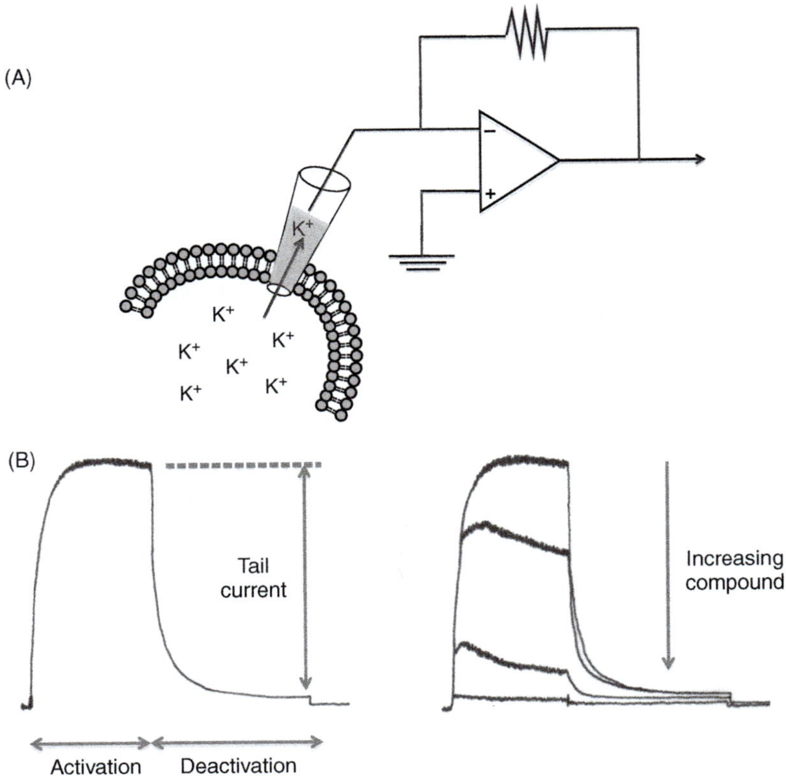

Figure 5.15 The patch-clamp assay of hERG function. (A) Schematic diagram of a single-cell patch-clamp setup. A glass micropipette is inserted through the cell membrane of a mammalian cell expressing recombinant hERG. A silver electrode is inserted into the micropipette and is used to apply a voltage step that caused potassium ions (K^+) to be effluxed via the hERG channel. A typical voltage protocol is as follows. The system at rest is poised at −80 mV. The voltage is then adjusted to +50 mV during the Activation phase and then returned to −80 mV during the Deactivation phase.
(B) Application of the voltage step results in an induced current as potassium ions efflux out of the cell. The current amplitude is measured as the vertical distance between the activation and deactivation plateaus (the tail current; left panel). This amplitude is diminished with increasing concentration of compounds that act as inhibitors of hERG ion channel function (right panel). *Source:* Copeland, 2013 / with permission of John Wiley & Sons.

5.5 Chemical Tractability

As we saw in the validation of hit compounds (Chapter 4), the purity, identity, aqueous solubility, and stability of compounds must continue to be ensured during the hit-to-lead stage of drug discovery. At this stage, however, compounds are being specifically designed and synthesized for SAR. Hence, identity and purity assessment are an integral part of the quality assurance typical of synthetic chemistry efforts. On the other hand, compound solubility (see above) and stability need to be assessed for each new compound.

5.5.1 Chemical Stability

For compounds approaching selection as development candidates destined for clinical studies, regulatory agencies, such as the FDA, have highly prescribed standards for the chemical stability of drug substances that form part of the systematic studies required for an IND application. These studies are summarized further in Appendix 2. At the hit-to-lead stage, one simply wants to assure that compounds are stable to storage in DMSO and when dissolved in aqueous solutions for assay uses. Typically, compounds are assessed for stability at different time points by LC/MS methods. Aqueous samples made from DMSO stock solutions need to be stable from decomposition for at least a week, and preferably a month, to facilitate their use in cellular and organismal studies. This is generally not an issue with compounds synthesized for medicinal chemistry purposes. However, on those occasions when a compound is not stable for the duration of an assay, this lack of stability can significantly confound the interpretation of assay results. Hence, caution must be exercised in establishing compound stability.

5.5.2 Synthetic Route(s)

When working within one or a few pharmacophore series, medicinal chemists can usually develop a clear synthetic route for members of the specific series, such that structural excursions from the founding compounds of the series can be readily accommodated. It is therefore critical to establish a clear synthetic route for each pharmacophore series at the outset of SAR studies. The details of the synthetic organic chemistry will vary from series to series and are beyond the scope of the present text. In general, however, the synthesis should be both scalable and reproducible. Additionally, later steps in the synthetic route, leading to final intermediates, should preclude the use of toxic metals, reactive species, and the like.

In favorable cases, the pharmacophore series may contain chemical moieties, often referred to as synthetic handles, that may make synthesis of a collection or library of compounds facile. For example, if the pharmacophore contains an amide linkage, this provides a convenient approach to generating a library of compounds in which the key pharmacophore elements are held constant on one side of the amide bond, and a wide variety of chemical diversity can be appended to the other side of the amide bond. This is done by synthesizing the constant portion of the molecule in bulk with a terminal carboxylate or amine group, and a library of variable structures with the alternative (carboxylate or amine) moiety, so that the carboxylate and amine groups on each half can be coupled to form the amide bond.

5.5.3 Patent Assessment

Typically, once validated hits are declared, a preliminary assessment of freedom to operate (FTO) is done by searching the patent literature. As one moves further into hit-to-lead activities, it is important to establish that there are no existing patent impediments to the pharmacophore structure(s) for the specific purpose of modulating the target of interest. Patent literature searches of this level of rigor are best performed by specialists in patent law. Large pharmaceutical companies and major universities may have such groups internal to their organizations. For smaller, biotechnology companies, external patent law firms can be hired to perform such searches on a case-by-case basis, or as part of a more comprehensive relationship between the law firm and biotechnology organization. Again, this is the purview of experts, and the details will vary among organizations, based on needs, costs, and other business considerations.

5.6 Lead Series

It is easy to think of a lead as a single compound that meets all the lead declaration criteria, such as those presented in Table 5.2. However, in practice, it is more likely that a collection of compounds within a pharmacophore series will, through SAR, begin to demonstrate sufficient target potency, solubility, lipophilicity, and cell permeability to elicit cellular PD and phenotypic responses. Whereas each of these compounds may not meet all the lead declaration criteria, they nevertheless constitute a *lead series*, of varying cellular potency, that will ultimately lead to one or more molecules meeting all the criteria. Thinking of such a collection of compounds as a lead series allows the researcher to test rigorously the veracity of the causal relationship between target occupancy (TO), PD, and phenotypic consequence that we expressed earlier in Scheme 5.1. Thus, one would expect that as the target-binding affinity increases, there should be a concomitant improvement in cellular PD potency, and this in turn should lead to improvement in cellular phenotypic potency. Of course, there can be exceptions to these general relationships. For example, two compounds of similar target potency in cell-free assays may display very different cellular activity if they differ dramatically in cell permeability. Nevertheless, across a large collection of compounds from a common pharmacophore series, one expects the relationships between target potency, PD, and phenotypic potency to be realized. In fact, Copeland and Boriack-Sjodin (2018) have codified the expectations for cellular activity of compounds based on modulation of a specific molecular target. They defined five key relationships that should be characteristic of the cellular activity of molecular targeted compounds, which they refer to as the elements of translational chemical biology. These elements are presented next.

5.6.1 The Elements of Translational Chemical Biology

The five elements of translational chemical biology, first published by Copeland and Boriack-Sjodin (2018), are as follows.

Element 1. Cellular PD, in the form of intracellular modulation of target function, should correlate quantitatively with target occupancy (TO).
Element 2. Phenotypic outcome should correlate quantitatively with the degree of intracellular modulation of target function (PD).
Element 3. Phenotypic outcome should be titratable and occur at reasonable compound concentrations, based on intracellular modulation of target function.
Element 4. Phenotypic outcome should be stratified by genotypic composition.
Element 5. Multiple pharmacophores modulating the same target should elicit the same phenotypic outcome.

Element 1 reflects the expectation that intracellular target engagement by the compound is the direct predicate of PD readout(s), and this should therefore be quantitatively related to the potency of target engagement seen in cell-free binding or functional assays (referred to below as "cell-free"). Hence, within a pharmacophore series, the EC_{50} for PD should be a linear function of the K_d of binding or EC_{50} of functional modulation of the target in cell-free assays. Figure 5.16A illustrates a typical example of a log–log scale plot of the PD EC_{50}s as a function of biochemical EC_{50}s for a hypothetical collection of compounds of a common pharmacophore series. Note that the linear relationship between PD and cell-free biochemical EC_{50} values holds only over a finite range of compound–target affinity. At low affinity (about 10–1000 μM in Figure 5.16A), the compounds are too weak to elicit a PD effect; this is a typical result in drug discovery – that a certain potency threshold must be overcome before cellular activity can be realized. Likewise, as compounds achieve very high target potency, we observe a second discontinuity in the plot. Here, the PD potency continues to improve, whereas the cell-free biochemical potency seems to have reached a constant limit. Again, this is a common observation in drug discovery that relates to exceeding the useful dynamic range of the cell-free assay. This limit is often due to the phenomenon of tight-binding modulator binding to the target, wherein the K_d of the target–compound complex approaches the concentration of target protein in the assay. Copeland (2013, 2023) discusses the concept of tight-binding effects in detail and provides useful suggestions for overcoming this limit. Despite these discontinuities at the high- and low-potency extremes, one should observe a linear relationship between biochemical/biophysical target potency (in cell-free assays) and PD potency over several orders of magnitude in EC_{50} values (Figure 5.16A).

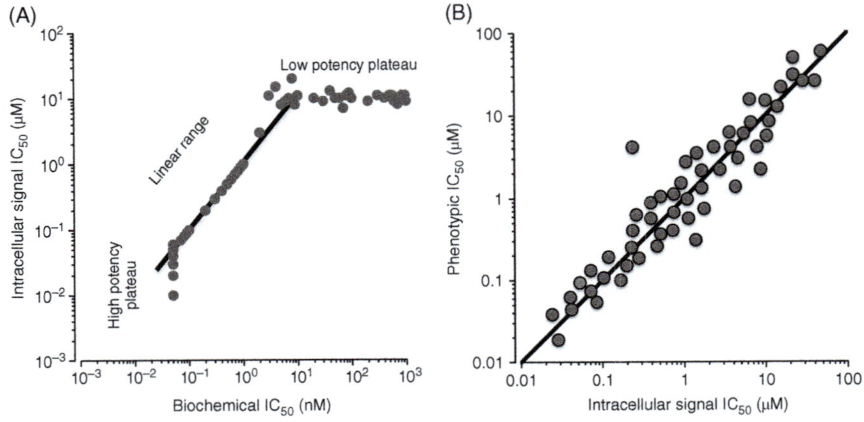

Figure 5.16 Correlation plots of (A) EC_{50} values for intracellular PD as a function of cell-free biochemical EC_{50} value. Here, the EC_{50} values are presented as IC_{50}s, reflecting a mechanism based on inhibition of target function and corresponding PD signal, and (B) EC_{50} values for phenotypic outcome as a function of PD EC_{50} value. *Source:* Copeland and Boriack-Sjodin (2018) / with permission of Elsevier.

Likewise, Element 2 predicts that a plot of phenotypic EC_{50} as a function of PD EC_{50} for a pharmacophore series should also yield a linear correlation, if intracellular TO (as measured by PD) is the direct predicate of phenotypic outcome (Figure 5.16B). Together, plots such as those illustrated in Figure 5.16 establish a clear, quantitative relationship between compound potency, intracellular TO (PD), and phenotypic outcome. Correlation plots of this type cannot prove causation in a rigorous mathematical sense. Nevertheless, the establishment of such a set of quantitative relationships between target binding, PD, and phenotypic outcome adds great confidence in the molecular targeted hypothesis upon which the drug discovery program was founded, especially when coupled with the prior target-validation studies. In this context, the lack of cellular activity for compound of low cell-free biochemical or biophysical potency (negative controls) further supports a causal relationship between target binding, PD, and phenotypic outcome. Whereas one cannot rigorously prove such causation, the opposite is true. That is, the lack of such correlative data, and especially phenotypic readout elicited by compounds of low cell-free potency, can be viewed as disqualifying of the molecular targeted hypothesis.

The slopes of these correlation lines also provide important information as to how challenging it may be to achieve intracellular PD and phenotypic outcome for a given pharmacophore series, and we will see that this relates to Element 3. The slope of the correlation plot of PD as a function of cell-free EC_{50}s (typically measured at $[S] = K_M$; see Chapter 2) is influenced by several factors, most notably compound solubility, PPB (when plasma proteins are present in cell culture media), and cell permeability (see above). For competitive modulators of target function, this slope will also be impacted by the cellular concentration of the competing natural ligand, relative to its K_d value for the target. If the cellular concentration of natural ligand is relatively constant among cells of the organism (i.e., humans, except for infectious disease targets or in the case of veterinary medicine), the slope of the PD/cell-free correlation plot will be steeper for targets displaying low K_d values for their natural ligand, as in this case the compound will be less able to effectively compete for target binding (according to the Cheng–Prusoff equation for competitive modulators, as presented in Chapter 2). Consider ATP-competitive inhibitors of the protein kinases as an illustrative example of this concept. The concentration of ATP within mammalian cells ranges from 0.5 to 5 mM, with most researchers considering 1 mM as a reasonable average value. The K_M value for ATP among protein kinases ranges from 10 to 100 μM. Let us say that kinase A displays a K_M value for ATP of 10 μM and kinase B displays a K_M value for ATP of 100 μM. If the cellular concentration of ATP is 1 mM (i.e., 1000 μM), the $[S]/K_M$ values would be 100 and 10 for kinases A and B, respectively. All other factors being equal, a pharmacophore series of ATP-competitive inhibitors of kinase A would thus be expected to display a PD/cell-free correlation plot with a slope of 51-fold, whereas an ATP-competitive pharmacophore series of inhibitors of kinase B would display a slope of only 5.5-fold for such a plot (based on the ratio of $(1 + [S]/K_M)$ between the cell-free assay $(1 + [S]/K_M) = 2$ and that in the cellular assay $(1 + [S]/K_M) = 101$ and 11, respectively).

These confounding variables notwithstanding, having established a quantitative relationship between PD and cell-free EC_{50} values for a pharmacophore series, we can treat the readout of a proximal PD signal as a measure of the degree of intracellular TO by a given compound concentration. Stated otherwise, the PD readout will follow a typical Hill–Langmuir isotherm, such that the concentration of compound that elicits a 50% change in PD signal (the EC_{50}) occurs at 50% TO. Based on the Hill–Langmuir equation (Chapter 2), this also means 80% TO is achieved at 4-fold above the EC_{50}, 90% TO is achieved at 9-fold above the EC_{50}, 95% TO is achieved at 19-fold above the EC_{50}, and 99% TO is achieved at 99-fold above the EC_{50} (Copeland 2013).

Now, the slope of the phenotypic EC_{50}s as a function of PD EC_{50}s (Figure 5.16B and Element 2) provides an estimate of the degree of TO required to elicit a specific percent modulation of phenotypic outcome. Another way to think about this is in terms of target vitality (Chapter 3, Section 3.3.5, and Copeland 2013), a concept that was first formulated for inhibitory compounds. The target vitality defines what degree of inhibition of target function is required to elicit a phenotypic modulation, or stated otherwise, what amount of residual target function is sufficient to maintain phenotypic integrity. To be effective as a therapeutic strategy, researchers prefer targets for which 50–99% TO is sufficient to modulate cellular phenotype. The slope of the phenotypic/PD EC_{50} correlation plot gives us a sense of target vitality. Accepting that the PD EC_{50} is a measure of intracellular TO, we can show from rearrangement of the Hill–Langmuir equation that the percent TO required to modulate phenotypic response by 50% (i.e., the phenotypic EC_{50}) is given by the following equation.

$$\%TO = \frac{100}{1 + \frac{EC_{50}^{PD}}{EC_{50}^{Phen}}} \tag{5.16}$$

where EC_{50}^{PD} is the EC_{50} for the PD signal and EC_{50}^{Phen} is that for the phenotypic outcome. But the ratio ($EC_{50}^{PD}/EC_{50}^{Phen}$) is simply the reciprocal of the slope of the phenotypic/PD EC_{50} correlation plot. Using this equality, we can recast Equation 5.16 as follows.

$$\%TO = \frac{100}{1 + \left(\frac{1}{\text{slope}}\right)} \tag{5.17}$$

In the most favorable cases, a slope of 1.0 indicates that the EC_{50} for phenotypic modulation occurs at 50% TO. As the slope increases, higher levels of TO are required to achieve 50% modulation of phenotype, as summarized in Table 5.5.

Element 3 was originally included by Copeland and Boriack-Sjodin (2018) to emphasize the needs for titration studies of phenotypic effect by compounds and for relating phenotypic effect to TO. This should be standard practice in drug discovery, as we have outlined here. However, Copeland and Boriack-Sjodin (2018) felt compelled to include this as a critical element of translational chemical biology because of the disturbing practice of testing phenotypic effects of compound at only two concentrations – zero and a supra-pharmacologic concentration – that

Table 5.5 Relationship between the slope of the phenotypic/PD EC_{50} correlation plot and the percent target occupancy (%TO) required to achieve 50% modulation of phenotypic outcome.

Slope value	%TO required at the phenotypic EC_{50}
1	50.0
5	83.0
10	90.0
50	98.0
100	99.0
500	99.8

Source: Data based on Equation 5.17.

had permeated the chemical biology literature. To illustrate this concept, consider a compound that modulates PD with an EC_{50} of 50 nM. Based on Hill–Langmuir behavior, one would expect to achieve 99% TO for this compound at a concentration of just under 5 μM. Hence, if one wishes to do a preliminary assessment of phenotypic response to this compound, it might make sense to compare phenotype at 0 and 5 μM compound. However, performing this experiment at 0 and 50 μM would be an unnecessarily excessive amount of compound, and at this supra-pharmacologic concentration, one would risk affecting nontarget proteins, membranes, nucleic acids, or other cellular components other than the target of interest (off-target effects). Hence, any phenotypic effect seen at 50 μM could not be directly and exclusively associated with target engagement. Nevertheless, the scientific literature is replete with examples of researchers making causal inferences about the phenotypic impact of specific target engagement by compounds from inappropriate data of this type.

Element 4 makes the point that for selectively essential targets (Chapter 3), the phenotypic (but not necessarily PD) effects of compounds should reflect the genotypic basis of the selective essentiality of the target. We have already discussed this concept at length, as illustrated in Figure 3.7 (Chapter 3) for the targets EZH2 and WRN, in Figure 3.8 (Chapter 3) for DHX9, and in Figure 5.12 (above) for $BRAF^{V600E}$.

Element 5 makes the point that compounds of different pharmacophore series that both selectively modulate the same target function should elicit the same phenotypic effect(s) on cells for which that target is selectively essential. This provides further evidence of a causal relationship between target engagement and phenotypic outcome. Let us illustrate this concept with the following example. Let us say that compound A is a selective, 5 nM inhibitor of a target of interest. When applied to appropriate cells, this compound blocks a proximal PD signal with an EC_{50} of 50 nM and elicits a phenotypic response with an EC_{50} of 150 nM. Thus, the ratio of EC_{50} values here is 3-fold, suggesting that 75% TO is required to achieve 50% modulation of phenotype (Equation 5.16). Suppose that the researcher has access to compound B that is also a selective, 100 nM inhibitor of the same target, but is structurally distinct from the pharmacophore series of compound A (for example, compound B might be a member of a distinct pharmacophore series representing a different validated screening hit, a competitor's compound, or a literature-derived probe compound). When compound B is applied to the same cell type, it blocks the proximal PD signal with an EC_{50} of 200 nM. For this compound, 75% intracellular TO would require 600 nM. Hence, we would expect this compound to display a phenotypic EC_{50} of 600 nM. This, of course, is a simplistic example and does not account for experimental error in the set of measurement. A better approach, if practical, would be to compare the EC_{50} values for PD and phenotypic effect for multiple compounds from each pharmacophore series. The overarching point here is that similar levels of intracellular modulation of target function (i.e., PD) should elicit similar levels of phenotypic modulation, if the phenotypic modulation is indeed a direct result of TO by the compounds. Element 5 is not always practical in the early stages of drug discovery. When the appropriate tools are available, however, this element provides additional, strong evidence of a causal relationship between TO and phenotypic outcome.

5.7 Case Studies

5.7.1 DOT1L Inhibitors for Mixed Lineage Leukemia

In Chapter 4, we illustrated some of the SAR for inhibition of DOT1L by a series of SAH mimetic inhibitors intended to treat patients with MLL-rearranged leukemias (see Table 4.3 and references therein). As described in Chapter 4, this SAR campaign began with compound EPZ0004 in which the homocysteine portion of SAH was replaced by a simple dimethyl nitrogen (see Table 4.3). From here, the team replaced one of the N-methyl groups of EPZ0004 by an isopropyl group to form EPZ2446, which was a weak inhibitor of DOT1L (Table 5.6) and therefore displayed no cellular activity. Likewise, the team replaced one of the N-methyl groups by an N-ethylene amide appended with a tricyclic ring structure, EPZ3344, which again yielded a weak DOT1L inhibitor without cellular activity. As described in Chapter 4, the first compound to demonstrate any cellular PD activity (reduction in the intracellular level of H3K79me2, the direct product

Table 5.6 SAR for SAH mimetic DOT1L inhibitors.

Compound	DOT1L K_i (nM)	PD (H3K79me2 reduction) EC_{50} (nM)[a]	Phenotypic (inhibition of cell proliferation) EC_{50} (nM)[a]
EPZ0004	38,000	>50,000	>50,000
EPZ2446	12,000	>50,000	>50,000
EPZ3144	20,000	>50,000	>50,000
EPZ3647	845	28,000	>50,000
EPZ3696	13	272	4900
EPZ4450	4	N.D.[b]	N.D.[b]
EPZ4777	0.3	11	150
EPZ5676	0.08	2.7	3.5

a) Measured in MV4–11 mixed lineage leukemia cells.
b) N.D., not determined.
Source: Adapted from Basavapathruni et al. (2012) and Daigle et al. (2013). See also Table 4.3.

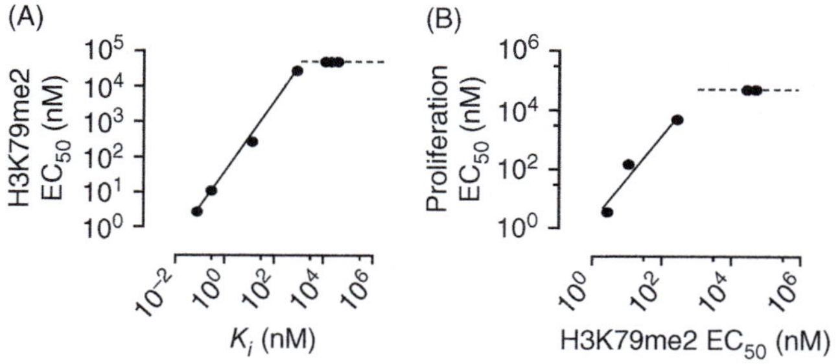

Figure 5.17 Correlation plots for the DOT1L inhibitor series of SAH mimetics (see text). (A) Correlation plot of intracellular PD (reduction of H3K79me2) EC_{50} as a function of inhibitor K_i as measured in a cell-free assay of DOT1L catalytic activity. (B) Correlation plot of phenotypic response (inhibition of cellular proliferation) EC_{50} as a function of PD EC_{50}. Data used to construct these plots were from Basavapathruni et al. (2012) and Daigle et al. (2013). The solid lines in each graph illustrate the best fit correlation between the x and y variables, whereas the dotted lines are meant to help visualize data regions where changes in the *x*-variable value did not lead to a change in the *y*-variable value.

of DOT1L enzymatic activity) was EPZ3647, which was an 845 nM inhibitor of DOT1L and displayed a PD EC_{50} of 28 μM; this compound was still too weak to elicit an antiproliferative phenotypic response (Table 5.6). Further SAR, as summarized in Tables 4.3 and 5.6, drove the potency of compounds as inhibitors of DOT1L with concomitant improvements in PD and phenotypic response efficacy.

Figure 5.17 illustrates the correlation plots for PD/cell-free biochemical EC_{50}s and phenotypic/PD EC_{50}s for the DOT1L series of inhibitors. As expected, the weak compounds display no cellular activity, resulting in flat, horizontal correlation lines (dashed lines). As the biochemical potency improved, the data for both PD and phenotypic response now conform to the expected linear relationships described above. These data fulfill the requirements of Elements 1–3 above for this SAR campaign. The slope of the phenotypic/PD EC_{50} correlation plot in Figure 5.17B is 11-fold, suggesting that low-nanomolar (or more potent) inhibitors of this enzyme are required to achieve potent inhibition of proliferation for MLL-rearranged leukemia cells.

Compound EPZ5676 was later declared a clinical development candidate (DC; see Chapter 6) and advanced into phase 1 clinical studies in leukemia patients. Consistent with the selective essentiality of DOT1L in MLL-rearranged leukemias, Daigle et al. (2013) demonstrated potent antiproliferative activity for EPZ5676 in MLL-rearranged leukemia cell lines, but not in non-rearranged cell lines (summarized in Figure 5.18), thus fulfilling the requirement of Element 4 above. In addition to modulating the proximal H3K79me2 PD mark and correlative modulation of cell proliferation, the cell-active compounds EPZ4777

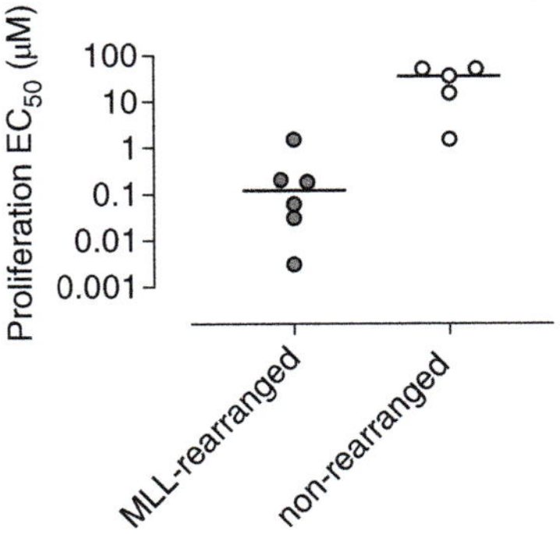

Figure 5.18 Selective antiproliferative activity in MLL-rearranged leukemia cell lines for the DOT1L inhibitor EPZ5676. Data used to construct this plot were estimated from that published by Daigle et al. (2013). *Source:* Adapted from Daigle et al. (2013).

and EPZ5676 also blocked intervening biological responses of DOT1L activity at ectopic gene locations, such as inhibition of HOXA9 and MEIS1 gene transcription (Daigel et al., 2011, 2013).

5.7.2 METTL3 Inhibitors for Acute Myeloid Leukemia (AML)

Not every validated target translates into a tractable drug discovery program. The mRNA methyltransferase METTL3/METTL14 program at Accent Therapeutics provides an example where, despite success in identifying sub-nanomolar, selective inhibitors of the target enzyme, the program was ultimately abandoned because of insurmountable challenges in translating potent enzyme inhibition into effective cellular and *in vivo* antitumor therapeutics.

Like DOT1L, METTL3/METTL14 is a SAM-dependent methyltransferase. However, rather than transferring the methyl group of SAM to a lysine or arginine side chain on a protein, METTL3/METTL14 is instead an RNA methyltransferase that selectively adds a methyl group to the *N*6 position of adenosine bases within messenger RNA (mRNA) to produce *N*6-methyl adenosine (m^6A). The catalytically active form of METTL3 occurs as a heterodimer with the related protein METTL14; in this, heterodimeric complex only METTL3 is catalytically active (Buker et al. 2020 and references therein). Multiple reports in the literature suggested that METTL3/METTL14 plays an oncogenic role in acute myeloid leukemia (AML). Hence, several groups began drug discovery programs against this target, including the team at Accent Therapeutics, Inc. The Accent team used mechanism-guided design (Chapter 4) to start the program with a compound initially designed as an inhibitor of another, undisclosed protein methyltransferase. This compound proved to be a modest inhibitor of METTL3, with a K_i value of ca. 5 μM. Despite this modest starting point, the team was able to use structure-guided optimization (Chapter 4), based on co-crystals of compounds bound to the METTL3 active site of the METTL3/METTL14 heterodimer, to

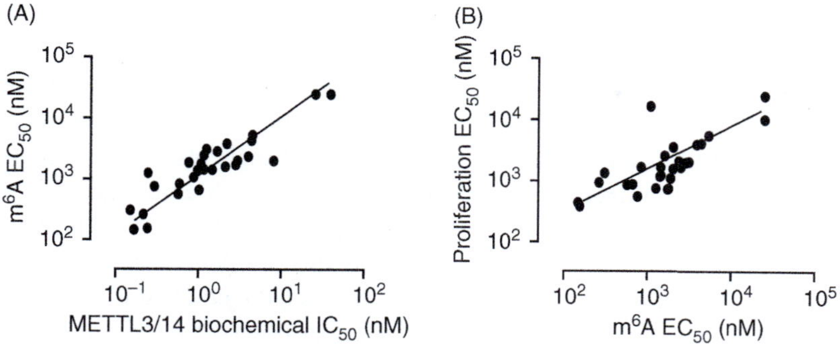

Figure 5.19 Correlation plots for the METTL3/METTL14 inhibitor series from Accent Therapeutics, Inc. (see text). (A) Correlation plot of intracellular PD (reduction of m^6A within mRNA) EC_{50} as a function of inhibitor IC_{50} as measured in a cell-free assay of METTL3/METTL14 catalytic activity. (B) Correlation plot of phenotypic response (inhibition of MOLM-13 cellular proliferation) EC_{50} as a function of PD EC_{50}. Data used to generate these plots were generously provided by Accent Therapeutics, Inc.

drive SAR. This allowed the team to drive the potency of the SAM-competitive pharmacophore series to IC_{50} values in the range of 100–0.1 nM when tested in a cell-free assay of METTL3/METTL14 enzymatic activity (Buker et al. 2020 and Figure 5.19A). Compounds within this pharmacophore series displayed excellent selectivity for METTL3/METTL4 when tested against other RNA methyltransferases, protein lysine and protein arginine methyltransferases, and a spectrum of other target classes such as kinases, receptors, and ion channels (Accent Therapeutics, unpublished data). As illustrated in Figure 5.19, the pharmacophore series reduced intracellular levels of m^6A in mRNA in a concentration-dependent manner that correlated with compound potency in the enzymatic assay. However, the slope of the correlation line in this case was about 700-fold. This large slope value may in part be due to the SAM-competitive modality of the pharmacophore series and the unusually low K_M value for SAM displayed by METTL3/METTL14 (ca. 100 nM; Buker et al. 2020). The reduction of intracellular m^6A in turn correlates with the antiproliferative activity of this pharmacophore series in AML cell lines, as illustrated for MOLM-13 cells in Figure 5.19B. Again, a strong correlation is observed in the phenotype/PD EC_{50} correlation plot (Figure 5.19B), but the slope of the correlation line is again very steep, with a value of ca. 410-fold. Consistent with the starting hypothesis, the Accent team found that potent members of this pharmacophore series displayed antiproliferative activity only in a subset of AML cell lines, with no METTL3/METTL14 inhibition-related antiproliferative activity in other cancers across a large sampling of solid tumor cell lines (unpublished data from Accent Therapeutics, Inc.).

The very high slope values in translating biochemical potency into intracellular PD and in translating this to antiproliferative activity made it unlikely that a robust clinical development candidate could emerge from these efforts. Hence, the Accent Therapeutics, Inc. team made the decision to abandon METTL3/METTL14 as a drug target to better marshal their resources on more productive targets. However, a team at Storm Therapeutics (Cambridge, United Kingdom) has advanced a METTL3/METTL14 inhibitor (STC-15) to a phase 1 clinic trial. Unlike other inhibitors of the enzyme, this compound shows activity in several solid tumor cell lines. As of November 2024, the compound had been used to treat 42 patients of varying cancer types at doses of 60–200 mg three times per week (TIW). Of these patients, 29 were evaluable for response, and three of these patients (ca. 10%) demonstrated a partial response (PR, a partial reduction in tumor size) on drug: one angiosarcoma patient treated at 60 mg, one immuno-oncology-refractory non–small cell lung cancer (NSCLC) patient treated at 100 mg, and one thymoma patient treated at 200 mg. All patients demonstrated reduction in m^6A within blood cells (Papadopoulos et al. 2024). Nevertheless, whether the mechanism of action of STC-15 in these three patients is indeed related to METTL3/METTL14 inhibition remains currently unclear.

5.7.3 EZH2 Inhibitors for Treatment of Non-Hodgkin's Lymphoma

As we discussed in Chapter 3, heterozygous point mutations of Tyr 641 within the EZH2 catalytic subunit of the PRC2 complex occur in about 20 to 30% of germinal center non-Hodgkin's lymphoma (NHL) patients. The work of Sneeringer et al. (2010) demonstrated that these mutations result in hypertrimethylation of histone H3 on lysine 27 (H3K27me3), leading to oncogenic repression of a program of gene transcription events.

Based on these data, several groups began drug discovery campaigns to identify potent, selective inhibitors of EZH2 enzymatic activity. The team at Epizyme, Inc. screened a 175,000-compound subset of a large diversity library and identified the 4,6-dimethyl pyridone-containing compound **1** (Figure 5.20) as a validated hit, displaying an IC_{50} of $3.4 \pm 0.9\,\mu M$ for inhibition of wild-type EZH2 enzymatic activity. Hit expansion was performed by first computationally scanning the unscreened portion of the diversity library for other 4,6-dimethyl pyridone-containing compounds. These compounds were then obtained and tested as EZH2 inhibitors. These efforts identified compound **2**, which displayed an IC_{50} of $0.62 \pm 0.2\,\mu M$. Compounds **1** and **2** both showed limited solubility in aqueous solution (<10 μM at pH 7), no PD activity (reduction of H3K27me3 levels) in cells and limited oral bioavailability (0.5%). Hence, the team began to systematically add polar groups at varying positions while leaving the core 4,6-dimethyl pyridine pharmacophore intact. They found that the 6-position of the 7-azaindazole of compound **2** tolerated polar groups. Basic amines with or

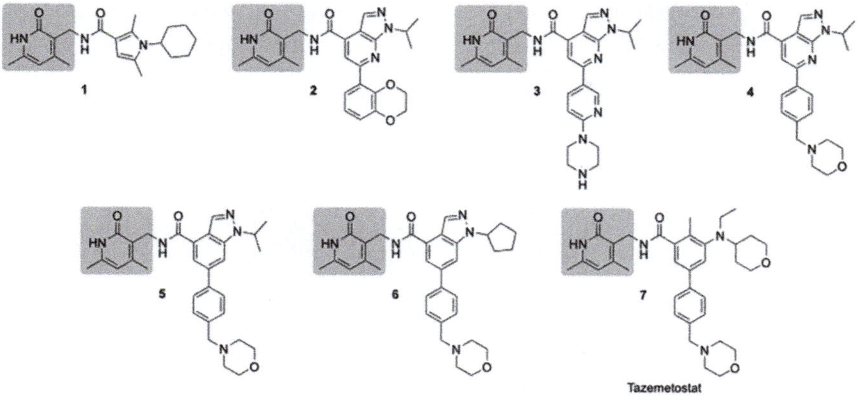

Figure 5.20 Chemical structures of EZH2 inhibitors within the 4,6-dimethyl pyridone pharmacophore series. The 4,6-dimethyl pyridone that is common to all these compounds is highlighted with gray shading. Data for this figure was derived from Knutson et al. (2012), Knutson et al. (2014), and Kuntz et al. (2016). Construction of this figure was completed with the aid of Dr. Brian Sparling.

without an aryl spacer appended to the 6-position improved solubility with minimal impact on target potency. This led to compounds **3** and **4**, with IC_{50} values of 0.2 ± 0.2 and $0.36 \pm 0.2\,\mu M$, respectively. Though both compounds **3** and **4** improved aqueous solubility, compound **4** was the first to demonstrate inhibition of intracellular H3K27me3 levels, with an EC_{50} of $10.5 \pm 2\,\mu M$. In preliminary *in vivo* mouse pharmacokinetics (PK) studies, compound **4** also proved to be 41% orally bioavailable and displayed a moderate clearance rate. Based on these data, compound **4** was declared a lead compound. Lead optimization explored a variety of substitutions at different locations on the molecule to further improve cellular potency and PK characteristics. For example, compound **5** demonstrated that the nitrogen at the 7-position of the 7-azaindazole group could be replaced by carbon. Increasing the size of the indazole N1–appended isopropyl to a cyclopentyl group further improved potency, resulting in the discovery of the *in vitro* tool compounds **6** (EPZ5687) and ultimately the clinical candidate **7** (EPZ6438, also known as tazemetostat). We shall discuss the characterization of tazemetostat further in Chapter 6. Table 5.7 summarizes the biochemical, PD, and phenotypic potencies of these EZH2 inhibitors.

Compound **6** was a potent, selective, SAM-competitive inhibitor of EZH2 (Knutson et al. 2012). As summarized in Table 5.7, this compound inhibited EZH2 with a K_i value of 24 nM (IC_{50} of 48 nM in an enzymatic assay run at $[S] = K_M$) and showed no inhibition of any other protein methyltransferase at concentrations up to 50 μM, except for the closely related enzyme EZH1, for which the selectivity of compound **6** was ca. 100-fold. When NHL cells were treated

Table 5.7 Biochemical and cellular potency of various EZH2 inhibitors from the group at Epizyme, Inc.

Compound	Biochemical IC_{50} (nM)	PD (H3K27me3 reduction) EC_{50} (nM)[a)]	Phenotypic (inhibition of cell proliferation) EC_{50} (nM)[b)]
1	3400	Inactive[c)]	Inactive
2	620	Inactive	Inactive
3	200	Inactive	Inactive
4	360	10,500	Inactive
5	160	Not reported[d)]	Not reported
6 (EPZ5687)	48	80	270
7 (EPZ6438; tazemetostat)	<5	28	4.3

a) Values represent the mean EC_{50}s across multiple NHL cell lines measured after four days of compound treatment.
b) Values represent the mean EC_{50}s across multiple NHL cell lines bearing oncogenic mutations of EZH2, measured after 11 days of compound treatment.
c) Inactive: no inhibition observed at any concentration of compound tested.
d) Not Reported: no data has been reported in the scientific literature.
Source: Data used to construct this table were taken from Kuntz et al. (2016), Knutson et al. (2012) and Knutson et al. (2014).

with 1 μM compound **6**, complete ablation of H3K27me3 was observed regardless of whether the cells contained homozygous wild-type EZH2 or heterozygous mutants at Tyr641. The compound-induced reduction in H3K27 methylation was accompanied by an increase in acetylation of H3K27, as would be expected since methylation and acetylation of H3K27 are mutually exclusive (Knutson et al. 2012). Apart from H3K27 methylation, no other histone methylation sites were impacted by compound **6** treatment, suggesting that the selectivity observed for EZH2 in cell-free enzymatic assays was recapitulated intracellularly. Inhibition of intracellular H3K27me3 by compound **6** in oncogenic mutant–bearing NHL cells resulted in the derepression of gene transcription for a program of tumor suppressor genes, consistent with an EZH2 inhibition mechanism of action (Knutson et al. 2012, 2014). The antiproliferative activity of compound **6** was tested on EZH2 wild-type and oncogenic mutant–bearing NHL cell lines. Consistent with the selective essential hypothesis, compound **6** titration had minimal impact on wild-type EZH2 NHL cells (e.g., OCI-LY19). In heterozygous EZH2 oncogenic mutant–bearing cells, such as WSU-DLCL2 cells (Y641F), titration of compound **6** showed little impact on cell proliferation over the first four days of treatment, but after that a robust time- and concentration-dependent inhibition of cell proliferation was observed, such that maximum inhibition of proliferation was realized after 11 days of treatment. Interestingly, this latency period prior to compound effect on cell proliferation is a common feature of epigenetic targeted drugs. Thus, the antiproliferative effects of compound **6** and other EZH2 inhibitors were, for the most part, limited to NHL cells bearing oncogenic mutations of EZH2, as previously illustrated for tazemetostat (compound **7**) in Figure 3.7A (Chapter 3).

5.8 Summary

In this chapter, we have described the systematic SAR that is performed in taking a validated hit and translating this into a compound (or compounds) that meet predetermined criteria for declaring a lead series. As stated previously, the specific criteria for lead declaration vary among different research groups. Generally, however, these criteria include potency, solubility, and permeability consistent with observing proximal PD, intervening biological, and perhaps phenotypic responses to compounds in appropriate cellular assays. Additional criteria include early assessments of target selectivity, metabolic stability, hERG inhibition, PPB, and chemical tractability. We also described a clear set of criteria by which one can test the assumed causal relationship between target binding, PD, and phenotypic impact of compounds that we refer to as "the elements of translational chemical biology" (Copeland and Boriack-Sjodin, 2018). Finally, we ended this chapter with several case studies of hit-to-lead activities against different targets for cancer indications. In the next chapter, we shall discuss the continuation of SAR-based optimization of compounds with the goal of declaring a clinical development candidate molecule(s). The intense scrutiny of compound target binding, *in vitro* microsomal stability, PPB, and the like, that we reviewed in this chapter, continues throughout lead optimization. These efforts are, however, augmented during lead optimization with *in vivo* assessments of PK, safety, PD, and efficacy.

References

Ackley, D.C., Rockich, K.T., Baker, T.R., Yan, Z. and Caldwell, G.W. (2004) Metabolic stability assessed by liver microsomes and hepatocyte. In *Optimization in Drug Discovery* (Z. Yan and G.W. Caldwell, Eds.). Humana Press, Totowa, NJ, pp. 151–162.

Aderniji, A. and Adejare, A. (2008) Chemical and physical characterizations of potential new chemical entity. In *Preclinical Development Handbook: ADME and Biopharmaceutical Properties* (S.C. Gad, Ed.). Wiley, Hoboken, NJ, pp. 211–225.

Basavapathruni, A., Jin, L., Daigle, S.R., Majer, C.R.A., Therkelsen, C.A., Wigle, T.J., Kuntz, K.W., Chesworth, R., Pollock, R.M., Scott, M.P., Moyer, M.P., Richon, V.M., Copeland, R.A. and Olhava, E.J. (2012) Conformational adaption drives potent, selective and durable inhibition of the human protein methyltransferase DOT1L. *Chem. Biol. Drug Des.* **80**, 971–980.

Blobaum, A.I. and Marnett, L.J. (2007) Structural and functional basis of cyclooxygenase inhibitors. *J. Med. Chem.* **50**, 1425–1441.

Buker, S.M., Gurard-Levin, Z.A., Wheeler, B.D., Scholle, M.D., Case, A.W., Hirsch, J.L., Ribich, S., Copeland, R.A. and Boriack-Sjodin, P.A. (2020) A mass spectroscopic assay of METTL3/METTL14 methyltransferase activity. *SLAS Discov.* **25**, 361–371.

Cohen, L.H., Yan, Z. and Galdwell, G.W. (2004) Plasma protein-binding methods in drug discovery. In *Optimization in Drug Discovery: in vitro Methods* (Z. Yan and G.W. Galdwell, Eds.). Humana Press, Totowa, NJ, pp. 111–122.

Copeland, R.A. (2013) *Evaluation of Enzyme Inhibitors in Drug Discovery: A Guide for Medicinal Chemists and Pharmacologists.* 2nd Edition. Wiley, Hoboken, NJ.

Copeland, R.A. (2021) Chance favors the perplexed mind: the critical role of mechanistic biochemistry in drug discovery. *Biochemistry* **60**, 2275–2284.

Copeland, R.A. (2023) *Enzymes: A Practical Introduction to Structure, Mechanism, and Data Analysis.* 3rd Edition. Wiley, Hoboken, NJ.

Copeland, R.A. and Boriack-Sjodin, P.A. (2018) The elements of translational chemical biology. *Cell Chem. Biol.* **25**, 128–133.

Copeland, R.A., Williams, J.M., Giannaras, J., Nurnberg, S., Covington, M., Pinto, D., Pick, S. and Trzaskos, J.M. (1994) Mechanism of selective inhibition of the inducible isoform of prostaglandin G/H synthase. *PNAS* **91**, 11202–11206.

Daigle, S.R., Olhava, E.J., Therkelsen, C.A., Majer, C.R. et al. (2011) Selective killing of mixed lineage leukemia cells by a potent small-molecule DOT1L inhibitor. *Cancer Cell* **20**, 53–65.

Daigle, S.R., Olhava, E.J., Therkelsen, C.A., Basavapathruni, A. et al. (2013) Potent inhibition of DOT1L as treatment of MLL-fusion leukemia. *Blood* **122**, 1017–1025.

Dubin, A.E., Nasser, N., Rohrbacher, J., Hermans, A.N., Marrannes, R., Grantham, C., van Rossem, K., Cik, M., Chaplan, S.R., Gallacher, D., Xu, J., Guia, A., Byrne, N.G. and Mathes, C. (2005) Identifying modulators of hERG channel activity using the PatchXpress planar patch clamp. *J. Biomol. Screening* **10**, 168–181.

Hafner, M., Niepel, M., Chung, M. and Sorgen, P.K. (2016) Grow rate inhibition metrics correct for confounders in measuring sensitivity to cancer drugs. *Nature Meth.* **13**, 521–527.

Harris, L.A., Frick, P.L., Garbett, S.P., Hardeman, K.N., Paudel, B.B., Lopez, C.F., Quaranta, V. and Tyson, D.R. (2016) An unbiased metric of antiproliferative drug effect *in vitro*. *Nature Meth.* **13**, 497–500.

Jarrett, A.M., Lima, E.A.B.F., Hormuth, D.A. 2nd,, McKenna, M.T., Feng, X., Ekrut, D.A., Resende, A.C.M., Brock, A. and Yankeelov, T.E. (2018) Mathematical models of tumor cell proliferation: a review of the literature. *Expert Rev. Anticancer Ther.* **18**, 1271–1286.

Knutson, S.K., Wigle, T.J., Warholic, N.M., Sneeringer, C.J., Allain, C.J., Klaus, C.R., Sacks, J.D., Raimondi, A., Majer, C.R., Song, J. and Scott, M.P. (2012) A selective inhibitor of EZH2 blocks H3K27 methylation and kills mutant lymphoma cells. *Nat. Chem. Biol.* **8**, 890–896.

Knutson, S.K., Kawano, S., Minoshima, Y., Warholic, N.M., Huang, K.C., Xiao, Y., Kadowaki, T., Uesugi, M., Kuznetsov, G., Kumar, N. and Wigle, T.J. (2014) Selective inhibition of EZH2 by EPZ-6438 leads to potent antitumor activity in EZH2-mutant non-Hodgkin lymphoma. *Mol. Cancer Therapeut.* **13**, 842–854.

Kuntz, K.W., Campbell, J.E., Keilhack, H., Pollock, R.M., Knutson, S.K., Porter-Scott, M., Richon, V.M., Sneeringer, C.J., Wigle, T.J., Allain, C.J., Majer, C.R., Moyer, M.P., Copeland, R.A. and Chesworth, R. (2016) The importance of being me: magic methyls, methyltransferase inhibitors, and the discovery of tazemetostat. *J. Med. Chem.* **59**, 1556–1564.

Luong, C., Miller, A., Barnett, J., Chow, J., Ramesha, C. and Browner, M.F. (1996) Flexibility of the NSAID binding site in the structure of human cyclooxygenase-2. *Nat. Struct. Biol.* **3**, 927–933.

Motulsky, H. and Christopoulous, A. (2004) *Fitting Models to Biological Data Using Linear and Nonlinear Regression: A Practical Guide to Curve Fitting*. Oxford University Press, Oxford, UK.

Papadopoulos, K.P., Moser, J.C., Rodon, J., Ofir-Rosenfeld, Y., Holz, J.B. and STC15-22101 Study Team (2024) Phase 1 dose escalation and cohort expansion study evaluating safety, PK, PD and clinical activity of STC-15, a METTL3 inhibitor in patients with advanced malignancies. *J. Immuno Therapy Cancer* **12**. DOI: 10.1136/jitc-2024-SITC2024.0676.

Paul, S.M., Mytelka, D.S., Dunwiddie, C.T., Persinger, C.C., Munos, B.H., Lindborg, S.R. and Schacht, A.L. (2010) How to improve R&D productivity: the pharmaceutical industry's grand challenge. *Nature Rev. Drug Discov.* **9**, 203–214.

Roberts, A.G. (2021) The structure and mechanism of drug transporters. *Methods Mol. Biol.* **2342**, 193–234.

Schurmann, M., Janning, P., Ziegler, S. and Waldmann, H. (2016) Small-molecule target engagement in cells. *Cell Chem. Biol.* **23**, 435–441.

Smith, W.L., Garavito, R.M. and DeWitt, D.L. (1996) Prostaglandin endoperoxide H synthases (cyclooxygenases)-1 and 2. *J. Biolumin. Chemilumin.* **271**, 33157–33160.

Sneeringer, C.J., Porter Scott, M., Kuntz, K.W., Knutson, S.K. et al. (2010) Coordinated activities of wild-type plus mutant EZH2 drive tumor-associated hypertrimethylation of lysine 27 on histone H3 (H3K27) in human B-cell lymphomas. *Proc. Natl. Acad. Sci. USA* **107**, 20980–20985.

Sou, T. and Bergstrom, C.A.S. (2018) Automated assays for thermodynamic (equilibrium) solubility determination. *Drug Discov. Today: Technologies* **27**, 11–19.

Stefaniak, J. and Huber, K.V.M. (2020) Importance of quantifying drug-target engagement in cells. *ACS Med. Chem. Lett.* **11**, 403–406.

Stein, E.M., Garcia-Manero, G., Rizzieri, D.A., Tibes, R. et al. (2018) The DOT1L inhibitor pinometostat reduces H3K79 methylation and has modest clinical activity in adult acute leukemia. *Blood* **131**, 2661–2669.

Strauss, O.H. and Goldstein, A. (1943) Zone behavior of enzymes: illustrated by the effect of dissociation constant and dilution on the system cholinesterase-physostigmine. *J. Gen. Physiol.* **26**, 559–585.

Wu, W.N. and McKown, L.A. (2004) in vitro metabolic profiling using hepatic S9 and human liver microsomes. In *Optimization in Drug Discovery* (Z. Yan and G.W. Caldwell, Eds.). Humana Press, Totowa, NJ, pp. 163–184.

Yang, H., Higgins, B., Kolinsky, K., Packman, K., Go, Z., Iyer, R., Kolis, S., Zhao, S., Lee, R., Grippo, J.F., Schostack, K., Simcox, M.E., Heimbrook, D., Bollag, G. and Su, F. (2010) RG7204 (PLX4032), a selective BRAFV600E inhibitor, displays potent antitumor activity in preclinical melanoma models. *Cancer Res.* **70**, 5518–5527.

6

Lead Optimization

Key Learning Points

- Having one or more lead series in hand, SAR continues with the goal of identifying compounds for clinical use.
- Increased stringency for target affinity, target selectivity, and *in vitro* ADME (absorption, distribution, metabolism, and excretion) criteria is applied during lead optimization, and these are augmented with additional criteria related to *in vivo* PK, efficacy, and safety.
- *In vitro* liver microsomes, hepatocytes, or related systems are used to identify the major metabolites of compounds across species. These data are used to identify one rodent and one non-rodent preclinical species within which all major human metabolites are represented, and that can be used to assess the *in vivo* safety of the compound(s).
- Preliminary *in vivo* safety assessment (i.e., toxicology) is performed in the two species that were selected based on metabolite profiling.
- Greater scrutiny of patent position, route of chemical synthesis, physical form (e.g., salt form), and formulation is applied to later-stage compounds to ensure the feasibility of progressing compounds to clinical studies and beyond.

With one or more lead series in hand, the continuum of preclinical drug discovery now transitions to a broader spectrum of multiparametric optimizations for each series. Medicinal chemistry structure–activity relationship (SAR) continues to strive toward greater target affinity, target selectivity, cellular phenotypic potency, and the other properties discussed in Chapter 5, but now also broadens to include much greater emphasis on *in vivo* assessment of compound safety, ADME, and pharmacologic action (efficacy). The goal of lead optimization is to produce one

Molecular Targeted Drug Discovery: A Guide to How Modern Medicines are Created,
First Edition. Robert A. Copeland.

or more clinical development candidate (DC)-quality compounds that fulfill a stringent set of predetermined criteria. Collectively, these criteria present a high bar for progression, such that compounds that fulfill all these criteria offer great confidence to the research team that the compounds are worthy of progression to the clinic. However, before clinical studies may begin, DC compounds are put through a rigorous and highly prescriptive battery of tests that are required by regulatory agencies as the basis of an investigational new drug (IND) application. These tests are referred to as IND-enabling studies and summarized in Appendix 2.

6.1 Typical Criteria for Development Candidate Declaration

As with the previous stages of drug discovery, lead optimization starts by defining a set of predetermined criteria for declaration of compounds as acceptable as clinical DCs, and by which the research team can gauge its progress along the way toward this goal. Again, as with milestone criteria for the other stages of drug discovery, there are no hard-and-fast rules for DC criteria, and the specific criteria will thus vary among research groups. Table 6.1, however, provides a typical set of DC criteria that have proved useful for several drug discovery organizations.

6.2 Continued Optimization of Target Potency, Selectivity, and Cellular Potency

The assays and analysis method described in Chapter 5 continue to drive further medicinal chemistry efforts toward optimization of target potency, target selectivity, and cellular potency. As summarized in Table 6.1, however, the stringency for these criteria is increased during lead optimization to ensure good translation of these *in vitro* measurements to *in vivo* efficacy. Often, considering these higher levels of stringency, certain adaptations must be made to how target potency, target selectivity, and cellular potency are measured. We shall discuss these potential adaptations in turn.

6.2.1 Measuring Target Potency Under Tight-Binding Conditions

In Chapter 5, we briefly discussed the concept of tight-binding drug–protein interactions in our discussion of plasma protein binding and of Element 1 of the elements of translational chemical biology. In the former case, the tight-binding conditions were a result of the very high concentration of albumin and other proteins in whole blood and plasma. However, as SAR continues to drive target potency, one can often reach conditions wherein the K_d values of compounds

Table 6.1 Typical criteria for the selection of compounds as clinical development candidates (DCs).

Property	Milestone criterion
Target potency (K_d)	≤10 nM (preferred); generally, potency consistent with an acceptable predicted human efficacious exposure (see below).
Target selectivity	≥50-fold against selected, related proteins.
Aqueous Solubility	≥100 μM (preferred); generally, sufficient to achieve linear dose–exposure relationship up to the maximum feasible human dose and to conduct *in vivo* safety assessment at high doses.
Patent position	Intellectual property (IP) established with freedom to operate, and composition of matter patent application(s) filed.
Cellular PD potency	≤100 nM. Correlated with cell-free biochemical/biophysical potency across series.
Cellular phenotypic potency	≤100 nM in target-sensitive cell lines (preferred). Potency (EC_{50} and/or EC_{90}) consistent with a reasonable predicted human efficacious exposure (see below).
General cytotoxicity	$EC_{50} \geq 10$ μM and not target mediated.
Safety panel screening of typical nontarget proteins	No off-target activity that limits exposure in the efficacious range. Generally, no EC_{50} < 30-fold over the plasma protein binding (PPB)-adjusted C_{max} at the efficacious exposure.
Perpetrator and victim drug–drug interactions (DDI)	Minimum potential for clinical DDI at the predicted human efficacious exposure. Generally, major CYP inhibition EC_{50}s ≥ 30 μM. CYP isoform(s) involved in compound clearance identified.
CYP time-dependent inhibition (TDI)	No meaningful TDI for any major CYP isoform at predicted human efficacious exposure.
CYP induction	No meaningful induction of any major CYP isoform at predicted human efficacious exposure.
hERG IC_{50}	>10 μM and/or 30-fold over the PPB-adjusted C_{max} at the predicted human efficacious exposure measured by patch-clamp electrophysiology, or evidence of compensatory cardiac ion channel activity.
Caco-2 permeability/efflux	$P_{app} \geq 1 \times 10^{-6}$ cm/s; efflux ratio ≤ 10.
Plasma protein binding (PPB)	Measurable and consistent with a reasonable predicted human efficacious exposure.
Whole blood or plasma stability	Half-life consistent with QD or BID dosing schedule for orally administered drugs (generally, $t_{1/2} \geq 8$ h)
Liver hepatocyte, microsome or S9 fraction stability	Scaled clearance (Cl) in multiple species <70% hepatic blood flow rate (Q_h).

Table 6.1 (Continued)

Property	**Milestone criterion**
Pharmacokinetics	Through understanding of PK/PD/efficacy relationship. *In vitro/in vivo* correlation (IV/IVC) between clearance relationship understood. Oral bioavailability >30%. *In vivo* clearance in one rodent and one non-rodent species suitable to achieve exposures needed for preliminary safety assessment studies.
Metabolic profiling	Identify major metabolites in humans and relevant preclinical safety assessment species (one rodent and one non-rodent) with *in vitro* evaluation of target modulation by major metabolites where required (see text).
Preliminary *in vitro* genotoxicity assessment (AMES test)	Negative or significance for disease indication clearly defined
Predicted human efficacious dose and dose schedule	Dose of ≤500 mg/day (preferred; ≤ 1 g/day acceptable) Fraction absorbed >30% Dose schedule QD or BID (preferred)
In vivo efficacy	Dose (and exposure)-dependent efficacy in relevant disease models at well-tolerated exposures. Extrapolated clinical exposure consistent with reasonable human dose.
PD biomarker	Suitable biomarker(s) identified and validated for *in vivo* studies.
Preliminary rodent 7–14 day repeat dose safety assessment	Acceptable safety profile at ≥2-fold the projected human PPB-adjusted C_{max} and AUC at the efficacious dose.
Preliminary non-rodent escalating single-dose safety assessment	Acceptable safety profile to enable subsequent non-rodent safety studies.
Chemistry	Synthetic route established consistent with scaling to >100 g. Evaluation of solid form(s) suitable for clinical development with acceptable chemical and physical stability. Initiation of formulation studies to enable clinical development.

for target-binding approach the concentration of target protein in the cell-free assay system. In this situation, the researcher's ability to assess continued SAR improvements in target affinity becomes compromised. Let us consider the following illustrative example. Suppose that we wish to optimize inhibitors of a specific enzyme target using an assay of enzymatic activity in which the

Table 6.2 Measured values of IC_{50} for a hypothetical series of enzyme inhibitors according to Equation 5.13, in a cell-free enzyme assay in which the enzyme concentration is 10 nM.

K_i (nM)	Measured IC_{50} (nM)
100	105
50	55
10	15
3	8
1	6
0.1	5.1

target enzyme concentration is fixed at 10 nM. Let us further say that within a lead pharmacophore series, the medicinal chemistry team has made structural changes leading to compounds with K_i values of 100, 50, 10, 3, 1, and 0.1 nM. Applying the Strauss and Goldstein equation (Equation 5.13) to these compounds, we would expect the measured IC_{50}s to be as shown in Table 6.2 and Figure 6.1.

Clearly, the researcher's ability to realize improvements in target affinity as the K_i value approaches the target protein concentration is severely compromised by the tight-binding conditions of the assay. If this situation is not recognized, it will lead to the type of discontinuity in PD/biochemical EC_{50} correlation plots that was illustrated in Figure 5.16A. The most obvious way to remedy this situation is to lower the concentration of target protein within the biochemical or biophysical

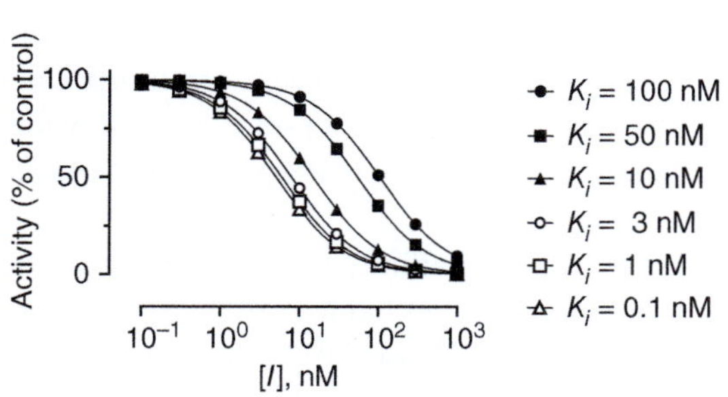

Figure 6.1 Concentration–response plots for enzyme inhibitors of varying target affinity (K_i) measured in an enzyme assay in which the enzyme concentration is 10 nM, illustrating the issue of accurately measuring the EC_{50} as the K_i value approaches the concentration of target protein in the assay. Note that as the K_i value approaches the enzyme concentration, the EC_{50} values converge to a value of ½ the enzyme concentration. See text for further details.

assay. However, that is not always a practical solution, due to the erosion of signal quality that may accompany lower protein concentrations in the assay. Copeland (2013), and Copeland (2023) have reviewed several approaches to dealing with tight-binding conditions that can extend the range of target affinity discrimination; such approaches include application of the quadratic equation shown in Equation 5.14, titration of target protein concentration and extrapolation of the resultant EC_{50} values to zero protein concentration, etc. Despite these varied approaches, there will nevertheless be a practical limit to the extent to which SAR for high-affinity compounds can be measured in cell-free assays. When this limit is reached, the researcher may need to use the cell-free biochemical or biophysical assay in a less quantitative and more categorical manner. That is, compounds with K_d values at or below the target protein concentration may need to be reported as $K_d \leq X$ nM, where X is the concentration of target protein in the assay. Further discrimination of SAR improvements in this case would rely on cellular assays of PD and/or phenotypic readout.

6.2.2 The Impact of Protein Conformational Dynamics on SAR

Until now, we have largely described drug-binding pockets as if they were static structures, for which compound affinity is related to complementarity between the fixed structure of the unoccupied binding pocket and that of the compound or other ligands. This view, however, is incomplete, as it does not consider the conformational dynamics of proteins that are common features of proteins and of protein–ligand interactions. Even in the absence of ligand binding, the most stable or "native" state of a protein is in fact a manifold of conformational substates that interconvert through vibrational and rotational motions of amino acid side chains and main chain structures. Additionally, unstructured loops can adopt different positions relative to the more structured protein core, and even elements of protein secondary structures, such as alpha helices, can translocate as units, to effect structural changes of the protein, which, in some cases, provide the basis for communication of allosteric ligand binding to distal orthosteric sites (Copeland, 2023).

Conformational dynamics can also attend high-affinity ligand binding by two principal mechanisms: conformational selection and induced-fit, both of which often lead to a slow, time dependence of ligand association and dissociation (Figure 6.2). In both cases, the ligand (L) structure is not optimally complementary to the most highly populated conformational state of the binding pocket (symbolized by R) but is instead optimally complementary to an alternative conformational state (symbolized by R*).

In the conformational selection mechanism (Figure 6.2C), conformations R and R* slowly interconvert in the absence of L, but the R state is more highly populated

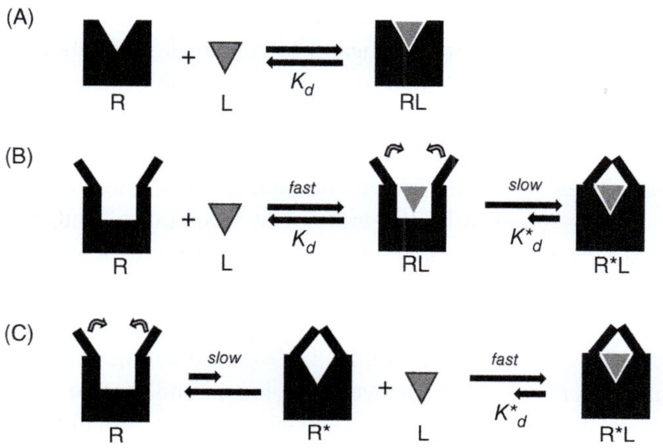

Figure 6.2 (A) Simple 1-step association and dissociation mechanism of receptor–ligand interactions. (B) The induced-fit model of receptor–ligand interactions. In this cartoon, the receptor (R) structure includes a ligand-binding pocket (in a square shape) that displays suboptimal structural complementarity for the ligand (L), and two disordered loops that fold over to occlude the reordered ligand-binding pocket (now a triangular shape) in the R*L conformational state. The small, curved, gray arrows associated with the RL state are intended to connote the motion of the disordered loops in transitioning between states RL and R*L. Note that in this model, the transition from RL to R*L and especially the reverse transition from R*L to RL are slow and rate-limiting to the overall binding kinetics. (C) The conformational selection model of receptor–ligand interactions. In this model, the free (i.e., unliganded) receptor equilibrates slowly between states R and R* (as in (**B**)) because the equilibrium greatly favors state R. Here, ligand (L) binds exclusively to the R* state of the free receptor. See text and Copeland (2013), Copeland and Knockenhauer (2023) for further details.

due to energetic (i.e., thermodynamic) differences between the two unliganded conformations of the receptor. L binds exclusively, and with high affinity, to the R* state in this model. Hence, the rate of ligand association is determined by the slow interconversion between the unliganded R and R* states. Because of the high affinity of ligand for the R* state, dissociation is also slow because it requires conversion from the R*L to the RL state prior to actual ligand escape from the complex.

In the induced-fit mechanism (Figure 6.2B), L binds initially to the R conformational state with modest affinity due to the suboptimal complementarity between the ligand structure and that of the binding pocket. After formation of the initial RL encounter complex, conformational rearrangements occur in the binding pocket to create state R*L, with optimal ligand complementarity; that is to say, the protein conformationally molds the binding pocket to fit the ligand optimally. The induced-fit model of ligand binding requires two distinct K_d values to describe the mechanism, one for state RL (K_d) and one for state R*L (K_d*).

There has been considerable debate in the academic community over which of these two models is most relevant to protein–ligand interactions. Experimental evidence for each model has been published in specific cases, and it has been argued that these two models merely represent the two poles of a continuum of conformational mechanisms that may attend protein–ligand interactions (Hammes et al., 2009). These academic arguments aside, most drug–protein binding kinetics can be best modeled by the induced-fit mechanism, and this mechanism seems consistent with crystallographic data comparing the R and the R*L states. In many cases of high-affinity, clinically active drugs, the conformational difference between R and R*L includes greater occlusion of the binding pocket from bulk solvent, thereby strengthening hydrogen bonding, van der Waals interactions, and other interactions between the ligand and protein. This greater occlusion is most often the result of loop movements of the protein that result in capping of the binding pocket by the loop(s). Certainly, it may be energetically possible for the free (i.e., unliganded) protein to access the loop-occluded conformation R*, but it is difficult to visualize how a ligand molecule could then bind to this occluded pocket. On the other hand, ligand dissociation can be visualized as a retrograde induced-fit mechanism, wherein the reverse transition from the R*L state to the RL state relieves the loop-mediated capping of the binding pocket, thus offering a straightforward escape trajectory for the ligand. A useful analogy for these aspects of the induced-fit mechanism is a Venus fly trap (as the receptor) interacting with a prey insect (as the ligand). When the trap is open, flying insects can enter and alight briefly on the trap's surface before flying off (dissociation). However, if the insect lingers, the trap closes. Now the insect's escape trajectory is blocked. The trapped insect can no longer fly away, nor can other insects enter the trap while it is closed. The only way that a trapped insect can escape is if the trap were to reopen prior to digestion of the prey (Figure 6.3).

The perturbed equilibrium that characterizes the 2-step mechanisms may at first glance seem like an exotic form of drug–target interactions. However, as reviewed by Swinney (2004, 2009), many FDA-approved drugs work through molecular mechanisms of action (MMOA) that deviate from the simple 1-step association/dissociation mechanism of Figure 6.2A. Indeed, it is quite common to encounter an induced-fit MMOA as medicinal chemistry SAR drives target potency, although this can go unrecognized by the research team. In fact, 2-step binding often occurs even for early compounds of more modest target affinity, but in these cases, the kinetics of the conformational changes are typically too rapid to be realized in commonly used assay formats and are only revealed by rapid kinetic methods, such as stopped-flow assays (see Copeland, 2023, for examples). Unfortunately, when 2-step binding mechanisms are not accounted for, discontinuities in SAR can be observed, and the true affinity of compounds

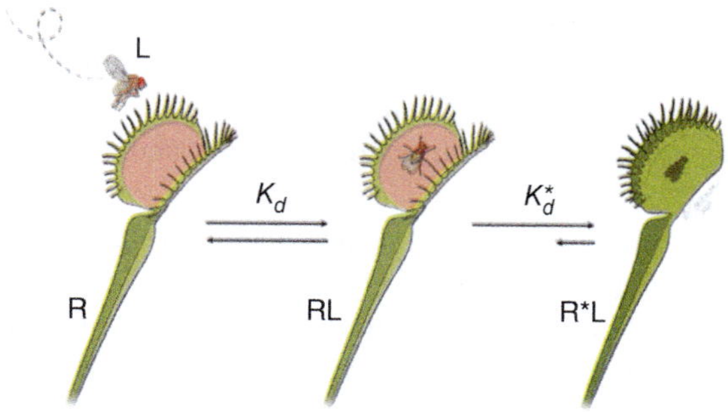

Figure 6.3 Cartoon of a Venus fly trap capturing a flying insect as a macroscopic analogy for the induce-fit model of receptor–ligand interactions. The insect encounters the plant in the open state of the trap. Once in contact with the plant surface, the insect can fly off (dissociation) or remain on the surface, triggering the trap to close. Once the trap is closed, the insect's escape trajectory is blocked, thus blocking dissociation, unless the trap were to reopen. Illustration by Jennifer E. Fairman, CMI, FAMI. © 2025 Fairman Studios, LLC.

for their targets can be greatly underestimated (see Copeland, 2013, for a detailed discussion of the proper analysis of such data).

The 2-step MMOAs presented in Figure 6.2 are generally rate-limited by slow dissociation of ligand from the R*L complex. In fact, surveys of small-molecule, peptide, and macromolecular (e.g., antibodies) ligands binding to target proteins have consistently shown that the SAR of binding affinity (i.e., K_d or K_d^*) is often driven almost exclusively by changes in the dissociation rate constant (k_{off}), with little change in the association rate constant (k_{on}). This has led Copeland and colleagues (Copeland et al., 2006; Copeland, 2016, 2021; Copeland and Knockenhauer, 2023) to propose the drug–target residence time model, which posits that cellular and *in vivo* drug PD and efficacy depend largely on the duration (i.e., lifetime or residence time) of the drug–target binary complex, rather than on drug–target affinity per se. These authors have recommended that preclinical lead-optimization efforts explore drug–target residence time (measured as the reciprocal of k_{off}) in addition to equilibrium binding affinity, as metrics for SAR; this recommendation has increasingly been taken up by many researchers throughout academia, biotechnology, and pharmaceutical companies. As noted by Copeland and colleagues, a long residence time is often favorable for drug efficacy and durability, but there are also clinical situations in which a short residence time is preferable. This is often the case when on-target toxicity is limiting, i.e., where toxicity and pharmacology are both driven by drug binding to the same target. In such cases, more intermittent or pulsatile target engagement may prove more tolerable (see Copeland, 2016). In previous chapters of this book, we have already encountered examples of target–drug complexes that form through induced-fit mechanisms and for which the optimized drug is characterized by a long drug–target residence time: the DOT1L inhibitor EPZ-5676 and the COX2-selective inhibitors Vioxx and Celebrex. Many additional examples can be found in the recent scientific literature (see Copeland, 2013, 2023, and references therein).

An interesting extension of the drug–target residence time model to a subset of heterobifunctional molecules and molecular glues has recently been reviewed by Deshaies and Potts (2025). These authors describe small molecules that elicit their pharmacological action by reducing the dissociation rate (measured as k_{off}; or equivalently increasing the residence time) of naturally interacting multi-protein complexes. That is, rather than forming novel protein–protein interactions, these compounds enhance the duration of naturally occurring dynamic protein complexes, thereby extending their biochemical and biological function. The compounds effectively "lock" the protein complex in place; hence, Deshaies and Potts have coined the term LOCKTACs to describe this emerging class of therapeutics. They go on to describe existing examples of LOCKTACs that can stabilize protein–protein complexes, leading either to the activation or inhibition of a biochemical mechanism, depending on the specific complex being stabilized. Examples of therapeutic applications for activating and inhibitory complexes are provided by these authors.

6.2.3 Extended Surveys of Target Selectivity

In Chapter 5, we introduced the concept of counter-screening of compounds for activity against collateral proteins during the hit-to-lead stage. In this early stage of SAR, counter-screening is usually limited to a few collateral proteins, with close structural and/or mechanistic relatedness to the target protein, and the hERG cardiac ion channel. As programs transition into lead optimization, it is common to expand the panel of counter-screens to include more distantly related proteins of the same target class. If, for example, the target is a kinase, or if the chemical structure of the lead series is related to kinase inhibitors, one may wish to screen compounds in lead optimization across an extended panel of other human kinases. Large panels of kinases and other target classes are available as fee-for-service screens at multiple contract research organizations (CROs). To illustrate this kind of extended target class selectivity survey, consider the EZH2 inhibitor EPZ6438 (tazemetostat), which was selected as a DC by the Epizyme team and ultimately became an FDA-approved drug for two distinct cancer indications (see below). During early SAR, compounds were tested against the wild-type (WT) EZH2, oncogenic mutants of EZH2 (see Chapter 5), and against the closely related enzyme EZH1. As lead optimization progressed, the team developed a panel of enzymatic assays for a large number of other protein lysine methyltransferases (PKMTs) and

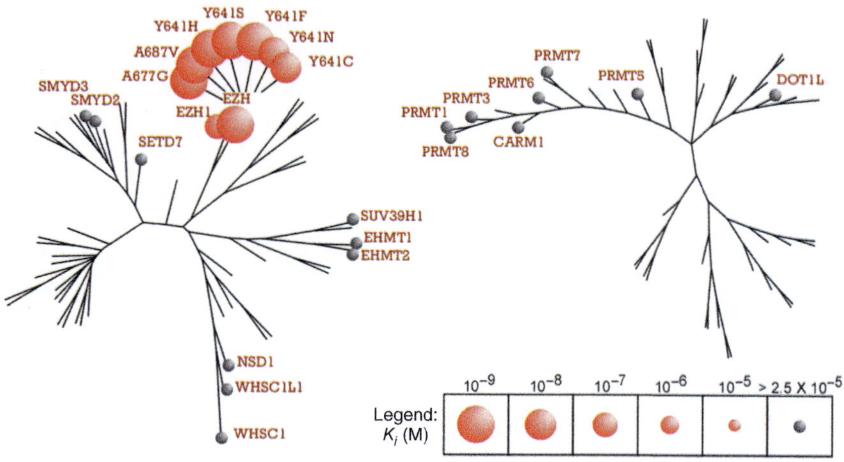

Figure 6.4 Selectivity of the EZH2 inhibitor tazemetostat (EPZ6438) mapped onto the dendrograms for the human PKMT and PRMTs. As indicated in the legend, the diameter of the circle is inversely proportional to the K_i value for inhibition of the indicated enzyme. Figure taken from Ho and Copeland (2024), based on data first reported by Knutson et al. (2013).

the related protein arginine methyltransferases (PRMTs), both as counter-screens for the DOT1L and EZH2 targets (see Chapter 5) and to create opportunities for serendipitous discovery of chemical starting points against other PKMT or PRMT enzymes that might be of therapeutic interest. Figure 6.4 illustrates the selectivity of tazemetostat for EZH2 mapped onto the dendrograms (family trees) for the human PKMTs and PRMTs. As can be gleaned from this figure, tazemetostat had a K_i of <2.5 nM for WT and oncogenic mutants of EZH2 and displayed ≥25,000-fold selectivity for all tested PKMTs and PRMTs except for EZH1, for which it displayed ca. 100-fold selectivity.

6.2.3.1 Secondary Pharmacology Screening

As a program continues to advance through lead optimization, the research team will also wish to begin counter-screening panels of proteins for which modulation of function could be problematic *in vivo*. Functional agonism or antagonism of certain receptors, enzymes, transporters, and ion channels is known to cause safety concerns (referred to as off-target toxicities) associated with specific organs or systems. Screening compounds for activity against these collateral proteins with *in vitro* binding and/or functional assays is known as secondary pharmacology or off-target pharmacology and is considered to be a cost- and time-efficient means of derisking the likelihood of off-target toxicities in the clinic (Scott et al., 2022; Bowes et al., 2012); secondary pharmacology counter-screening has more recently become a routine part of preclinical data packages that are submitted to regulatory agencies in IND applications (Appendix 2), and the inclusion of such data has been recommended by the International Council of Harmonization Guidance (S7A).

Typically, compounds are first screened against collateral proteins at a fixed concentration of 10 μM for their ability to inhibit the binding of a well-known ligand for each protein (specific controls). Compounds are scored as hits if they exhibit ≥50% inhibition of specific control binding under these conditions. Hits from such screening are then screened in a concentration-dependent manner in cell-based assays of collateral target function to determine if they act as agonists or antagonists of specific collateral proteins and to calculate the EC_{50}. The use of cell-based assays is preferred for this purpose to ensure that the measured EC_{50}s faithfully reflect compound affinity for the full-length protein(s) under the physiological conditions (e.g., pH, ionic strength, natural ligand, metabolite, nucleic acid, and partner protein binding) at which they function within the cell.

What collateral proteins are used for counter-screening varies among organizations and can depend on the specific target and disease indication. Scott et al. (2022) analyzed 2,701 IND applications for which secondary pharmacology data were included during the period between 1999 and 2020. They found that among these IND applications a total of 747 distinct proteins had been tested. These 747 proteins were categorized by target class as follows: 192 G-protein–coupled receptors, 80 ion channels, 15 transporters, 375 enzymes, and 85 belonging to other classes. The average number of targets tested in an IND application was 74, with mean and median hit rates of 5.1% and 0%, respectively. Bowes et al. (2012) made more of an attempt to define a common set of collateral proteins for secondary pharmacology screening. They recommended a panel of 44 proteins, as summarized in Table 6.3. They also categorized the hit rates for these targets into four broad categories: very high (hit rate ≥ 20%), high (5–20%), medium (1–5%), or low (0–1%). Although not universally adopted, there is a growing trend of researchers using the recommendations of Bowes et al. (2012) for secondary pharmacology assessment (Scott et al., 2022).

6.2.3.2 Assessing Affinity for Target Orthologues

When the target of drug discovery is a human protein (i.e., not an infectious disease target), it is important to understand the extent to which compounds also bind and modulate target orthologues of other mammalian species that are to be used for *in vivo* efficacy and/or safety studies. This is an important consideration because if the affinity for a target orthologue is significantly diminished relative to that for the human target, the interpretation of *in vivo* pharmacology and toxicity can be severely compromised. Research teams typically begin by sequence alignments for the human, mouse, rat, dog, and monkey orthologues of the target to determine the degree of homology within the drug-binding pocket. Even when there is a high degree of sequence identity within the binding pockets across species, this is often

Table 6.3 Recommended counter-screening proteins to provide early assessment of potential off-target toxicity for a compound or chemical series.

Target (gene)	Hit rate		Main organ class or system affected
	Binding	Functional	
G-protein–coupled receptors			
Adenosine receptor A_{2A} (ADORA2A)	High	Low (agonist)	CVS, CNS
α_{1A}-adrenergic receptor (ADRA1A)	High	Low (agonist); high (antagonist)	cardiovascular system (CVS), gastrointestinal (GI), CNS
α_{2A}-adrenergic receptor (ADRA2A)	High	Low (agonist); medium (antagonist)	CVS, CNS
β_1-adrenergic receptor (ADRB1)	High	N/A	CVS, GI
β_2-adrenergic receptor (ADRB1)	High	Medium (agonist); medium (antagonist)	Pulmonary, CVS
Cannabinoid receptor CB_1 (CNR1)	Medium/high	Medium (antagonist)	CNS
Cannabinoid receptor CB_2 (CNR2)	Medium	Medium (agonist)	Immune
Cholecystokinin A receptor (CCKAR)	Low/medium	N/A	GI
Dopamine receptor D_1 (DRD1)	Medium/high	Medium (antagonist)	CVS, CNS
Dopamine receptor D_2 (DRD2)	Medium/high	Medium/high (agonist); medium (antagonist)	CVS, CNS, endocrine
Endothelin receptor A (EDNRA)	Low	N/A	CVS, development
Histamine H_1 receptor (HRN1)	High	Very high (antagonist)	CVS, immune
Histamine H_2 receptor (HRN2)	High	Low (agonist)	GI, CVS
δ-type opioid receptor (OPRD1)	Medium/high	N/A	CNS, CVS
κ-type opioid receptor (OPRK1)	High	Medium (agonist and antagonist)	GI, CNS, CVS

Table 6.3 (Continued)

Target (gene)	Hit rate		Main organ class or system affected
	Binding	Functional	
μ-type opioid receptor (OPRM1)	High	Medium (agonist and antagonist)	CNS, GI, CVS
Muscarinic acetylcholine receptor M_1 (CHRM1)	High	Low (agonist); high (antagonist)	CNS, GI, CVS
Muscarinic acetylcholine receptor M_2 (CHRM2)	High	Low (agonist); medium (antagonist)	CVS
Muscarinic acetylcholine receptor M_3 (CHRM3)	High	N/A	GI, pulmonary
5-HT_{1A} (HTR1A)	Medium/high	Low (agonist); medium (antagonist)	CNS, endocrine
5-HT_{1B} (HTR1B)	High	High (agonist); medium (antagonist)	CVS, CNS
5-HT_{2A} (HTR2A)	Very high	Low/medium (agonist); medium/high (antagonist)	CVS, CNS
5-HT_{2B} (HTR2B)	High/very high	Low (agonist); high (antagonist)	CVS, pulmonary, development
Vasopressin V_{1A} receptor (AVPR1A)	Medium	High	Renal, CVS
Ion channels			
Acetylcholine receptor α1 (CHRNA1) or α4 (CHRNA4)	Medium/high	Low (opener); very high (blocker)	CNS, CVS, GI, pulmonary
Voltage-gated calcium channel subunit α Cav1.2 (CACNA1C)	N/A	Medium/high (blocker)	CVS
Gamma-aminobutyric acid (GABA) receptor α1 (rat cortex) benzodiazepine (BZD) site (GABRA1)	Medium	N/A	CNS

Table 6.3 (Continued)

Target (gene)	Hit rate		Main organ class or system affected
	Binding	Functional	
Potassium voltage-gated channel subfamily H member 2; hERG (KCNH2)	High	High	CVS
Potassium voltage-gated channel KQT-like member 1 (KCNQ1) and minimal potassium channel MinK (KCNE1)	N/A	Low	CVS
NDMA receptor subunit NR1 (GRIN1)	Low/medium	Medium (blocker)	CNS
5-HT_3 (HTR3A)	Medium	Very high	GI, endocrine
Voltage-gated sodium channel subunit α Nav.1.5 (SCN5A)	N/A	High	CVS
Enzymes			
Acetylcholinesterase (ACHE)	N/A	High	CVS, GI, pulmonary
Cyclooxygenase 1; COX1 (PTGS1)	N/A	Medium	GI, pulmonary, renal
Cyclooxygenase 2; COX2 (PTGS2)	N/A	Medium/high	Immune, CVS
Monoamine oxidase A (MAOA)	N/A	Medium	CVS, CNS
Phosphodiesterase 3A (PDE3A)	N/A	High	CVS
Phosphodiesterase 4D (PDE4D)	N/A	Very high	CNS, immune
Lymphocyte-specific protein tyrosine kinase (LCK)	N/A	Medium/high	Immune
Transporters			
Dopamine transporter (SLC6A3)	High/very high	N/A	CNS
Noradrenaline transporter (SLC6A2)	High/very high	N/A	CNS, CVS

Table 6.3 (Continued)

Target (gene)	Hit rate		Main organ class or system affected
	Binding	Functional	
Serotonin transporter (SLC6A4)	High	N/A	CNS, CVS
Nuclear receptors			
Androgen receptor (AR)	Medium	Medium	Endocrine
Glucocorticoid receptor (NR3C1)	Medium	Medium	Endocrine, immune

Source: Adapted from Bowes et al. (2012).

followed by recombinant protein production and assay development for the orthologues to experimentally determine any differences in affinity across species. If the target orthologue of a particular species displays greatly reduced compound affinity, this species may be deemed inappropriate for use as an *in vivo* model of efficacy and/or safety assessment.

For many drug discovery programs, this turns out to not be a major issue, as the sequence variation among mammals is such that similar compound affinity is often observed among the species typically used in preclinical drug discovery and development. However, in unfavorable cases, the research team may need to alter plans for *in vivo* efficacy and/or safety assessment to ensure that such studies reflect reasonable levels of *in vivo* target occupancy.

6.2.4 Optimization of Potency and Durability of Cellular Pharmacology

As discussed in Chapter 5, compound potency for cellular PD and phenotypic outcome should correlate with target affinity measured in cell-free biochemical and biophysical assays. As medicinal chemistry efforts continue to drive target potency to lower K_d or EC_{50} values, discontinuities in the PD/biochemical EC_{50} plot can occur due to tight-binding and long residence time, as described in Section 6.2.1 and 6.2.2. In addition to their impact on cell-free assays of target affinity, these same phenomena can also impact the kinetics of proximal PD, intervening biology, and phenotypic readout of cellular activity. Hence, during lead optimization, it is important to test for time-dependent cellular effects with sentinel compounds over the course of SAR campaigns. Not accounting for any time dependence of cellular effects of compound can result in underestimating compound potency. We already encountered an example of this in Chapter 5 in the case of inhibitors of epigenetic targets such as EZH2, where a latency period of about four days preceded the observation of time- and concentration-dependent cell killing of mutant-bearing non-Hodgkin's lymphoma (NHL) cell lines by compounds. Similarly, a time dependence of proximal PD effects is often seen in cellular studies of compounds displaying high target affinity, especially when this is associated with a long drug–target residence time. For example, this is seen with potent EZH2 inhibitors, DOT1L inhibitors, and various high-affinity kinase inhibitors. Figure 6.5 illustrates some examples of such time-dependent PD effects. This figure shows the time dependence of reduction in intracellular H3K79me2 induced by administration of the DOT1L inhibitor EPZ5676 (pinometostat) to MV4–11 leukemia cells. These data conform to a single-phase exponential

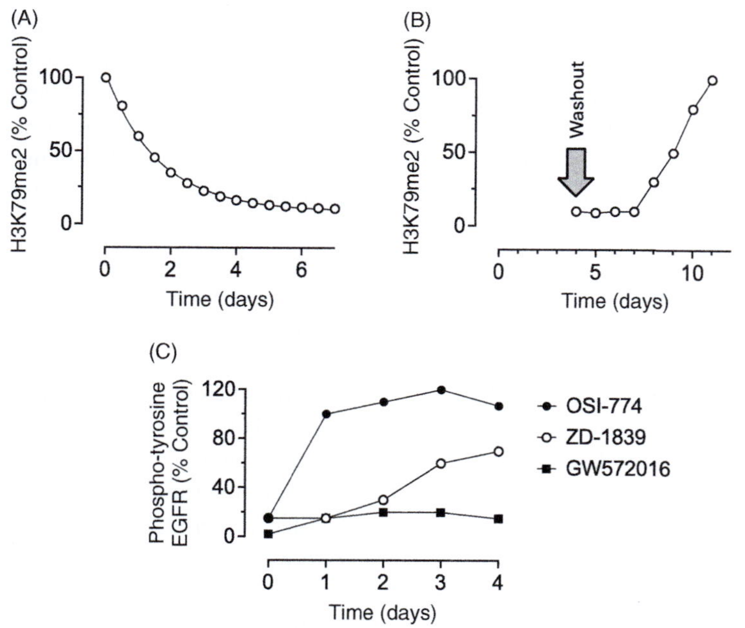

Figure 6.5 Examples of the time dependence of cellular PD readouts. (A) Time-dependent diminution of intracellular H3K79me2 levels following treatment of MV4–11 leukemia cells with the DOT1L inhibitor EPZ5676 (administered at day zero). (B) Time course of recovery of H3K79me2 levels in MV4–11 cells after treatment of cells with EPZ5676 for four days, then washout of compound (by exchange of culture media). (C) Comparative rates of recovery of EGFR tyrosine autophosphorylation upon washout of different EGFR kinase inhibitors: OSI-774, ZD-1839, and GW572016 (washout at day zero on the *y*-axis of this graph, following four hours of treatment with each compound at 1 μM). Data used to construct this figure (panels **A** and **B**) were estimated from that published by Daigle et al. (2025). Data used to construct this figure (panel **C**) were estimated from that published by Wood et al. (2004).

decay with a half-life of 1.0 days, which corresponds well with the natural rate of H3K27me2 decay in cells (Daigle et al., 2013). Panel B of Figure 6.5 illustrates the rate of recovery of intracellular H3K79me2 levels after compound washout following four days of treatment of MV4–11 cells with EPZ5676. In this case, the PD signal lags for several days before a quasi-linear recovery phase (Daigle et al., 2025). Panel C of Figure 6.5 compares the recovery rates for epidermal growth factor receptor (EGFR) tyrosine autophosphorylation after washout of three EGFR kinase inhibitors (Wood et al., 2004). It is clear from these data that the three compounds display widely differing PD recovery rates, which correlate well with their drug–target residence times on the EGFR enzyme (Wood et al., 2004).

Assuming that phenotypic outcome is predicated on robust PD activity (as an intracellular surrogate of target occupancy), the kinetics of PD onset and recovery (after compound washout) provide important information with respect to the timing and durability of phenotypic effects, *in vitro* and *in vivo*. Hence, if the half-life for onset of PD is prolonged (e.g., ≥1 day), it is likely that one will observe a corresponding latency period before phenotypic outcome is manifested in cells and *in vivo*. Similarly, the half-life of PD recovery helps to set expectations for the durability of phenotypic effects that may be observed in animal models and humans. As we shall discuss below, this can help inform appropriate dosing schedule(s) for realizing the full efficacy and safety of therapeutics *in vivo*.

In a like manner, it is important to measure the kinetics of phenotypic effects of compound treatment in cellular studies to ensure that one is reporting the true potency of each compound. As with PD effects, in some cases, the EC_{50} for phenotypic readout can change over treatment time, asymptotically approaching a minimum value (as seen for PD in Figure 6.5A). Thus, if sufficient treatment time has not elapsed, the measurement of phenotypic EC_{50} will underestimate the true potency of the compound. In some cases, the kinetics of phenotypic outcome may vary among different cell types. Hence, one should determine the time course of phenotypic outcome for each cell type to be tested with compound, to ensure that potency and selectivity of effect is not compromised by temporal differences in phenotypic response.

Having determined the kinetics of PD, intervening biology, and phenotypic outcome, one should carefully determine the EC_{50} values of each relevant metric for multiple cell lines in which the selective essentiality hypothesis (Chapter 3) suggests the compound should be active, as well as for a collection of cell lines in which the hypothesis would suggest the compound should be inactive. For each cell line tested, both the EC_{50} and EC_{90} should be determined. When the Hill coefficient is unity, the Hill–Langmuir equation (Chapter 2) predicts that the EC_{90} occurs at $\sim 9 \times EC_{50}$ value (Figure 6.6).

However, in cellular assays, the apparent Hill coefficient often varies, so the EC_{50} and EC_{90} values should both be determined directly from curve fitting. The values of EC_{50} and EC_{90} for cellular phenotypic readout, after correction for

Figure 6.6 Typical concentration–response plot for inhibition of cellular phenotypic readout (in this case, cell proliferation) fit to the Hill–Langmuir equation, identifying the EC_{50} and EC_{90} concentrations of compound.

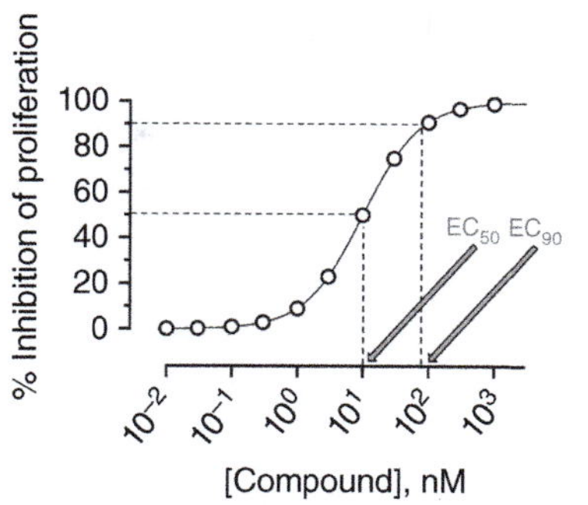

the effects of plasma protein binding, are commonly predictive of the systemic exposure (concentration) of drug required to observe *in vivo* efficacy (see below). Hence, it is common to design *in vivo* efficacy studies at compound doses to bracket the predicted exposure levels around the cellular phenotypic EC_{50} and/or EC_{90} values (again, after correcting these values for plasma protein binding). As discussed further below, it is best to use the median EC_{50} and EC_{90} values from survey of multiple sensitive cell lines for predicting *in vivo* efficacious exposure, rather than the mean values or the values for the most sensitive cell line tested. The median is often the best measure of central tendency for these purposes, as the mean can be disproportionately influenced by extreme values (i.e., outliers). We will have more to say on this point below in our discussion of *in vivo* studies.

6.2.5 Expanded Survey of Additional Cell Lines

As a program approaches identification of DC quality compounds, it is often valuable to perform more expansive surveys of cell lines for compound phenotypic modulation effects. Despite the selective essentiality hypothesis that often forms the basis of target validation (see Chapter 3), broad cell line screening with highly potent, target-selective compounds can unveil unanticipated sensitivities that can expand or redefine one's understanding of the genetic or molecular basis of sensitivity to compound.

An illustrative example of this comes from preclinical studies of the EZH2 compound tazemetostat. As we have presented in earlier chapters, tazemetostat was originally discovered as a treatment for germinal center NHL patients bearing oncogenic mutations in EZH2. As tazemetostat was approaching DC and IND-enabling studies, the team at Epizyme tested the compound across other cancer cell lines. They found that tazemetostat also displayed potent antiproliferative activity in some EZH2 WT NHL cell lines, in multiple myeloma cell lines, and in a subset of soft tissue sarcomas. The activity of tazemetostat in WT NHL and multiple myeloma pointed to a broader dependence on EZH2 enzymatic activity in B-cell malignancies (Brach et al., 2017; Drew et al., 2018).

The activity of tazemetostat in certain soft tissue sarcomas was, at first, perplexing to the researchers. However, data in the literature had shown that H3K27me3 production by EZH2 catalysis led to transcriptional repression of a set of tumor suppressor genes. This action of EZH2 was in direct opposition to that of another chromatin-modifying enzyme complex known as the switch/sucrose non-fermentable (SWI/SNF) complex. Using the energy derived from ATP hydrolysis, SWI/SNF alters histone–DNA interactions within chromatin in a manner that promotes transcription of the same set of tumor suppressor genes. Hence, SWI/SNF and EZH2 share a mutually antagonistic relationship with one another. The ATP-hydrolyzing activity of SWI/SNF requires the full complement of protein subunits of the complex. In certain soft tissue sarcomas, inactivating mutations are commonly found in several subunits of SWI/SNF. For example, both malignant rhabdoid tumor (MRT) and epithelioid sarcoma (ES) are characterized by biallelic deletions or mutations of the INI1 (also known as SNF5, BAF47, and SMARCB1) subunit of SWI/SNF, resulting in complete loss of INI1 expression. In fact, INI1 negativity – as determined by immunohistochemical staining of tumor biopsy samples – is a defining clinical hallmark of these cancers. Both MRT and ES are highly aggressive cancers that respond so poorly to current therapeutics that they have been designated as "chemotherapy refractory tumors". Based on the antagonistic relationship between EZH2 and SWI/SNF, Wilson et al. (2010) studied the functional relationship between these enzymes in human oncogenic transformation. They found that loss of INI1 led to elevated expression of EZH2, broad H3K27me3, and resultant transcriptional repression of target genes. In conditional mouse models, loss of INI1 was sufficient for tumorigenesis; genetic ablation of EZH2 was able to block INI1 loss–driven tumor formation in these models (Wilson et al., 2010).

The work of Wilson et al. (2010) suggested that soft tissue sarcomas with INI1 loss would be particularly sensitive to inhibitors of EZH2, and this seemed to be consistent with the pattern of sensitivity observed for tazemetostat. To test this hypothesis further, Knutson et al. (2013) compared the impact of tazemetostat on H3K27me3 levels and proliferation on a collection of INI1-negative MRT cell lines and INI1-replete rhabdomyosarcoma cell lines. In all these cells, H3K27me3 levels were diminished by tazemetostat treatment with similar EC_{50}s (1–5 nM) independent of INI1 status. In stark contrast, however, cells that were INI1-deficient displayed selective sensitivity to tazemetostat in terms of cell proliferation (Figure 6.7).

These and other preclinical data (see below) suggested that tazemetostat might be beneficial to patients with solid tumors that were deficient in SWI/SNF function. This inference was also consistent with the phase 1 experience with tazemetostat, wherein patients with loss of the SWI/SNF subunits INI1 or SMARCA4 demonstrated a clinical benefit rate of 31% (Ho and Copeland, 2024). Based on these data, a phase 2 trial in patients with INI1-negative MRT and

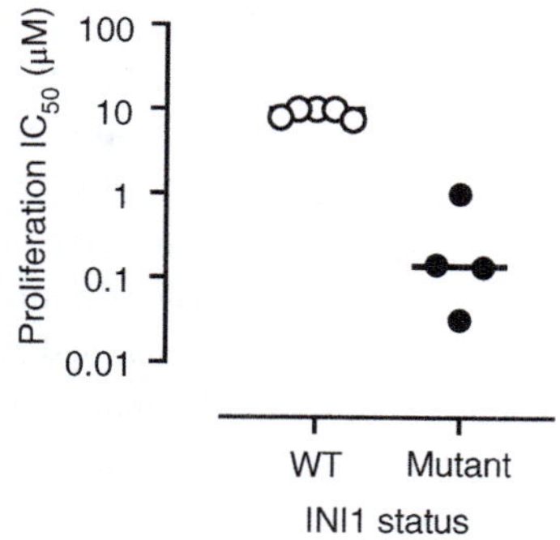

Figure 6.7 Comparison of the IC_{50} values for inhibition of cell proliferation by tazemetostat in INI1-replete rhabdomyosarcoma cell lines (WT) and INI1-deficient MRT cell lines (mutant). *Source:* Adapted from Knutson et al. (2013).

ES was conducted, eventually leading to the FDA approval of tazemetostat as a treatment for ES patients (Ho and Copeland, 2024).

Similarly, Castro et al. (2025) reported that the RNA helicase, DHX9, was a selectively essential target for microsatellite instability (MSI)-high cancers, such as colorectal cancer. They also reported the discovery of a small-molecule inhibitor of DHX9, ATX968, and showed that treatment of colorectal cancer cell lines with this compound led to MSI-high–selective antiproliferative effects both *in vitro* and *in vivo* (see also Chapter 5). Screening of a broad panel of cancer cell lines subsequently demonstrated that ATX968 also displayed potent antiproliferative activity in cancers bearing loss-of-function mutations in the DNA damage repair proteins BRCA1 and BRCA2 (e.g., in triple-negative breast cancer) and in other cancer indications as well (Castro et al., 2024; Copeland et al., 2024). These data and follow-up experiments led Castro et al. (2024) to a broader view of the mechanism of selective essentiality for DHX9 in cancer. They hypothesized that MSI-high status and disruption of DNA damage repair are but examples of a spectrum of ways in which the set point (or basal level) of cellular replication stress and genomic instability can be elevated in certain cancers. In these cases, inhibition of DHX9 by ATX968, and the resultant accumulation of R-loops (Castro et al., 2025, and Chapter 5), increases the level of replication stress beyond the survival threshold for the cell, leading to cell cycle arrest and subsequent apoptotic cell death. In other cancers and normal cells, the basal level of replication stress is far below the survival threshold, so that inhibition of DHX9 is tolerated. This hypothesis continues to be tested and refined through preclinical and clinical studies today.

6.3 *In Vitro* ADME Properties

During lead optimization, the ADME properties of compounds are assessed through both *in vitro* and *in vivo* studies to support SAR. The *in vitro* studies consist mainly of continued assessment of overall hepatic clearance using liver microsomes, S9 fractions, or intact hepatocytes (as described in Chapter 5). In lead optimization, these studies are augmented with assessment of compound transformation rates by individual cytochrome P450s (CYP450s).

The CYP450s are a family of heme-containing monooxygenases that constitute major drug-metabolizing enzymes of the liver. Understanding which member(s) of this enzyme family is dominant in hepatic drug metabolism is important in helping to guide SAR efforts aimed at optimizing hepatic clearance rate. These studies also help to define the major products of hepatic metabolism in humans and in preclinical species. Because the CYP450 family is responsible for much of hepatic metabolism, it is also important to understand if a compound acts as an inhibitor or inducer of any CYP450 function, as either could impact the metabolism of the compound under study and of other drugs that the patient may be taking.

6.3.1 CYP450-Mediated Metabolism

The human genome contains 57 genes that encode distinct CYP450 isozymes, which are categorized into 18 families of enzymes. Of these, three families of CYP450 account for most drug metabolism reactions: the 3A family (accounting for 50% of marketed drug metabolism), the 2C family (accounting for 15% of marketed drug metabolism), and the 2D family (accounting for 25% of marketed drug metabolism). Recombinant CYP450 isozymes from human, rat, mouse, dog, and monkey are commercially available for studies of *in vitro* drug metabolism. These enzymes generally follow Michaelis–Menten kinetics (see Chapter 2), so that under steady-state conditions the initial velocity (v_i) is given by the following equation.

$$v_i = \frac{V_{max}[S]}{K_M + [S]} \tag{6.1}$$

where the substrate concentration ($[S]$) refers to the compound being tested. The intrinsic metabolic clearance (Cl_{int}) due to a particular CYP450 enzyme is defined as v_i divided by $[S]$. However, typically *in vivo*, the free fraction of compound that reaches the liver is far less than the K_M of most CYP450, so the $[S]$ term in the denominator of Equation 6.1 can be ignored (Copeland, 2013). Under these conditions, the intrinsic metabolic clearance then becomes

$$Cl_{int} \sim \frac{V_{max}}{K_M} \tag{6.2}$$

Using Equation 6.2, the Cl_{int} of each CYP450 isozyme can be compared to determine the isozyme(s) that contribute greatest to hepatic metabolism of the compound.

The total intrinsic metabolic clearance (${Cl_{int}}^T$) within liver would then be the sum of the individual isozyme V_{max}/K_M values. This is most typically determined by measuring the apparent V_{max} and K_M values using liver microsomes (or hepatocyte or S9 fraction). Thus,

$$\mathrm{Cl}_{\mathrm{int}}^{T} = \left(\frac{V_{\mathrm{max}}}{K_M}\right)^{T} \tag{6.3}$$

In favorable cases, when *in vivo* drug metabolism is driven largely by hepatic metabolism, ${Cl_{int}}^T$ can be used to predict the intrinsic clearance *in vivo* from Equation 6.4.

$$\mathrm{Cl}_{int} = \left(\frac{V_{\mathrm{max}}}{K_M}\right)^{T} \times \frac{\mathrm{mL}_{\mathrm{reaction}}}{\mathrm{mg}_{\mathrm{microsomes}}} \times \frac{\mathrm{mg}_{\mathrm{protein}}}{\mathrm{g}_{\mathrm{liver}}} \times \frac{\mathrm{Mass}_{\mathrm{liver}}}{\mathrm{kg}_{\mathrm{body}}} \tag{6.4}$$

where $mL_{reaction}/mg_{microsomes}$ is the volume of the reaction divided by the mass of microsomal protein (mg) used, $mg_{protein}/g_{liver}$ is the microsomal protein per gram of liver, and $Mass_{liver}/kg_{body}$ is the mass of the liver divided by the total body mass of the organism. Values of $mg_{protein}/g_{liver}$ and $Mass_{liver}/kg_{body}$ for humans and different preclinical organisms have been compiled in Copeland (2013) and can thereby be readily looked up. Values of Cl_{int} can be used to rank compounds for their relative rates of clearance as a metric for medicinal chemistry SAR during lead optimization. The company Cyprotex (an Evotec company) suggests categorizing compounds into one of three classes with respect to their *in vitro* clearance rates using microsomes or hepatocytes for determination of Cl_{int} (Cyprotex, 2017, Everything you need to know about ADME, 3rd Edition, www.cyprotex.com), as summarized in Table 6.4.

Finally, the overall hepatic clearance of a compound *in vivo* (Cl_H) depends on the intrinsic clearance, the fraction of unbound drug (f_u), and the rate of blood flow through the liver (Q_H) as follows.

$$\mathrm{Cl}_H = \frac{Q_H \times f_u \times \mathrm{Cl}_{\mathrm{int}}}{Q_H + (f_u \times \mathrm{Cl}_{\mathrm{int}})} \tag{6.5}$$

Table 6.4 Categories of intrinsic clearance (Cl_{int}) rates as measured by *in vitro* assays using microsomes or hepatocytes.

Clearance category	Cl_{int} from *in vitro* microsomes (μL/min/mg protein)		Cl_{int} from *in vitro* hepatocytes (μL/min/10^6 cells)	
	Human	**Rat**	**Human**	**Rat**
Low	≤8.6	≤13	≤3.5	≤5.1
Medium	8.6–47	13–72	3.5–19	5.1–28
High	≥47	≥72	≥19	≥28

Values of Q_H for different organisms can also be found in Copeland (2013). Equation 6.5 provides a useful means of estimating *in vivo* hepatic clearance. However, the precision of such estimates can vary among pharmacophore series and from organism to organism. Hence, the calculations described in this section should be viewed as a useful *in vitro* guidepost by which to follow SAR among compounds of a common pharmacophore series (Smith et al., 2019). The actual *in vivo* hepatic clearance, however, must be measured in the organism, as will be discussed below. For many pharmacophore series, the *in vitro* intrinsic clearance obtained from microsomes, or hepatocytes correlates well with the *in vivo* hepatic clearance. Examples of linear correlations between the *in vitro* and *in vivo* clearance values are available on the Cyprotex website (see above). In cases where there is a discordance between the *in vitro* and *in vivo* clearance values, common causes include extrahepatic metabolism being a significant contribution to overall *in vivo* clearance, slow equilibration between systemic circulation and liver hepatocytes, active transport of compound through the sinusoidal membrane, and the like. More information on causes of discordance between *in vitro* and *in vivo* clearance can be found on the Cyprotex website (see above), and in the review by Wienkers and Heath (2005).

6.3.2 CYP450 Inhibition and Drug–Drug Interactions

Not only can compounds act as substrates for CYP450s (i.e., be metabolized by these enzymes), but in some cases, they can also act as inhibitors of specific CYP450s. If a compound inhibits a specific CYP450, it may interfere with its own metabolism and/or with the metabolism of other drugs that a patient may be taking. The diminished metabolism of a second drug by a compound that acts as a CYP450 inhibitor will lead to elevated concentrations of the second drug, which could in turn lead to adverse events; this situation is referred to as a drug–drug interaction and can be a serious concern for the clinical use of the compound. A drug that inhibits a CYP450 and thereby affects the metabolism of a second drug is referred to as a "perpetrator" of drug–drug interactions, and a drug that is thus impacted by the perpetrator of drug–drug interactions is referred to as the "victim". Table 6.5 provides some clinical examples of drug–drug interactions.

CYP450 inhibition is generally measured with isolated, recombinant CYP450 isozymes using selective fluorescent substrates of the specific isozyme. Copeland (2013) provides examples of selective fluorescent substrates for different CYP450 isozymes and more detail on assay methods. Typically, compounds within a lead series are tested for CYP450 inhibition at a fixed concentration (e.g., 1 or 10 μM), and the percent inhibition of fluorescent substrate metabolism is determined as a metric for rank-ordering compounds. Compounds that display inhibition can be subsequently tested in a concentration–response manner to determine the

Table 6.5 Some examples of clinical drug–drug interactions due to CYP450 inhibition.

Inhibitor (perpetrator drug)	CYP450 isoform affected	Drug for which metabolism is affected (victim drug)	Clinical impact
Amiodarone Metronidazole	2C9	Warfarin	Increased bleeding
Clarithromycin Erythromycin Telithromycin	3A4	Simvastatin	Myopathy or rhabdomyolysis
Clarithromycin Erythromycin Telithromycin	3A4	Verapamil	Hypotension and QT prolongation
Cimetidine	2D6	Doxepin	Increased anticholinergic side effects
Diltiazem	3A4	Prednisone	Immunosuppression
Fluoxetine	2D6	Risperidone	Increased extrapyramidal effects
Ritonavir	3A4	Erythromycin	Nausea, vomiting
Terbinafine	2D6	Amitriptyline	Dizziness, cardiac toxicity

Data used to construct this table were taken from Copeland (2013) with permission of John Wiley & Sons.

IC_{50} and from that the K_i of the compound (using the appropriate Cheng–Prusoff equation: Copeland, 2013, and Chapter 2). In practice, one finds that most CYP450 inhibitors function as competitive inhibitors of the enzyme(s).

As with any enzyme inhibitor, the degree of inhibition realized will be a function of the concentration of compound relative to its K_i value (Chapter 2 and Copeland, 2013, 2023), and this will determine the degree to which the systemic exposure (i.e., concentration) of the victim drug may be elevated in patients. As we shall discuss below, the systemic exposure of a drug over a specific time interval is often measured as the area under the curve (AUC) of a plot of total or free drug (i.e., corrected for plasma protein binding) concentration as a function of time. The AUC for a victim drug in the presence of a CYP450 inhibitor (AUC_i) relative to its AUC in the absence of the inhibitor (AUC_0) is given by the following equation for competitive inhibition.

$$\frac{AUC_i}{AUC_0} = 1 + \frac{[I]}{K_i} \tag{6.6}$$

Hence, the impact of CYP450 inhibition will depend on the dose and exposure of the perpetrator drug, its K_i, and the therapeutic index of the victim drug. All these factors must be considered in determining the degree of CYP450 inhibition (if any) that may be acceptable in a specific clinical indication. Note that Equation 6.6 does not explicitly account for plasma protein binding for the perpetrator or victim drug and only accounts for victim metabolism by a single CYP450 isozyme that is inhibited by the perpetrator drug. Nevertheless, this simple equation suffices to make the pedagogic point here. A more complete treatment of the impact of perpetrator drug concentration on the exposure of victim drug can be found in Mao et al. (2011).

Ultimately, the risk of drug–drug interactions (by any mechanism) depends on the degree to which an efficacious dose (or exposure) of the perpetrator drug results in increased exposure of the victim drug (measured as its AUC), relative to its therapeutic index. The FDA has categorized the relative risk for drug–drug interactions as weak, moderate, or strong, depending on the level to which the perpetrator drug diminishes the clearance or augments the AUC of the victim drug; the FDA guidance on these categories is summarized in Table 6.6.

6.3.2.1 Time-Dependent CYP450 Inhibition (TDI)

A particularly problematic form of CYP450 inhibition, with respect to drug–drug interactions, is time-dependent inhibition (TDI). As discussed above, and in greater detail elsewhere (Copeland, 2013, 2023), TDI refers to a situation in which the apparent affinity of an inhibitor–enzyme complex evolves over the course of incubation time. TDI is problematic for two reasons. First, it is possible to miss the true potency of inhibition by this mechanism in standard assays of CYP450 inhibition if the incubation time with compound is too brief. Second, and more critically, TDI mechanisms often result in irreversible or slowly reversible inhibition of the CYP450 isozyme, making the likelihood of drug–drug interactions that much greater. Thus, compounds can act directly as rapidly reversible (not time-dependent), slowly reversible, or covalent (irreversible)

Table 6.6 FDA guidelines for relative risk of drug–drug interactions based on observed diminution of clearance or augmentation of AUC for the victim drug in the presence of the perpetrator drug.

Category of relative risk	Observed change in clearance or AUC of victim drug
Strong inhibitor	≥5-fold increase in AUC or >80% decrease in clearance
Moderate inhibitor	>2- but <5-fold increase in AUC or 50–80% decrease in clearance
Weak inhibitor	>1.25- but <2-fold increase in AUC or 20–50% decrease in clearance

Data used to construct this table were taken from Copeland (2013) with permission of John Wiley & Sons.

inhibitors of CYP450s. Additionally, a compound can be a CYP450 substrate, and the enzymatic activity of the CYP450 may produce a reactive product that then acts as a covalent inhibitor of the same or a different CYP450 isozyme – an inhibition process known as mechanism-based inactivation (Copeland, 2013). Testing for CYP450 TDI requires specialized assay designs that are best performed by researchers with expertise in enzymology and allied fields. The interested reader can find additional details on TDI in the texts by Copeland (2013, 2023) and references therein.

6.3.3 CYP450 Induction

Compounds can also cause increases in CYP450 activity that can alter the metabolism of the compound itself (autoinduction) or of other drugs that a patient may be taking. This creates the potential for drug–drug interactions in which the systemic exposure of drug at a given dose is diminished due to the increased activity of specific CYP450 isozyme(s). Drug-induced increases in CYP450 activity are referred to as "induction" and can occur by several mechanisms; the two most common mechanisms are (1) activation of gene transcription, leading to increased message and enzyme levels, and (2) stabilization of CYP450 mRNAs or enzymes leading to reduced message or protein degradation. Nuclear receptor–mediated induction is the most common mechanism of CYP450 induction. In this mechanism, compounds interact with nuclear receptors such as the aryl hydrocarbon receptor (AhR), constitutive androstane receptor (CAR), or peroxisome proliferator-activated receptor (PPAR). This results in increased transcription of mRNAs for multiple phase 1 and phase II enzymes and transporters, including CYP450 isozymes such as CYP1A2, CYP2B6, and CYP3A4 (Kwon, 2001).

CYP450 induction is measured *in vitro* with commercially available human hepatocytes (use of hepatocytes from three separate donors is recommended). Both enzymatic activity and mRNA levels can be measured in human hepatocyte cultures, although measurement of mRNA levels by quantitative reverse transcription polymerase chain reaction (qRT-PCR) is the more common approach. Assays of CYP450 induction by test compounds are typically performed at a minimum of three concentrations and should also include known inducers of specific CYP450 isozymes as positive controls; Table 6.7 provides some examples of compounds known to induce specific isozymes. Flumazenil, a compound that has been clinically shown to be a non-inducer is also often included as a negative control.

Test compounds are titrated in the assay and scored as *in vitro* inducers if they cause an increase in mRNA level >2-fold relative to vehicle control or >20% of positive control in at least one donor sample. Any compounds that score as inducers in this assay are then further evaluated *in vivo*. The US Food and Drug Administration (2020) has provided detailed guidance to industry on the *in vitro*

Table 6.7 Some examples of known inducers of CYP450 isozymes in humans.

CYP450 isozyme	Inducer
1A1, 1A2	Omeprazole
2A6	Rifampicin, dexamethasone
2B6	Rifampicin, phenobarbital
2C8, 2C9, 2C19	Rifampicin
2E1	Isoniazid
3A4	Rifampicin, dexamethasone

evaluation of drug interactions, including CYP450 induction. Also, the Evotec company website (www.evotec.com) provides further details on their human hepatocyte assays of CYP450 induction.

6.3.4 Major Metabolite Identification

In addition to understanding the efficacy, safety, and pharmacokinetics (PK) of a specific compound (referred to in the present context as the "parent compound"), it is also critical to understand the role of major metabolites of that compound that may contribute to pharmacology and toxicology in human subjects. The US Food and Drug Administration defines a major metabolite as any metabolite that accounts for plasma levels ≥10% of the total drug-related exposure, measured as the AUC at steady state.

Although drug biotransformation occurs via both phase 1 and phase 2 metabolic pathways (Copeland, 2013), the FDA acknowledges that phase 1 reactions are more likely to produce species that may retain pharmacological or toxicological activity. Hence, *in vitro* assays using microsomes are generally acceptable for initial studies to identify and quantify major metabolites, although these studies can also be performed with hepatocytes, S9 fractions, or liver slices. Microsomes (or other sources) from human and preclinical species (e.g., mouse, rat, dog, and monkey) are studied in parallel with two aims: (1) first, to identify all major metabolites generated by human microsomes, and (2) to identify one rodent and one non-rodent species for which the sum of their metabolite profiles subsumes all major human metabolites at concentrations similar to humans. Metabolite identification and quantification are performed by several analytical methods, but initial studies are commonly analyzed by liquid chromatography with mass spectroscopic detection (LC/MS), and sometimes nuclear magnetic resonance (NMR) spectroscopy follow-up. *In vivo* studies in the appropriate preclinical species are then performed to validate the *in vitro* results.

For any major human metabolite, the research team must determine whether the molecule is pharmacologically or toxicologically active. This can be a laborious, costly, and time-consuming effort, as the chemical structure of each major metabolite must be deduced and the compound must then be synthesized in sufficient quantity and purity for a battery of tests. Fortunately, if a rodent and non-rodent species can be identified for which the metabolic profiles represent all major human metabolites at exposures similar to those expected in humans, toxicology studies in these two species are sufficient for risk mitigation without the need for separate studies. In its guidance to industry, the FDA provides a decision tree for determining what preclinical studies are required for major human metabolites; this decision tree is summarized in Figure 6.8 (https://www.fda.gov/media/72279/download). However, in unfavorable cases, the exposure

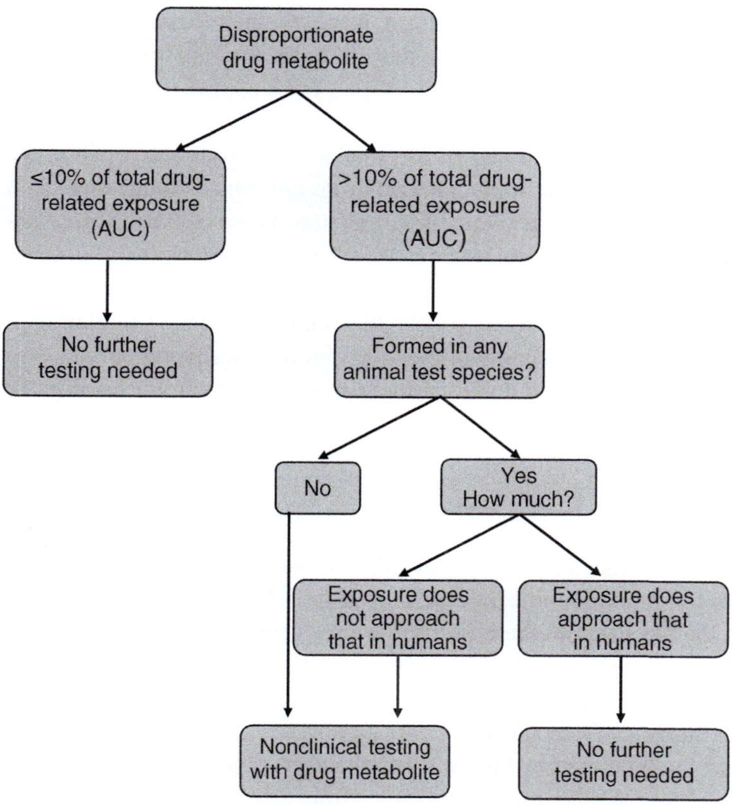

Figure 6.8 Decision tree for determining testing needs for major human metabolites of a compound. *Source:* U.S. FDA Safety Testing of Drug Metabolites Guidance for Industry (2020).

of a human metabolite cannot be recapitulated in any preclinical species; such metabolites are referred to as "disproportionate drug metabolites" and in this case detailed profiling of the metabolite is required per the FDA guidance (see above).

6.4 *In Vivo* ADME Properties

6.4.1 Oral Bioavailability

Human therapeutics are administered by a variety of parenteral (e.g., direct injection or infusion into the vasculature) and nonparenteral (e.g., extravascular such as oral, topical, nasal, ocular, vaginal, and rectal suppository) routes. For most disease indications, oral administration is the most convenient dosing route for adult patients, especially for long-term, chronic therapy. When a compound is swallowed as a pill, capsule, or other form, it must transit through the gastric compartment without significant degradation to enter the small intestines. In some cases, drug is absorbed directly within the gastric compartment, but most drugs are absorbed through the small intestines. The drug then enters the liver through the portal vein, where it may be metabolized and eliminated or continue to systemic circulation. Hence, orally ingested drugs are subject to metabolism in the gut and liver before they enter systemic circulation. These metabolic processes in the gut and liver are collectively referred to as "first pass metabolism". Because of first pass metabolism, only a portion of the drug mass that is ingested reaches circulation. The fraction of ingested drug that enters systemic circulation is referred to as the oral bioavailability of the drug and is typically measured as the ratio of systemic AUC for drug dosed orally to that of drug administered by IV injection, as described next.

6.4.2 Pharmacokinetic Profile of Orally Bioavailable Drugs

When a compound is injected directly into the vasculature, the entire mass of drug enters systemic circulation immediately (i.e., it is 100% bioavailable), and its concentration in the bloodstream then begins to wane as compound is distributed to bodily tissues and is metabolized and eliminated. The PK profile of an IV-dosed drug is thus characterized by a pseudo-first-order decay over time from a maximum value at the time of injection (time zero on a typical graph), as we have seen previously and as illustrated in Figure 6.9A (see also Appendix 1); the half-life ($t_{1/2}$) for such a PK profile is usually determined from direct fitting of the data to a first-order decay equation (see Chapter 1). The area under the PK curve is determined by the summed trapezoid method, as described in Copeland (2013), and illustrated by the gray area in Figure 6.9A.

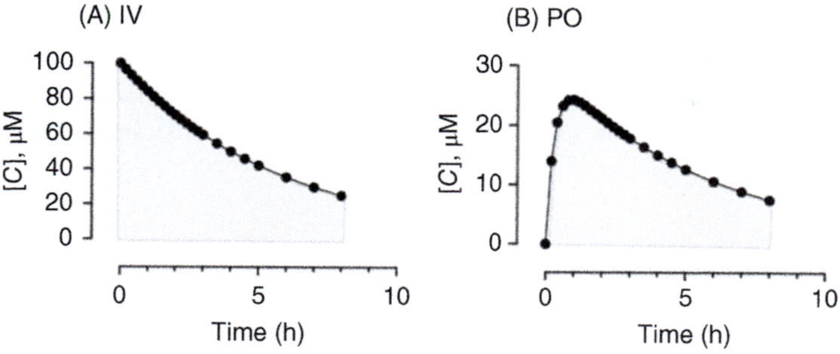

Figure 6.9 PK profiles for a compound administered as a single dose to an animal by different routes. (A) PK profile for a compound administered by intravenous (IV) injection. (B) PK profile for the same compound administered orally (PO). In both panels, the PK profile is presented as black circles for each data point, connected by a line of best fit to an appropriate nonlinear equation (see Appendix 1). The gray area under each curve represents the summed area under the curve (AUC). Note that these graphs represent idealized data over many time points. In practice, far fewer time points are typically measured. For example, a common practice would be to measure compound concentration at 0.25, 0.5, 1.0, 2.0, 4.0, 7.0, 12.0, and 24.0 hours.

The PK profile for an orally administered drug is quite distinct from that of an IV-administered drug, as illustrated in Figure 6.9B. In the case of oral administration, the PK profile displays several distinct phases. First, one observes an ascending phase of systemic drug concentration ($[C]$) as the compound is absorbed from the gastrointestinal tract into the bloodstream. The systemic concentration reaches a maximum value (C_{max}) at a well-defined time point (t_{max}), based on the rate of absorption of the compound. Next, one typically observes an initial phase of $[C]$ declination due to tissue distribution, followed by a final phase of terminal elimination that results in a minimum concentration of drug (C_{min}) in circulation at the end of the observation period (typically 12–24 hours after dosing). The terminal half-life for an orally administered drug is calculated by fitting this terminal phase of the PK profile. The AUC for an orally administered drug is also calculated by the summed trapezoid method and is displayed in Figure 6.9B as the gray-shaded area.

The percent oral bioavailability ($F\%$) of the compound is then calculated as the ratio of AUC values for oral (per os, PO) administration over IV administration, multiplied by 100, assuming the doses used for IV and PO administration are identical. If different doses are used for IV and PO administration, then the AUC values must be corrected for dose (D) given in each case, as follows.

$$F\% = 100 \times \frac{\text{AUC}_{\text{PO}} \times D_{\text{IV}}}{\text{AUC}_{\text{IV}} \times D_{\text{PO}}} \tag{6.7}$$

The above treatments refer to single doses administered to animals of a specific species. The PK parameters (e.g., C_{max}, C_{min}, $t_{1/2}$, and AUC) will vary among species and must be determined empirically for each species studied. Also, repeat dosing at some defined dosing interval (e.g., once per day [QD] or twice per day [BID]) will eventually lead to a steady state of drug exposure. The AUC over a specified period can thus also be calculated once the drug has reached steady state (Copeland, 2013).

6.4.3 Dose–Exposure Proportionality

One may think that increasing the dose of drug administered orally to an animal would lead to an increase in systemic exposure of the drug, and in general terms, this is what is observed experimentally. However, several factors can affect the range of doses over which one observes a linear relationship between dose and total systemic exposure. Hence, it would be beneficial to experimentally determine the dose–exposure relationship for a compound before embarking on *in vivo* efficacy studies to ensure that adequate exposures can be achieved at higher doses.

One of the factors that impacts dose–exposure proportionality is drug absorption from the gastrointestinal tract. Both active and passive transport mechanisms can contribute to intestinal absorption of drugs, and some of these mechanisms can become saturated at higher dose levels, leading to a decline in the ratio of exposure to dose. Conversely, saturation of CYP450 and other metabolic processes can result in greater than expected systemic exposure at higher doses. These are just two examples of myriad mechanisms that can lead to discontinuities in the dose–exposure relationship. Greater detail on such mechanisms can be found in texts devoted to drug PK. For our present purpose, it is critical to realize that a linear dose–exposure relationship will often occur only over a finite range of drug doses, and the limits of this relationship must be determined experimentally.

6.4.4 Relating Plasma Protein–Binding–Adjusted Exposure to Cellular EC_{50} or EC_{90}

The determination of the compound dose(s) that will elicit a robust pharmacological response in an animal model is based on experimental studies at multiple doses of compound, as we shall discuss in more detail below. However, animal studies are more costly and more time- and compound-consuming than typical *in vitro* studies; hence, researchers typically limit *in vivo* studies to a finite number of doses over a range for which they expect to see compound efficacy.

A general rule of thumb is to select a range of doses, based on compound bioavailability (see above), that will yield systemic exposures that bracket the cellular phenotypic EC_{50} or EC_{90} values. One must recall, however, that total systemic concentration of compound is not what leads to pharmacological action. Rather, it is only the free or unbound fraction of compound (i.e., that fraction not bound to plasma proteins) that is thought to be available for distribution from the bloodstream to the tissues and organs in which the target protein resides. Hence, it is the free drug exposure and not the total drug exposure that should approach or exceed the EC_{50} or EC_{90} value. To illustrate this point, consider the following example. Suppose that a compound displayed a median phenotypic response EC_{50} of 10 nM across a range of sensitive cell lines. Based on a standard Hill–Langmuir concentration–response relationship, that would mean that the EC_{90} for this compound would be 90 nM ($9 \times EC_{50}$). Suppose also that the fractional PPB for this compound was 0.98 (i.e., 98% bound), which defines the free fraction of compound as 0.02 (i.e., 2% free). To achieve a free compound exposure equal to the EC_{50}, we would need a total compound exposure in systemic circulation of 500 nM:

$$[C]_{\text{total}} = \frac{EC_{50}^{\text{phenotypic}}}{f_u} = \frac{10\,\text{nM}}{0.02} = 500\,\text{nM} \tag{6.8}$$

where $[C]_{\text{total}}$ is the total systemic concentration of drug required, $EC_{50}{}^{\text{phenotypic}}$ is the cellular phenotypic EC_{50} value, and f_u is the fraction of unbound (free) drug. If we replace the cellular phenotypic EC_{50} value in Equation 6.8 by the corresponding EC_{90} value, we find that the $[C]_{\text{total}}$ in this case would be 4,500 nM (i.e., 4.5 μM). Thus, for an initial exploration of *in vivo* efficacy, we might decide to test compound at doses that would yield total systemic exposures of 0 (vehicle control), 300, 3,000, and 10 000 nM to bracket the EC_{50} and EC_{90} values in this example.

Let us assume that the minimum total systemic concentration of drug at which pharmacological activity may be observed is one at which the free concentration of drug is equal to its cellular phenotypic EC_{50} or EC_{90}. This would define our expected minimum effective concentration (MEC) of drug. Returning to the PK profile for an orally bioavailable drug, we can thus define several important PK parameters for the compound, as illustrated in Figure 6.10: the C_{max}, the t_{max}, the duration of time over which total systemic concentration reaches or exceeds the MEC, the terminal half-life ($t_{1/2}$), and the C_{min}.

It is worth noting that there are cases where the free concentration of drug in systemic circulation is not what drives pharmacology, but rather it is the free concentration of drug in a specific tissue or anatomical compartment, such as the central nervous system (CNS). In such cases, the free fraction in systemic circulation and that in the relevant tissue may be quite different, requiring the researcher to measure free and bound drug concentrations in the relevant tissue from preclinical animal models

6.4.5 Relating PK and PD Time Courses

In reviewing the data provided by Figure 6.10, one might naturally assume that the duration period for which systemic drug concentration meets or exceeds

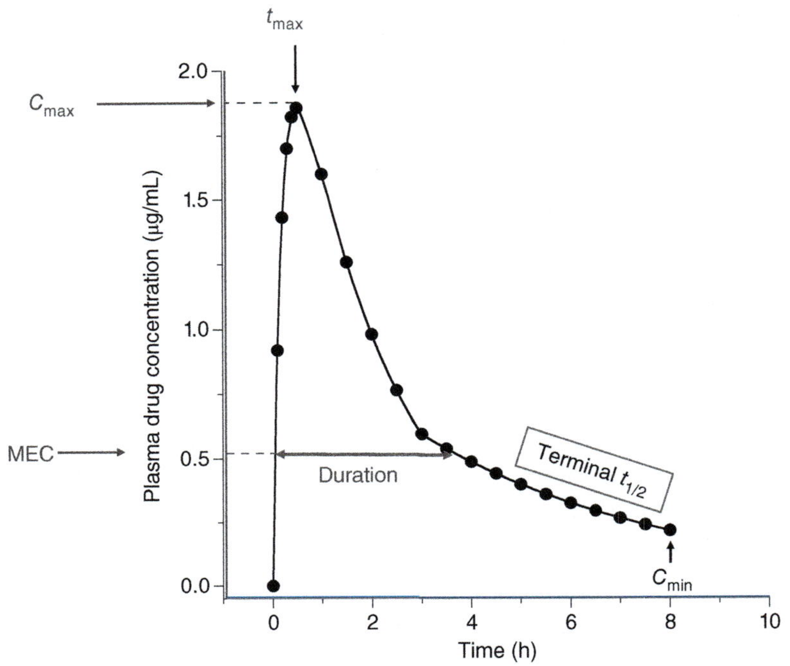

Figure 6.10 Idealized PK profile for an orally administered drug, illustrating several important PK parameters: C_{max}, C_{min}, t_{max}, MEC, duration, and terminal half-life ($t_{1/2}$). *Source:* Adapted from Copeland (2013) with permission of John Wiley & Sons.

the MEC would define the period of pharmacological action of the drug. This assumes rapid and full equilibration of drug concentration between the bloodstream and the target-bearing tissue, and that pharmacological action depends on sustained drug concentration above the MEC. Another way to express this same concept is that we expect to see a temporal correlation between PK and PD, such that PD (and attendant pharmacological action) endures only as long as the systemic concentration of drug is above the MEC. For many compounds, this assumption holds true, and one observes strong concordance between the PK and PD temporal profiles, such that drug efficacy depends on the AUC meeting or exceeding the MEC. However, for other compounds, one can observe a disparity between the time courses of PK and PD. This is especially true for covalent modulators and compounds with drug–target residence times far longer than the PK half-life of the drug. In such cases, the PD and pharmacological activity can endure long after the systemic drug concentration has declined below the MEC, due to the sustained target occupancy of such drugs. Consider, for example, the drug finasteride, which is used to treat benign prostate hyperplasia.

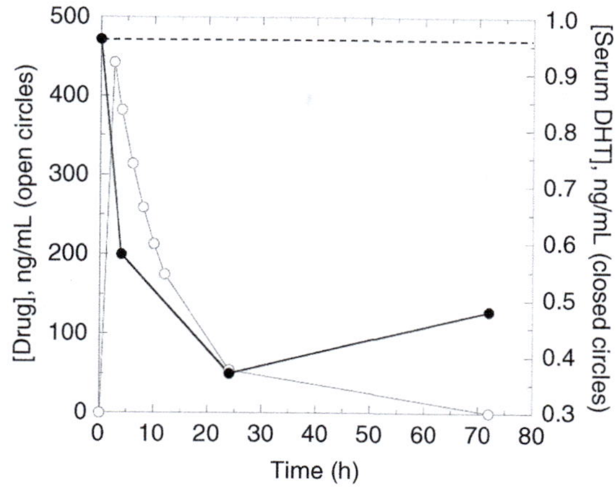

Figure 6.11 Plasma drug concentration (open circles) and serum DHT levels (closed circles) after a single oral dose (50 mg) of finasteride in healthy volunteers. The dashed, horizontal line indicates the initial serum level of DHT prior to drug absorption. Figure taken from Copeland (2013) based on data reported by Tian (1996).

Finasteride inhibits the enzyme 5α-reductase, thereby blocking the reduction of testosterone to dihydrotestosterone (DHT). Tian (1996) reported PK and PD (serum DHT levels) for healthy volunteers after a single dose of 5–100 mg of finasteride. Figure 6.11 illustrates the data for patients given a 50 mg dose of the drug. Over the entire 5–100 mg dose range, the PK half-life was between 4.7 and 7.1 hours, such that at all doses the compound reached a very low trough level by 24 hours and continued to decline further over time. In stark contrast, the serum concentration of DHT remained significantly diminished (ca. 52% inhibition of pre-dose DHT level) even 70 hours after dosing, a point at which the systemic concentration of drug was undetectable. This topic, and its relationship to drug-target residence time, has been discussed extensively by Copeland and colleagues (Copeland et al., 2006; Tummino and Copeland, 2008; Copeland, 2013; Copeland, 2016; Copeland, 2021; Knockenhauer and Copeland, 2024) as well as other researchers.

Likewise, compounds that act as degraders of a target protein, such as proteolysis targeting chimera (PROTACs) and glues (see Chapter 2), can display a disparity between the PK and PD time courses when the half-life for new target protein synthesis is long compared to the degrader PK half-life. Thus, the temporal relationship between PK, PD, and efficacy can be driven by the C_{max}, the AUC above the MEC, and, in some cases, by the C_{min} or trough value of drug concentration in systemic circulation.

6.4.6 Compartmental Modeling of Pharmacokinetics

To understand the PK profile of a drug, the body (human or preclinical species) can be considered as made up of a system of compartments among which bodily fluids, containing the drug molecule, are reversibly exchanged.

The simplest of these models is a one-compartment model, which is useful in describing the PK for some drugs administered as an intravenous (I.V.) bolus (Figure 6.12A). In this model, the entire body is viewed as a single, large container; this implies that bodily fluids and tissues equilibrate rapidly (i.e., instantaneously), so that there is a uniform distribution of drug throughout the body. Drug is eliminated from the body compartment by a combination of metabolic and excretion processes that are quantified by the composite elimination rate constant k_e, which is assumed to behave as a first-order rate constant. However, for many drugs, the one-compartment model is insufficient to describe the PK profile because there may be a slow equilibration between the central compartment (compartment 1) and a second compartment (compartment 2). For example, the central compartment may represent drug in systemic circulation, and compartment 2 may represent drug within specific tissues of the body. This is modeled as a two-compartment system, as illustrated in Figure 6.12B for I.V. administration and in Figure 6.11C for extravascular (e.g., oral) administration. In two-compartment models, the drug rapidly equilibrates within each compartment, but the forward and reverse exchange rates between compartments (k_{12} and k_{21}, respectively) are slower than the intra-compartmental equilibrations. Drug is again eliminated from the central compartment with composite rate constant k_e. For extravascular administration, the rate of drug absorption from the gut to the central compartment (k_a) must also be accounted for in the two-compartment model (Figure 6.12C). Note that when equilibration between compartments 1 and 2 is very rapid (based on the values of k_{12} and k_{21} in Figure 6.12C), this reduces to a one-compartment model for a drug administered by an extravascular route (e.g., oral).

All the models depicted in Figure 6.12 are relatively simple. Nevertheless, they are useful in modeling the PK profiles of many drugs. More complex models can also be required to describe the PK behavior of some drugs, and detailed descriptions of these models can be found in the literature and in textbooks devoted to PK. For our pedagogic purposes, it is sufficient to discuss the one-compartment model in more detail below. The interested reader can learn more about multi-compartmental PK modeling in texts such as Kwon (2001), Shargel and Yu (1999), and Riviere (1999).

Returning to the one-compartment model, the PK profile for a drug administered as an I.V. bolus (all the dose in a single injection) should reflect a single, exponential decay of total plasma concentration of drug ($[C_p]$) as a function of time after injection, as illustrated in Figure 6.13.

Such data can today be fit by nonlinear curve fitting to determine the extrapolated starting concentration of drug at time zero, $[C_p^0]$, and the rate constant for elimination, k_e.

$$[C_p^t] = [C_p^0]\,e^{-k_e t} \tag{6.9}$$

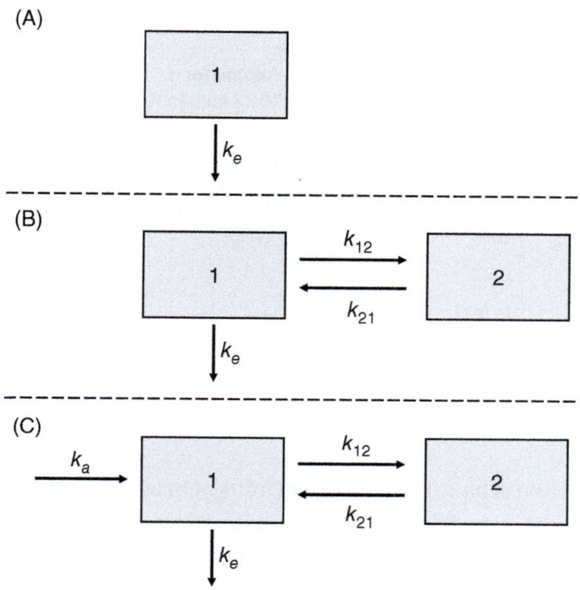

Figure 6.12 Schematic illustrations of common compartmental models of drug distribution. (A) The one-compartment model for I.V. bolus injection. (B) The two-compartment model for I.V. dosing and (C) the two-compartment model for extravascular (e.g., oral) administration. For all models, compartment 1 refers to the central compartment, and compartment 2 refers to the tissue compartment. Rate constants in this figure are as follows: k_a = rate constant for absorption from the gastrointestinal tract, k_e = rate constant for drug elimination, k_{12} = forward rate constant for distribution from compartment 1 to compartment 2, k_{21} = reverse rate constant for redistribution from compartment 2 back to compartment 1. Figure taken from Copeland (2013) with permission of John Wiley & Sons.

Figure 6.13 PK profile for a hypothetical drug measuring the total concentration of drug in plasma ([C_p]) as a function of time after an I.V. bolus injection. The line drawn through the data and extrapolated to time = 0 represents the nonlinear best fit to a single exponential decay (Equation 6.9).

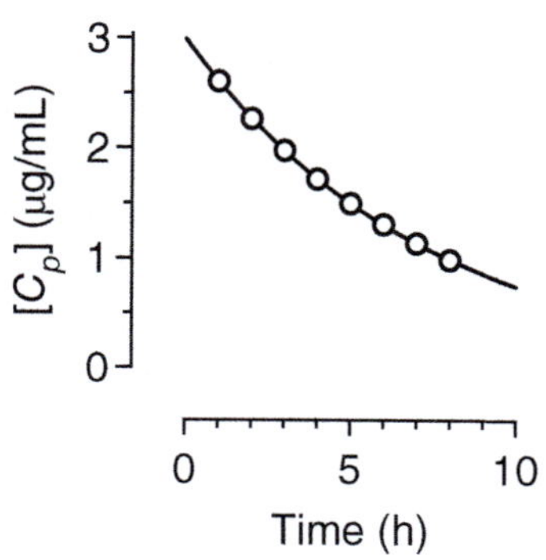

Table 6.8 Physiological volumes of bodily fluids normalized by body mass.

Physiological compartment	Volume per kg body mass (L)	Volume for a 70 kg human (L)
Total body water	0.60	42.0
Extravascular fluid	0.56	39.2
Blood	0.08	5.6
Plasma	0.04	2.8
Interstitial fluid	0.16	11.2
Tissue	0.60	42.0

Source: Adapted from Smith et al. (2015).

The total plasma concentration is typically reported in units of mass per volume, such as ng/mL, μg/mL, and the like. The dose of drug administered is typically reported in units of mass, such as mg. Knowing the mass of drug that was administered (D) and the initial total plasma concentration ($[C_p^0]$) from fitting of the PK data, one can calculate a hypothetical volume into which the drug is distributed within the body at time zero, referred to as the initial volume of distribution (V_D^0), with units of volume (mL, L, etc.). This volume can be divided by the mass of the organism being studied to give a relative value in units of L/kg, which is often useful in relating volume of distribution among different bodily fluids and species (see Table 6.8).

$$V_D^0 = \frac{D}{C_p^0} \tag{6.10}$$

The volume of distribution is an *apparent* volume into which drug is dispersed within the body; it should not be equated with any specific anatomical space. Nevertheless, it is sometimes useful to compare the volume of distribution for a drug molecule with various fluid volumes within the body. Smith et al. (2015) provide values of the volumes of various bodily fluids normalized to body mass, and these are summarized in Table 6.8.

For example, the typical volume of whole blood for a 70 kg human is about 5.6 L, whereas the total volume of bodily fluids is about 42 L. Thus, if the volume of distribution for a drug in humans was around 5–6 L, it would be tempting to assume that this reflects a compound that is only distributed within the bloodstream. However, we cannot make such an inference from these data. On the other hand, if the volume of distribution were >> 5.6 L, it would be reasonable to assume that the drug is distributed to extravascular tissues. In some cases, the volume of distribution can exceed the total fluid volume of the body; this generally indicates that the

Table 6.9 Categories of V_{ss} values for all species.

Category	V_{ss} (L/kg)
Low	<0.6
Moderate	0.6–5
High	5–100
Very high	>100

drug is being accumulated (i.e., concentrated) within a specific tissue or organ of the body. Equation 6.10 defines the initial volume of distribution of the drug. It is also possible to calculate the volume of distribution at other points along the PK profile (e.g., for the distribution and elimination phases) of the drug and at steady state. According to Smith et al. (2015), the volume of distribution at steady state (V_{ss}) is the most used value for comparative studies. They categorize compounds based on the V_{ss} value as summarized in Table 6.9.

6.5 *In Vivo* Drug Efficacy

Nonhuman animals have been used as surrogates for testing the effectiveness of therapeutics since the earliest days of systematic drug discovery. Many of the early understandings of digestion and other physiological processes came from 18th and 19th century studies in birds, fish, amphibians, and small mammals (see Copeland, 2023, for some examples), and early studies of the effects of salicylates and morphine derivatives were likewise observed in animal models (e.g., dogs). We also saw in Chapter 1 that in 1910, Ehrlich and Hata tested compounds for their ability to cure syphilis in rabbits infected with the bacterium that causes this venereal disease (*Treponema pallidum*). Thus, there is a rich history of the use of animal models for testing drug efficacy.

Today, a broad range of animal models continue to be used in preclinical testing of compounds for drug efficacy. Figure 6.14 illustrates the broad range of species that are presently used for efficacy testing and compares these species for their general genetic relatedness to humans and their ease of use.

General criteria for the use of a particular species for testing drug efficacy are as follows.

1) Reasonable relationship to human physiology with respect to the disease or condition to be treated.
2) Relatively short time to an outcome reading of efficacy.

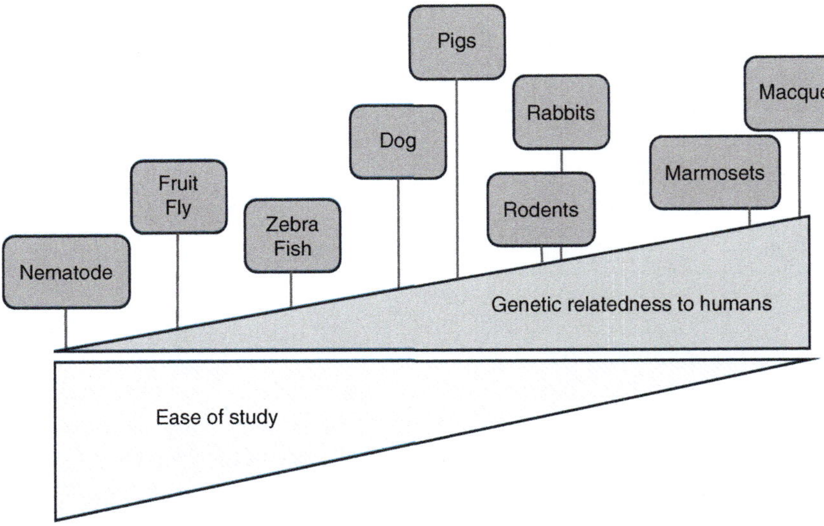

Figure 6.14 Diagram of some species currently used to test compounds in preclinical drug discovery according to their genetic relatedness to humans (Graphodatsky et al., 2011) and their ease of study. The scales used for genetic relatedness and ease of study are arbitrary and not meant to be quantitatively accurate. Rather, they are merely meant to give a general sense of these two opposing assay criteria. Macaques refers to two common monkey species used in preclinical studies: cynomolgus macaque (also known as the crab-eating macaque) and rhesus macaque.

3) Ease of breeding and other aspects of animal husbandry.
4) Small mass or surface area to minimize compound quantity needed for testing.
5) Ease of assessing disease-affected tissue and/or organs.
6) Ethical considerations.

Taking all these criteria into consideration, one finds that the most used animals for efficacy studies across all therapeutic areas are rodents, particularly mice. The predominance of mouse models in drug discovery is based largely on the balance of genetic and physiological relatedness to humans; the ease of breeding, care, and handling of mice; and the availability of different strains of mice to meet different therapeutic area needs. Because of their small size, less compound is needed for conducting efficacy studies in mice, relative to other mammalian species. Mouse models are used in analgesic, cardiovascular, inflammatory, neurological, respiratory, and oncology disease studies. Space does not permit a comprehensive discussion of mouse and other animal models across all therapeutic areas. Instead, we will discuss the role of mouse models in cancer drug discovery as an illustrative example of how such models are used to assess drug efficacy.

6.5.1 Mouse Models of Efficacy in Cancer Drug Discovery

The first mouse models of cancer were developed in the late 1940s and 1950s and were based on growth of leukemic cells as ascites after intraperitoneal (I.P.) inoculation of mice with mouse leukemia cells, such as L1210 or P388 cells. Early researchers noted that the exponential growth of the implanted cells resembled that of bacterial cell growth in culture; therefore, the simple, mathematical models of growth kinetics that had been established for bacterial cell culturing could be adapted for quantifying cancer cell growth *in vivo*. The ability of a compound to slow or halt the proliferation of tumor cells in these models was taken as evidence of compound efficacy, measured either as tumor growth inhibition or tumor growth delay (Teicher, 2006).

Since that time, a wide variety of hematologic and solid tumor mouse models have been established for preclinical drug testing. Today, preclinical testing of potential cancer drugs is most often performed in mouse models that are based on the ability to implant cultured tumor cells in mice, where the cells proliferate to form tumors. Table 6.10 summarizes the most used mouse models for cancer drug discovery and compares the relative advantages and limitations of each model.

6.5.1.1 Subcutaneous (Ectopic) Xenografts

By far the most common animal model for cancer drug discovery is the subcutaneous xenograft model in nude mice; this model has truly become a workhorse of cancer drug discovery because of the many advantages summarized in Table 6.10. A xenograft is any surgical grafting of tissue (or organs) from one species to another.

In the case of cancer models, the most common form of xenograft is the engrafting of cultured human tumor cells or tumor tissue under the skin (i.e., subcutaneously) of a mouse. A universal issue with all xenografts is rejection of the graft due to an immune response by the host. To circumvent this issue, immunocompromised mice are used as the host organism for the tumor cell xenograft. Nude mice are most used in this context, but other immunocompromised mouse strains are also used. The nude mouse refers to a strain of mice that are homozygous for a recessive mutant form of the gene Foxn1. These mice are athymic, meaning that they are born without a thymus and therefore lack T cells. Hence, the mice are unable to mount a proper immune response and cannot reject the foreign graft. In addition to being born athymic, the mice are also born hairless, hence the moniker "nude". Female nude mice that are homozygous for the mutant form of Foxn1 also display poorly developed mammary glands and therefore cannot nurse their young properly. Hence, nude mice are bred by mating heterozygous females with homozygous males to produce litters enriched for homozygous offspring. Although the athymic nude mouse is immunocompromised due to T-cell deficiency, these mice retain intact humoral immunity,

Table 6.10 Comparison of common mouse models of cancer used for preclinical drug discovery. A + sign indicates that the descriptor is an advantage of the model type. Likewise, a – indicates that this description is not an intrinsic property of the model type and thus a limitation.

	Subcutaneous xenograft	Orthotopic xenograft	Syngeneic, subcutaneous models	Syngeneic, orthotopic models	Xenograft disseminated disease models	Syngeneic disseminated disease models	Transgenic tumor models
Easily established and/or straightforward implantation/injection of tumor cells.	+	–	+	–	–	–	–
Tumor of human origin.	+	+	–	–	+	–	–
Tumor of same origin as host (i.e., mouse or rat); hence, proper tumor–stroma interactions.	–	–	+	+	–	+	+
Tumor is visible and can be quantified by volume measurements; model does not rely on survival or PD endpoints alone.	+	–	+	–	–	–	–
Animal is immunocompetent, thus allowing evaluation of immune-oncology drugs and combinations.	–	–	+	+	–	+	+
Tumor grows in tissue of origin environment.	–	+	–	+	–	+	+
Large number of cell lines available for inoculation.	+	+	–	–	+	–	–
Allows for determination of metastatic potential; pattern of disease often more accurately reflects clinical course of disease.	–	–	–	–	+	+	+
Ability to assess distribution, size, and number of tumors at study conclusion.	–	–	–	–	+	+	–

which can limit tumor formation. Mouse strains with additional immune defects also exist and are sometimes used for xenograft studies where greater efficiency of tumor formation is required. These additional mouse strains include SCID (Severe Combined ImmunoDeficient) mice, which lack both T cells and B cells, and the nonobese diabetic strain (NOD)/SCID Il2rg−/−, which also lacks natural killer cells (Kim et al., 2009).

Xenografts can be established by engrafting cultured human cancer cells into a mouse host (referred to as cell line–derived xenograft or CDX), engrafting primary tumor tissue directly from a patient sample (referred to as a direct xenograft or patient-derived xenograft, PDX), or engrafting patient tumor tissue that has previously been serially passaged in mice (referred to as an indirect xenograft). A wide range of human tumor cell lines have been found to grow well in the context of mouse subcutaneous xenografts.

Once the tumor cells are implanted, they proliferate within a confined subcutaneous space to form a tumor of human origin. Because the tumor grows directly under the skin, it is visible to the researcher without the need for invasive procedures. More important, the volume of the tumor can be measured directly using calipers, thereby allowing the researcher to track the course of tumor growth over time (Figure 6.15).

Typically, xenograft models are used to measure compound efficacy, where the tumor is allowed to grow to a certain size before compound dosing begins. For example, the CRO CrownBio (www.crownbio.com) provides a study timeline for

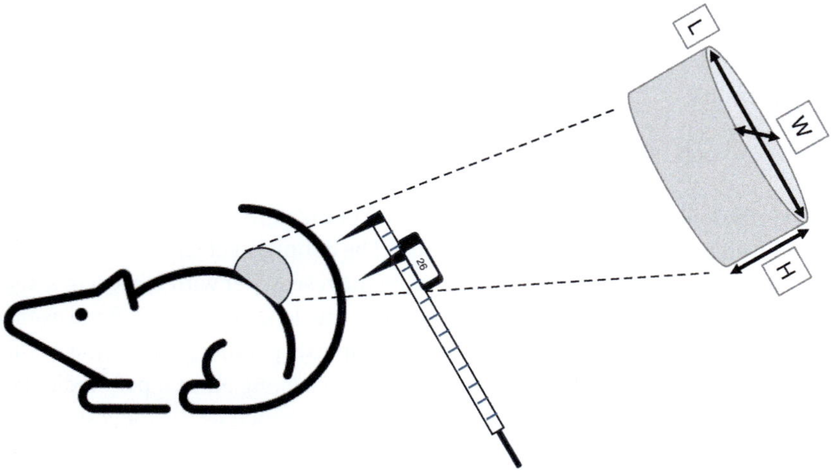

Figure 6.15 Cartoon of a mouse bearing a subcutaneous human tumor xenograft. Because the tumor is visible just below the skin surface, its length (L), width (W), and height (H) can be measured with a caliper. From these measurements, the volume of the tumor can be calculated using Equations 6.11 or 6.12.

LNCaP (prostate cancer cell line) xenografts. After inoculation, the tumors are allowed to grow to a volume of 100–200 mm^3, and the mice are then divided into groups for each study arm (typically 3–10 mice per arm). Different tumor types grow at different rates, so the amount of time required to reach the specified tumor volume for dose initiation will vary. For LNCaP cells, it typically takes about a week to reach the 100–200 mm^3 size after initial inoculation. Once the specified tumor volume is reached, mice in the test group(s) are dosed with compound, while mice in the control group are dosed with the appropriate vehicle. P.O., I.V., or I.P. dosing continues at regular intervals according to the study protocol (e.g., QD, BID), which is based on the mouse PK properties of the compound(s) under testing (see above). Usually, tumor volume is assessed by caliper reading daily over the course of the study. Additionally, it is also typical to assess total body mass of the animals daily as an acute measure of any gross toxicity.

With large enough numbers of animals per arm, a subset of animals can be sacrificed at defined time points and/or at the end of the study, and remaining tumor tissue can be excised. This tumor material can be used to measure drug exposure and PD markers intratumorally. Additionally, but less commonly, residual tumor size can be accurately determined as mass by gravimetric analysis, or as volume by water displacement. Although these are very accurate measures, they can only be performed once, after sacrifice of the individual animals.

As stated above, most researchers instead estimate tumor volume *in situ* during the study period (Euhus et al., 1986). In this way, the volume of the tumor for each animal can be followed longitudinally. To do this, calipers are used to measure the length, width, and height of the tumor at defined time points during the study (e.g., once per day). The tumor volume is then computed using a modified ellipsoid volume formula:

$$V = \frac{\pi}{6} \times (l \times w \times h) = 0.5236 \times (l \times w \times h) \tag{6.11}$$

where l is the length of the tumor, w is the width of the tumor, and h is the height of the tumor, all measured in millimeters (mm). The definition of length and width is somewhat arbitrary but is typically defined by the researcher with respect to some anatomical feature of the organism. For example, the length may be defined as the distance from edge to edge of the tumor along an axis parallel to the spine of the mouse, while the width is the edge-to-edge distance along an axis perpendicular to the length axis. Alternatively, the length may be defined as the longest axis of the tumor, and the width as the axis perpendicular to the length. Of the three measures, height is often the measurement most prone to error. This is because height is measured as the distance above the normal skin surface, and that surface is quite pliable. Recognizing the inherent error in height measurements, a common practice is to use a simplified calculation that relies only on caliper measurements of tumor length and width to approximate tumor volume.

$$V = \frac{1}{2} \times (l \times w^2) \tag{6.12}$$

This much simpler equation replaces $\pi/6$ with 0.5 and uses the width of the tumor as an approximation for the height; that is, w^2 as an approximation of $(w \times h)$. Length is defined here as the longest axis of the tumor, and width is defined as the axis perpendicular to length. Although Equation 6.12 only approximates true volume, researchers have found that this method gives more reproducible values of tumor volume because it is simple to calculate and relies only on the more precise measures of tumor length and width.

In the context of lead optimization, one would typically divide the animals into groups based on different doses of a single compound, plus a vehicle control group. Figure 6.16 illustrates an example of such a dose–response experiment for three different doses of the EZH2 inhibitor tazemetostat, tested in a mouse xenograft model bearing human KARPAS-422 lymphoma cells. It is readily obvious from this figure that the compound displays a potent, time- and dose-dependent antitumor effect in this model. In fact, at doses ≥161 mg/kg BID, the tumor volumes drop below the starting tumor volume and eventually result in no quantifiable tumor remaining. Thus, at sufficient doses, tazemetostat leads to complete tumor regression. While a significant inhibition of tumor growth may be sufficient for advancement of a cancer therapeutic to clinical testing, tumor regression, as seen in Figure 6.16, is the optimal desired outcome of xenograft studies.

To study more generally compounds of varying efficacy in xenograft models, a systematic method is needed for calculating the degree of tumor growth inhibition exerted by different doses of compound. Several different approaches have been suggested to quantify the effect of drug from data such as in Figure 6.16. The most accurate reflection of effect on the course of tumor growth would be derived from measuring the slope of the growth curve during exponential growth. The percent residual growth at a particular dose of drug would then be calculated as 100 multiplied by the ratio of the treated group (T) average slope over the vehicle control group (C) average slope. This calculation could be performed at multiple doses of drug to generate a dose–response curve and thus calculate an EC_{50} for tumor growth inhibition, much like what was presented earlier for cellular growth curves.

However, in practice, few researchers go to the trouble of measuring slope values for individual animals. More commonly, researchers analyze the treated and control tumors at the initiation of study (day 0) and at a single, specified day during the tumor growth experiment. Some researchers use the last day of study as the fixed time point for calculating tumor growth inhibition, while others select a time point for each treated animal that represents the nadir (i.e., lowest observed value) of tumor volume to report the Best Response achieved

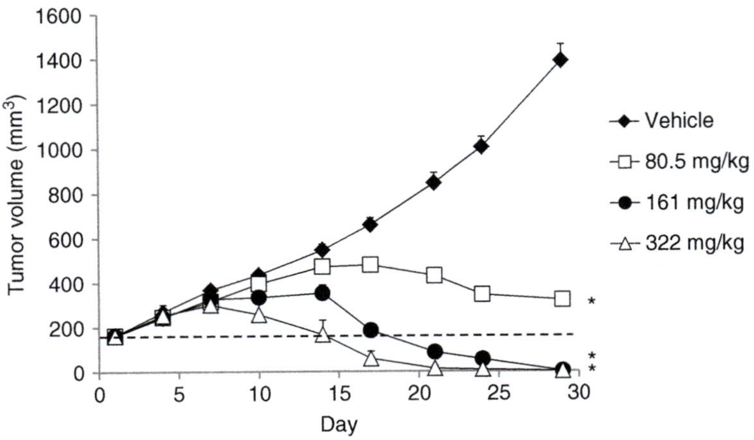

Figure 6.16 Dose–response experiment for tazemetostat, dosed orally at the stated BID doses, in a mouse xenograft model bearing KARPAS-422 human lymphoma cells. Each data point represents the mean value of nine mice per treatment group. The dashed horizontal line represents the starting volume of tumors at the beginning of the study. *Source:* Adapted from Knutson et al. (2014).

(Gao et al., 2015). Regardless of which time point is selected, two methods are commonly used to calculate the effect of treatment on tumor volume. The first method involves measuring the change in volume of the treated tumors between day 0 and day t, the selected day of observation. The percent tumor volume change for each treated animal (%ΔV) is then calculated as follows (Gao et al., 2015).

$$\%\Delta V = 100 \times \left(\frac{V_t - V_0}{V_0} \right) \tag{6.13}$$

where V_t is the tumor volume on day t and V_0 is the initial tumor volume on day 0. The advantage of this approach is that it mirrors how the effect of drug treatment on solid tumor growth is assessed clinically. When a patient is about to start drug treatment (especially in a clinical trial setting), a radiometric or other assessment of tumor volume is made just prior to the start of drug treatment. Tumor size is then similarly assessed at predefined time points during drug treatment. The response of the patient to treatment is then categorized according to the guidelines established by the National Cancer Institute for Response Evaluation Criteria in Solid Tumors, version 1.1 (RECIST 1.1) as having a complete response (CR), partial response (PR), stable disease (SD), or progressive disease (PD), depending on the tumor volume at assessment, relative to the initial tumor volume (Nishino, 2018). Similarly, Gao et al. (2015) have suggested categorizing Best Response achieved in xenograft studies into RECIST-like categories. They define Best Response as the minimum value of %ΔV for $t \geq 10$ days for each treated animal

Table 6.11 Categories of response to compound treatment in xenograft models based on Best Response and Best Average Response criteria as recommended by Gao et al. (2015).

Category	Best response criteria		Best average response criteria
Complete response (CR)	< −95%	and	< −40%
Partial response (PR)	< −50%	and	< −20%
Stable disease (SD)	< +35%	and	< +30%
Progressive disease (PD)	Not otherwise categorized		

Source: Adapted from Gao et al. (2015).

and the Best Average Response as the average value of Best Response over all animals in a treatment group. The criteria established by Gao et al. (2015) for each category of response is summarized in Table 6.11.

A disadvantage of the method of Gao et al. (2015) is that it measures compound effect relative to the starting tumor volume for each treatment group but does not reference tumor volume in the treatment groups to that of the vehicle control group. To account for the growth of the untreated tumors in the vehicle control group, some researchers use a different metric of compound effect. These researchers calculate the difference in volume for treatment group animals on day 0 (T_0) and day t (T_t) and the difference in volume for vehicle control group animals on day 0 (C_0) and day t (C_t). They then use these values to calculate a percent change in volume for the treatment group relative to the control group as follows.

$$\%\frac{\Delta T}{\Delta C} = 100 \times \left(\frac{T_t - T_0}{C_t - C_0}\right) \tag{6.14}$$

The advantages of the ectopic, subcutaneous xenograft model have made it the most used model of cancer drug efficacy in preclinical compound evaluation, especially during the lead-optimization phase. Yet, critics of this model system are keen to point out its limitations (see above) and the fact that there are several examples of compounds that have proved effective in mouse xenograft models but have failed to demonstrate efficacy in human clinical trials. While no preclinical model in a nonhuman species will ever perfectly predict outcome in human clinical trials, the ectopic, subcutaneous xenograft model has served the cancer drug discovery community well. Essentially all drugs in clinical use today for human cancers have been tested and shown efficacy in this preclinical model (Kerbel, 2003; Peterson and Houghton, 2004). Further, the correlation between efficacy in preclinical xenograft models and human clinical studies of conventional cytotoxic agents is quite reasonable. For example, Peterson and Houghton (2004) summarized data for six cytotoxic agents in rhabdomyosarcoma and colon

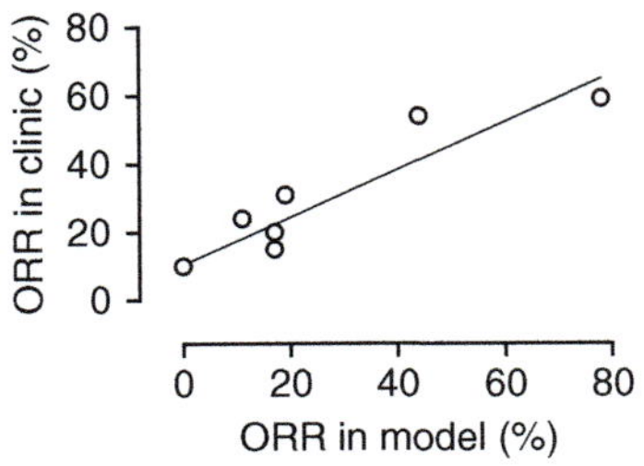

Figure 6.17 Objective response rate (ORR) for cancer cytotoxic agents in clinical studies as a function of ORR observed in preclinical xenograft models. ORR is presented as percent response in both cases. *Source:* Adapted from Peterson and Houghton (2004).

adenocarcinoma, comparing objective response rates in mouse xenografts and human clinical data. The data from this study have been used to generate the correlation plot illustrated in Figure 6.17.

As can be seen from this figure, there is a strong correlation between efficacy in the preclinical model and human clinical experience. For molecular targeted therapies, it is likely that a good correlation between preclinical animal and human efficacy will also be realized, especially if frank observations of efficacy in the preclinical model(s) are coupled with quantitative assessments of pharmacodynamics and/or molecular target engagement.

Historically, ectopic, subcutaneous xenografts have been performed with human cancer cell lines that have been passaged many times *in vitro* (CDX models). This is because such cultured cell lines have been most readily available to the research community. There is always a concern, however, that the multiple *in vitro* passages of such cell lines may induce genetic changes that make these cells less faithfully reflect the genetic background of patient tumors. More recently, however, the availability of tumor material directly from patients, or after a minimal number of *in vitro* passages, has become more widely available to the research community. Hence, xenografts in which patient-derived tumor material is engrafted (PDX models) are becoming more common and are considered more reflective of the clinical situation (Williams, 2018). Today, it is still common for research teams to begin *in vivo* assessment of potential cancer drugs using CDX models during lead optimization. As the team narrows the field to a single or small number of clinical candidate molecules, however, it is common to reassess this smaller number of compounds in appropriate PDX models. Finally, for molecular targeted drugs based on a selective essential hypothesis, it is also important to run xenograft models bearing cells that are expected to respond to the targeted drug and, as a negative control, models bearing cells that are not expected to be responsive to the targeted modality.

As we have seen in this section, ectopic, subcutaneous xenograft models are the workhorse approach to *in vivo* assessment of efficacy for small-molecule, targeted cancer drugs. However, because the mice used in xenograft studies are immunocompromised, these assays are not useful for the assessment of immuno-oncology approaches, such as anti-PD1 and anti-PD-L1 antibodies and other mechanisms of action that rely on cancer cell–immune cell interactions or other aspects of the tumor microenvironment. For approaches targeting these aspects of the tumor microenvironment, more specialized approaches must be used, such as coculturing of cancer cells and immune cells, three-dimensional organoid models, syngeneic animal models, and the like. The CrownBio website (www.crownbio.com) provides a good starting point for exploring these types of assay formats. The review article by Franklin et al. (2022) also provides a useful overview of preclinical discovery tools for immuno-oncology approaches.

6.5.1.2 Relating Tumor Growth Inhibition to Systemic and Intratumoral Drug Exposure and PD

During *in vivo* efficacy studies, the researcher will typically take blood samples at the beginning, at the end (e.g., by cardiac puncture after animal sacrifice), and at one or two time points during the study. The intermediate blood draws are typically limited for mouse models because of the small size of the animal and the attendant small volume of blood; blood sampling during the in-life portion of the study is commonly obtained by tail vein puncture, retro-orbital bleeding, or submandibular bleeding. It has been recommended that in aggregate no more than about 200 μL of blood should be taken over the course of a 28-day study in mice. These blood samples provide the ability to determine total and PPB-free (i.e., unbound) compound concentration in systemic circulation. Studies of compound PK in larger animals, such as rats, dogs, etc., also allow for total and free compound concentrations to be determined and additionally open the potential to collect peripheral blood mononuclear cells (PBMCs) that may be useful in measuring PD effects (see below). At the end of an *in vivo* efficacy study, and at intermediate points at which a subset of animals may be sacrificed, the researcher can excise residual tumor material and use this to determine intratumoral compound concentration and measures of PD. We stated earlier that the cellular phenotypic EC_{50} and EC_{90} serve as good guideposts for the free compound systemic exposure required to see efficacy *in vivo*. The blood and tumor samples collected during efficacy studies in animals provide a means of testing this assumption and understanding the quantitative relationship between free systemic exposure, tissue (e.g., tumor) exposure, PD, and efficacy.

6.5.1.3 Case Studies

Tazemetostat as a Treatment for Non-Hodgkin's Lymphoma Let us return to tazemetostat as a treatment for mutant-bearing NHL to provide an illustrative example of assessing *in vivo* exposure, efficacy, and PD (Knutson et al., 2014; Kuntz et al., 2016). Prior to efficacy testing in mouse xenografts, the PK of tazemetostat (a.k.a., EPZ6438) was determined in mouse and rat. In mice, the compound

Table 6.12 Summary of EZH2 mutant-bearing NHL cell line sensitivity to tazemetostat. LCC is the lowest cytotoxic concentrations, as defined in Chapter 5. The EC_{90} values were calculated as $9XEC_{50}$.

Cell Line	EC_{50} (nM)	EC_{90} (nM)	LLC (nM)
Pfeiffer	0.49	4.41	0.5
KARPAS-422	1.80	16.20	120
WSU-DLCL2	8.60	77.40	170

Source: Adapted from Knutson et al. (2014).

was rapidly absorbed after oral dosing, displaying a t_{max} of 15–18 minutes and an oral bioavailability of 55%. The volume of distribution at steady state was about 1.5 × total body water, suggesting that the compound was rapidly and well distributed throughout tissue. The terminal half-life in mice for tazemetostat was about four hours. Similar PK properties were observed in the rat. Knutson et al. (2014) tested tazemetostat at multiple doses and dosing schedules (e.g., QD, BID, TID, with and without dosing holidays) in three xenograft models of NHL: Pfeiffer, KARPAS-422, and WSU-DLCL cell lines. These three cell lines were all sensitive to tazemetostat treatment *in vitro*, but with variable potencies, as summarized in Table 6.12.

Exposures in mice were determined after 7 and 28 days of dosing. Exposure exceeded the EC_{90} for Pfeiffer and KARPAS-422 cells at doses above 80 mg/kg TID and exceeded the LLC for all three cell lines at doses at or above 160 mg/kg TID and at or above 213 mg/kg BID. A direct comparison of efficacy among the three xenografts is difficult because they were treated at different dose ranges and dose schedules; the WSU-DLCL2 xenograft was dosed at 40, 80, and 160 mg/kg TID, the KARPAS-422 xenograft was dosed at 80.5, 161, and 322 mg/kg BID and the Pfeiffer xenograft was dosed at 34.2, 114, 342, and 1140 mg/kg QD. Nevertheless, we can assess the best responses seen for each of these xenografts, as summarized in Table 6.13.

The best responses seen in these three models categorically align with their relative sensitivity to tazemetostat from *in vitro* cellular assays of antiproliferative activity (Table 6.12). Focusing on the KARPAS-422 xenograft, Knutson et al. (2014) excised tumors after seven days of compound treatment with QD and BID dosing schedules and at the end of the BID study period (28 days of treatment) and measured intratumoral H3K27me3 levels, relative to total histone H3 protein levels, as a proximal measure of PD. Figure 6.18A illustrates the dose-dependent reductions in intratumoral H3K27me3 levels observed after seven days of tazemetostat treatment. In this graph, the dose labeled 0 refers to the vehicle control animals. The data fit well to Hill–Langmuir dose–response models and yield EC_{50} doses of

Table 6.13 Summary of xenograft responses to treatment with tazemetostat, based on data reported.

Tumor type	Dosing schedule	Dose assessed	Best outcome
Pfeiffer	QD × 28 days	114 mg/kg	Complete regression[a]
KARPAS-422	BID × 28 days	161 mg/kg	Complete regression[a]
WSU-DLCL2	TID × 28 days	160 mg/kg (highest dose tested)	54% TGI[b]

a) Complete regression; no quantifiable tumor remaining at the end of the study period.
b) TGI; tumor growth inhibition.
Source: Adapted from Knutson et al. (2014).

431 and 147 mg/kg for QD and BID schedules, respectively. Knustson et al. (2014) also measured the intratumoral H3K27me3 levels at the end of the BID study and plotted these for each animal as a function of the trough exposure level of drug in systemic circulation; this is presented in Figure 6.18B. Again, the data fit well to a Hill–Langmuir model and yield an EC_{50} of 23 nM.

Knutson et al. (2014) also dosed rats with tazemetostat at doses from 100 to 1000 mg/kg, QD for 28 days, to determine if drug-mediated PD (reductions in H3K27me3 levels) could be measured in readily accessible normal tissues. They demonstrated dose-dependent reductions in H3K27me3 levels in PBMCs and in

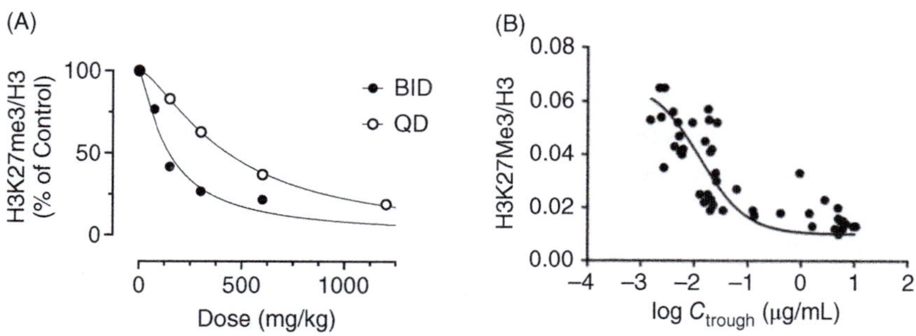

Figure 6.18 Tazemetostat treatment leads to dose-dependent reductions in intratumoral H3K27me3 levels. (A) Intratumoral H3K27me3 level as a function of dose of tazemetostat for KARPAS-422 xenografts treated on QD (open circles) and BID (closed circles) schedules for seven days. Each data point represents the mean value of all animals in the dosed group. (B) Intratumoral H3K27me3 levels as a function of systemic trough concentrations of tazemetostat after 28 days of BID dosing. *Source:* Adapted from Knutson et al. (2014). Note that in panel (A) the term "dose" refers to the amount of drug give per treatment (i.e., given once [QD] or twice [BID] each day). The dose labeled 0 refers to the vehicle control animals.

bone marrow and dose-dependent reductions in H3K27me3 positive skin cells using an H2K27me3-specific antibody for IHC (immunohistochemistry) staining. These data were in part used in the validation of skin punch biopsy samples and IHC staining as a PD assay during clinical trials of tazemetostat (Italiano et al., 2018; Ho and Copeland, 2024).

ATX968 as a Tool Compound DHX9 Inhibitor In Chapter 3, we covered the target validation for the RNA helicase DHX9 as a selectively essential target for MSI-high colorectal cancer and other solid tumors with high baseline levels of replication stress. Based on the excellent target validation, Castro et al. (2025) initiated high-throughput screening for inhibitors of DHX9 (see Chapter 3). A validated hit from screening, compound **1** (Figure 6.19A), was selected for optimization. Recently Daniels et al. (2025) have provided a detailed account of the hit-to-lead and lead-optimization SAR that led to ATX968 (Figure 6.19B), a potent, selective, orally bioavailable tool compound DHX9 inhibitor. The SAR for this series of DHX9 inhibitors was greatly facilitated by the availability of co-crystal structures of compounds bound to the feline (domestic cat) and human DHX9 orthologues (Lee et al., 2023; Figure 6.19C and D).

Compound **1** and its analogues partially inhibited the ATP hydrolysis activity of DHX9 (compound **1** EC_{50} = 2.9 μM, maximum inhibition = ca. 70%). Generally, partial inhibitors are not considered pharmacologically tractable (Chapter 2 and Copeland, 2013). However, Daniels et al. (2025) found that these compounds were full, dead-end inhibitors (Chapter 2 and Copeland, 2013; Copeland, 2023) of the RNA unwinding activity of the enzyme (compound **1** EC_{50} = 21.4 μM). The partial inhibition of ATP hydrolysis but complete inhibition of helicase activity likely reflects some energetic uncoupling between ATP hydrolysis and the conformational changes that engender nucleic acid unwinding by the enzyme.

Titration of the enzyme reaction with ATP or RNA substrates indicated that the compounds were noncompetitive with respect to both substrates, and these data were concordant with surface plasmon resonance (SPR)-binding data that suggested no change in compound K_d value (compound **1** K_d = 0.33 μM) with increasing concentrations of either substrate. The crystal structure of compound **1** and ADP bound to feline (2.27 Å resolution) and human (3.26 Å resolution) DHX9 showed that the compound bound in an allosteric pocket at the predicted exit site of the single-stranded RNA-binding channel formed between the lower RecA1 and RecA2 domains and the upper OB and HA2 domains (Figure 6.19C and D). This binding site is 20 Å away from the catalytic active site, to which ADP was bound in the structures. The binding pocket amino acids were identical between feline and human DHX9, so Daniels et al. (2025) relied on the higher-resolution cat enzyme for further SAR and attendant structural biology studies.

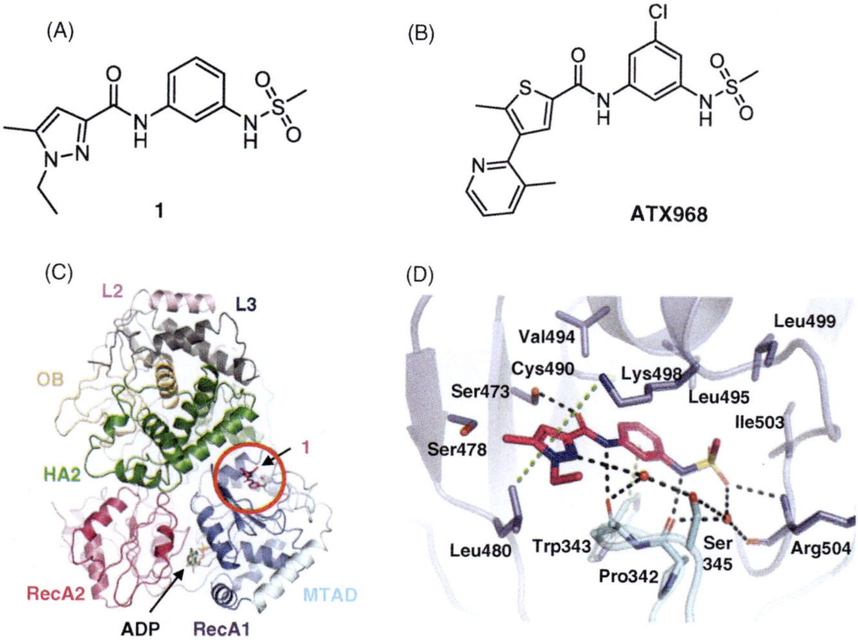

Figure 6.19 Inhibitors of the RNA helicase DHX9 as reported in Castro et al. (2025) and Daniels et al. (2025). (A) Chemical structure of the validated screening hit compound **1**. (B) Chemical structure of the optimized tool compound ATX968. (C) Crystal structure of feline DHX9 with ADP bound in the enzyme active site and compound **1** bound in an allosteric site highlighted by the red circle. (D) Close-up view of the allosteric binding pocket with compound **1** bound, highlighting the key interactions between the compound and recognition elements within the binding pocket. Panels C and D reproduced from Daniels et al. (2025; https://pubs.acs.org/doi/10.1021/acs.jmedchem.5c00252) with permission of American Chemical Society.

The structural details of compound **1** interactions with the allosteric binding pocket of DHX9 are shown in Figure 6.19D. The pyrazole ring of compound **1** is inserted between Lys 498 and Leu 480, interacting with the side chains of these two amino acids through π–cation and π–hydrogen contacts. The side chain of Trp343 interacts with the central benzene ring of compound **1** through an edge-to-face π–π interaction. The compound is further stabilized within the binding pocket through a network of hydrogen bonds to amino acid side chains and two bound water molecules. Additionally, van der Waals interactions with hydrophobic amino acids lining the top and back of the pocket also strengthen compound **1** binding. This allosteric binding pocket appears to be unique to DHX9 and is not observed or predicted to be part of the structure of other human nucleic acid helicases (Daniels et al., 2025).

Table 6.14 Some preclinical characteristics of ATX968.

Parameter	Value
EC_{50} DHX9 ATPase activity	3 nM
EC_{50} DHX9 RNA unwinding activity	8 nM
SPR K_d	1.3 nM
EC_{50} for other enzymes[a)]	>10 μM
LS411N (MSI-high) PD EC_{50} (circBRIP1 induction)	60 nM
LS411N (MSI-high) antiproliferative EC_{50}	69 nM
H747 (MSS) antiproliferative EC_{50}	>10 μM
Solubility (fasted simulated intestinal fluid)	62.5 μM
Solubility (simulated gastric fluid)	264 μM
Caco-2 permeability A-to-B rate	7.6×10^{-6} cm/s
Caco-2 efflux ratio	3.5
Intrinsic clearance (mouse hepatocytes)	1250 mL/min/kg
Clearance (mouse, unbound compound)[b)]	3661 mL/min/kg
$AUC_{0\text{-}24h}$ (mouse, unbound)[b)]	39.6 h•nM

a) Tested against helicase DHX36, SMARCA2, WRN, and a panel of 97 human kinases.
b) From a single 10 mg/kg PO dose study.

Compound **1** required significant optimization along multiple vectors including target potency, cell permeability, solubility, and PK parameters. The report by Daniels et al. (2025) systematically describes the SAR optimization of these parameters that ultimately led to ATX968 and its related clinical candidate molecule. Some of the preclinical characteristics of ATX968 are summarized in Table 6.14. Co-crystal structures of ATX968 bound to ADP-replete feline and human DHX9 not only verified that the compound bound to the same pocket as compound **1**, but also indicated additional interactions with ATX968, as detailed in Daniels et al. (2025) and shown in Figure 6.20.

As summarized in Table 6.14, ATX968 is a potent, selective inhibitor of DHX9 enzymatic activity that induces circular RNA (e.g., circBRIP1) as a proximal cellular PD marker and selectively kills MSI-high colorectal cancer cells. The PD induction and selective cell killing exhibited by ATX968 recapitulates the effects of siRNA knockdown of DHX9, as summarized in Chapter 3 and in Castro et al. (2025). In cellular assays, there was a reasonable correlation between the enzymatic EC_{50} values, or SPR K_d values, in cell-free assays and intracellular PD across the pharmacophore series ($R^2 = 0.49$ and 0.55, respectively). However, the correlation between drug–target residence time (measured from SPR off-rates)

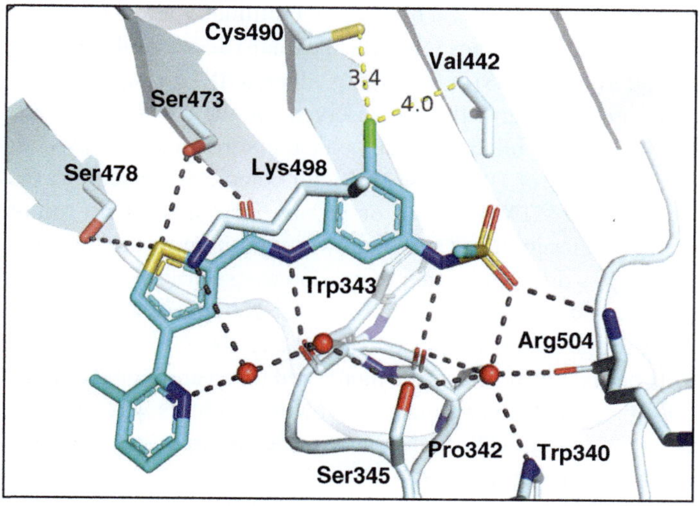

Figure 6.20 Crystal structure of ATX968 bound to feline DHX9 showing details of compound-binding pocket interactions. Figure kindly provided by Dr. Brian Sparling of Accent Therapeutics, based on data from Daniels et al. (2025) with permission of American Chemical Society/CC BY NC ND.

and intracellular PD was far superior, displaying an R^2 value of 0.84. These data suggest that the durability of the drug–target complex, as measured by residence time, is a critical factor in cellular PD (Daniels et al., 2025). A strong correlation was likewise observed between intracellular PD (circBRIP1 induction) and antiproliferative activity in MSI-high cells, such as LS411N (Castro et al., 2025).

To determine the tumor growth inhibition caused by ATX968 treatment in mouse xenografts, the research team first explored the PK properties of ATX968 in mice (Daniels et al., 2025). Initial PK studies found that the compound was well absorbed upon oral dosing (C_{max} [unbound] of 19.9 nM), and that the unbound compound clearance in mice was within 3-fold of the intrinsic clearance determined *in vitro* with mouse hepatocytes (Table 6.14). A preliminary assessment of *in vivo* target engagement suggested that exposure would need to be at or above the circBRIP1 EC_{90} (corrected for PPB) for 24 h, thus requiring a C_{min} (i.e., trough) value of 3964 ng/mL. From this, a dose range of 10–300 mg/kg was selected for dose escalation studies. Over this dose range, the C_{max} value was found to be dose-proportional, but the $AUC_{0\text{-}24h}$, the oral bioavailability, and the terminal half-life all demonstrated greater than dose proportionality, suggesting saturation of *in vivo* clearance mechanism(s), hepatic recirculation, and/or other factors. For example, the terminal half-life increased from 1.42 to 14.7 h over the 10–300 mg/kg dose range (Daniels et al., 2025).

Castro et al. (2025) reported the effects of ATX968 on tumor volume in mouse xenografts of colorectal cancer cell lines LS411N (MSI-high) dosed orally between 30 and 300 mg/kg, and SW480 (MSS) at 300 mg/kg, all on a BID dosing schedule. No significant effect on tumor growth was observed for the SW480 xenograft dosed with 300 mg/kg ATX968 relative to the vehicle control group, consistent with the selective essentiality of DHX9 for MSI-high colorectal cancers. In the LS411N xenograft, doses of ATX968 of 30 or 100 mg/kg displayed no tumor growth inhibition. However, significant tumor growth inhibition was observed in this model at the 200 mg/kg dose, and tumor regression was seen at the 300 mg/kg dose (Figure 6.21A). As illustrated in Figure 6.21A, at 200 and 300 mg/kg, the LS411N tumors were suppressed over the course of the 28-day dosing period. At the end of the dosing period, the tumors were monitored for an additional 28 days. During this post-dosing period, some modest regrowth of tumors was observed for the mice dosed at 200 mg/kg ATX968, but no significant tumor regrowth was seen for mice dosed at 300 mg/kg ATX968.

Castro et al. (2025) also measured intratumoral PD (as the induction of circBRIP1 RNA relative to vehicle-control–treated tumors). As seen in Figure 6.21B, the level of circBRIP1 increased with increasing dose of ATX968; the dose-dependent circBRIP1 level increases according to a Hill–Langmuir model, yielding an EC_{50} of ca. 98 mg/kg. As expected for this proximal PD marker, the level of circBRIP1 induction was similar in LS411N (MSI-high) and SW480 (MSS) xenografts at the 300 mg/kg BID dose (Castro et al., 2025). Likewise, the intratumoral circBRIP1 level was dependent on the intratumoral exposure of compound, as illustrated in Figure 6.21C, and in turn the reduction in tumor volume correlated well with the intratumoral circBRIP1 levels, as seen in Figure 6.21D.

Ultimately, this work also resulted in the identification of a clinical candidate from the same pharmacophore series, ATX-559 (structure not disclosed), that is currently in phase 1 clinical studies (clinicaltrials.gov, NCT06625515); this trial represents the first human clinical study of a DHX9 inhibitor.

EPZ5676 as a Treatment for Mixed Lineage Leukemia In Chapter 5, we reviewed the SAR for the SAM-competitive pharmacophore series of DOT1L inhibitors, leading up to the clinical candidate EPZ5676 (pinometostat). EPZ5676 was found to be a very potent inhibitor of DOT1L ($K_i = 0.08$ nM, or 80 pM) that was selective relative to other human lysine and arginine protein methyltransferases; the selectivity for DOT1L was >37,000-fold for all protein methyltransferases tested (Daigle et al., 2025). As previously discussed, EPZ5676 displayed the expected selective essentiality for mixed lineage leukemia cells with chromosomal translocations affecting the 11q23 locus (MML gene). In MV4–11 mixed lineage leukemia cells, for example, EPZ5676 displayed potent antiproliferative activity with an EC_{50} value of 3.5 ± 0.7 nM (Daigle et al., 2013). Based on these favorable *in vitro* properties, Daigle et al. (2025) studied the PK properties of EPZ5676 in mice and

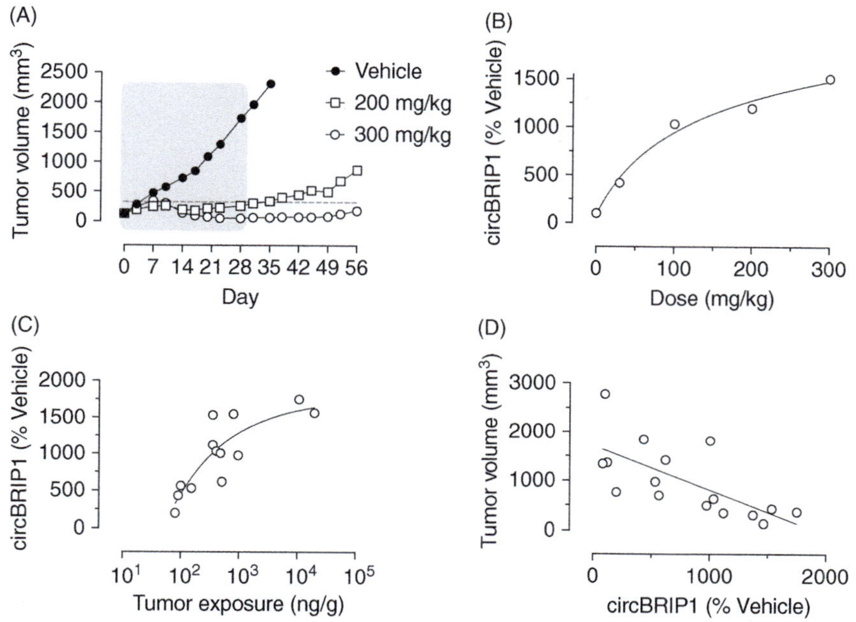

Figure 6.21 Tumor growth inhibition and intratumoral PD resulting from treatment of LS411N xenografts with ATX968 by oral dosing (P.O.). (A) Tumor growth curves for vehicle and ATX968 at 200 and 300 mg/kg doses. The horizontal dashed line indicates the starting volume of tumors in this study. The compound was dosed PO, BID for 28 days, as indicated by the gray shaded area of the plot. Tumor volume was evaluated during this dosing period and for an additional 28 days after dosing was discontinued. (B) circBRIP1 induction as a proximal PD marker of DHX9 inhibition as a function of dose. The line through the data points is the best fit to a Hill–Langmuir dose–response model and yields an EC_{50} of 98 mg/kg. (C) Intratumoral PD (circBRIP1 induction) as a function of intratumoral compound exposure. The line drawn through the data represents the best fit to a Hill–Langmuir concentration–response model. (D) Tumor volume as a function of intratumoral PD (circBRIP1 induction), demonstrating a linear relationship between intratumoral PD and phenotypic outcome. The line drawn through the data represents the best fit to a linear equation. *Source:* Castro et al. (2025).

rats as a prelude to xenograft studies of tumor growth inhibition. Single-dose studies by I.V. bolus in mice and rats resulted in the data presented in Table 6.15. A single P.O. dose study was also conducted in mice and showed that the compound had oral bioavailability of <1%. The lack of oral bioavailability, low volume of distribution at steady state, and the high clearance of the compound in mouse and rat dictated that the compound be dosed by continuous I.V. infusion to ensure adequate exposures for efficacy. The small size of blood vessels in mice made this species untenable for continuous I.V. infusion studies. Instead, Daigle et al. (2025)

Table 6.15 PK parameters for EPZ5676 upon single-dose I.V. bolus injection in mouse and rat.

Parameter	Mouse	Rat
Dose (mg/kg)	5	1
C_{max} (μM)	7.99	2.04
t_{max} (h)	0.05	0.05
$AUC_{0\text{-}t}$ (μM•h)	1.95	0.43
$AUC_{0\text{-infinity}}$ (μM•h)	1.96	0.44
$t_{1/2}$ (h)	1.14	3.73
Clearance (L/h/kg)	4.6	4.07
V_{ss} (L/kg)	1.58	1.66
Oral bioavailability (%)	<1	N.D.[a)]

a) N.D., not determined.

chose to study the efficacy of EPZ5676 in nude (i.e., immune-compromised) rat xenografts, using implanted pumps to deliver compound continuously.

Figure 6.22A illustrates tumor growth curves for rat xenografts of the MV4–11 mixed lineage leukemia cell line, treated with vehicle and with 35 mg/kg or 70 mg/kg of EPZ5676 administered by continuous I.V. infusion. Infusion was continued for 21 days as indicated by the gray shaded area of the plot. Tumor measurements were continued after the end of the infusion period for a total of 53 days. A significant tumor growth delay was observed at the 35 mg/kg dose and complete tumor regression was seen at the 70 mg/kg dose. A subset of rats from each group was sacrificed after 14 days of drug treatment, and the reduction of H3K79me2 levels was measured in tumor, bone marrow, and PBMCs as a proximal PD marker. As illustrated in Figure 6.22B, a dose-dependent diminution of H3K79me2 level was observed in all three tissues, suggesting that the compound was well distributed to these tissues. For all three tissues, the reduction in H3K79me2 level followed a Hill–Langmuir dose–response relationship, with EC_{50} values of 12, 27, and 12 mg/kg for tumor, bone marrow, and PBMCs, respectively (note that these dose–response curves are based on a minimal number of doses, so the veracity of these EC_{50} values must be viewed with caution). EPZ5676 was eventually taken into phase 1 clinical studies in mixed lineage leukemia patients. Despite the challenges of administering a drug by continuous I.V. infusion, Stein et al. (2018) found that the proximal PD biomarker, diminished H3K79 methylation, was observed after dosing in most of the patients treated. Table 6.16 summarizes the clinical activity seen with EPZ5676 at all doses. Although leukemia is primarily a disease of the bone marrow, in some patients', leukemia can also manifest as skin lesions;

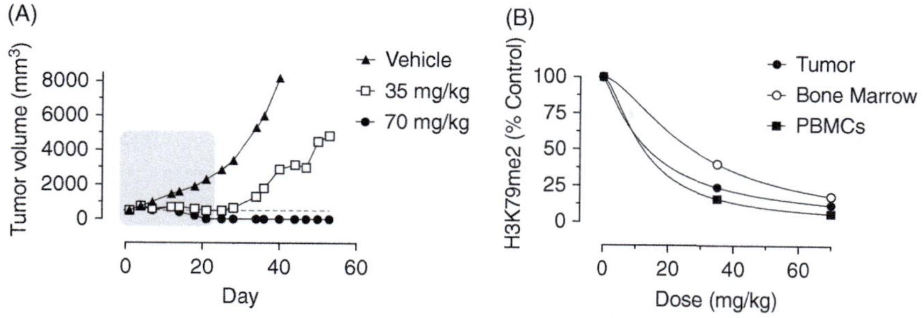

Figure 6.22 Tumor growth inhibition and tissue PD due to treatment with EPZ5676 (pinometostat) in MV4–11 rat xenografts. (A) Tumor growth curves for vehicle and EPZ5676 treatment at 35 mg/kg and 70 mg/kg dosed by continuous I.V. infusion. The dashed horizontal line indicates the starting volume of tumors in this study. The gray shaded area indicates the dosing period for rats in this study. Tumor volumes were evaluated during and after the dosing period for a total of 53 days. (B) H3K79me2 level in tumor, bone marrow, and PBMCs as a function of EPZ5676 dose. The lines through each data set represent the best fit to a Hill–Langmuir dose–response model. Both panels were redrawn from estimates of the data presented in Daigle et al. (2025).

this condition is referred to as leukemia cutis. As seen in Table 6.16, nine patients at various dose levels demonstrated leukocytosis/differentiation and three of these patients also demonstrated resolution of leukemia cutis. Two patients, both with *t*(11;19) translocations (resulting in MLL–ENL fusion protein), treated at a dose of 54 mg/m^2 per day by continuous I.V. infusion achieved CRs. Although the response rate on this trial was modest, the drug-induced PD, clinical responses, and accompanying cytogenetic changes provided proof of concept that targeting DOT1L can lead to clinically meaningful responses in mixed lineage leukemia patients (Stein et al., 2018). Further work on orally bioavailable compounds targeting DOT1L is currently ongoing (see, for example, Chen et al., 2016).

While this chapter was being finalized, an exciting new report was published by Nguyen et al. (2025) showing synergistic activity and a broader range of responsive B-cell cancer cell lines and in xenografts when pinometostat (EPZ5676) is combined with tazemetostat. This exciting and unanticipated combination effect of two epigenetic targeted drugs may lead to clinical benefit for a broader range of B-cell lymphoma patients.

6.6 Allometric Scaling and Predicted Human Dose

Despite the diversity of animal life on earth, nature tends to use common solutions to provide the structural, physiological, and biochemical needs of living organisms. In the early 20th century, a plethora of data were collected across various

Table 6.16 Clinical activity of EPZ5676 (pinometostat) by dose. Data from Stein et al. (2018).

Dose (mg/m^2 per day)	Number of patients	Leukemia cutis resolution	Leukocytosis/ differentiation	Overall response
12	1	0	0	
24	5	0	1	
36	4	1	2	
54	6	1	1	1 CR[a)]
54 (28-D CIV)	8	1	1	1 CR[a)]
80	3	0	2	
90 (28-D CIV)	24	0	2	

a) CR = complete response.

animal species relating physiological processes to organism size in terms of body mass (kg) and/or body surface area (m^2). Kleiber (1932) synthesized a large body of these data and found that the relationship between body mass and physiological functions (e.g., heart rate, respiration, and metabolic rate) could be expressed as a simple, power function as follows.

$$P = A \times M^b \tag{6.15}$$

where P is some physiological function, M is the body mass of the organism, A is a normalization constant referred to as the allometric coefficient, and b is the allometric exponent. Because body mass varies so greatly between humans and the species used in drug discovery, Equation 6.15 is usually presented in a log–log form as in Equation 6.16. This form of the equation is useful for fitting data plotted on a log–log scale.

$$\log(P) = \log(A) + b \times \log(M) \tag{6.16}$$

Similarly, many PK properties of biologics and small-molecule therapeutics scale with organism body mass in an allometric manner (Huang and Riviere, 2014). Hence, by studying PK properties in preclinical animal species, one can often develop a form of Equations 6.15 and 6.16 to fit the data and thereby predict the corresponding PK property in other species, including humans. For example, Jairam et al. (2019) study the PK properties of fomepizole, a drug used as a treatment for methanol poisoning, in multiple preclinical species (BALB/C mice, Sprague Dawley rats, New Zealand white rabbits, and beagle dogs). From these data, they were able to develop allometric scaling equations relating $AUC_{0\text{-}\infty}$, clearance, half-life, and volume of distribution at steady state (V_{ss}). For example, Figure 6.23 illustrates the V_{ss} measured by Jairam et al. (2019) as a function of

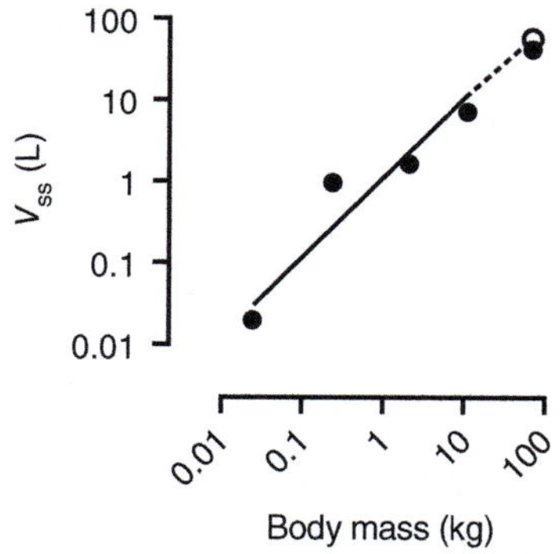

Figure 6.23 Allometric scaling of V_{ss} for the drug fomepizole. Closed circles represent the experimentally determined V_{ss} values in BALB/C mice (mean body mass, 0.025 kg), Sprague Dawley rats (mean body mass, 0.24 kg), New Zealand white rabbits (mean body mass, 2.10 kg), beagle dogs (mean body mass, 11.0 kg), and humans (mean body mass, 70 kg). The open circle represents the predicted value for humans based on extrapolation using Equation 6.17. The line drawn through the data represents the best fit linear relationship defined by Equation 6.17. The dashed line represents the extrapolation of this fit to predict the value in humans. Data used to construct this graph were taken from Jairam et al. (2019).

body mass. The data from mouse, rat, rabbit, and dog track linearly with body mass on a log–log scale and were best fit by the following equation (a modified form of Equation 6.16).

$$V_{ss} = 0.9307 \log(M) + 0.0429 \tag{6.17}$$

Using this equation, Jairam et al. (2019) calculated the predicted V_{ss} in humans to be 57.55 L, which matched well with the reported value of 41.17 L determined clinically.

Allometric scaling is typically combined with compartmental PK modeling to predict human PK properties such as fractional bioavailability (F), absorption rate constant (k_a), clearance (Cl), and volume of distribution at steady state (V_{ss}). These predicted PK values can then be combined with the PPB-adjusted MEC of drug in preclinical models to provide a prediction of the dose needed in humans and other organisms (see next section) to achieve efficacy.

$$\text{Dose} = \frac{\text{MEC} \times [k_a - k_e]}{F \times k_a \times \left[\dfrac{1}{1 - e^{-k_e}} - \dfrac{1}{1 - e^{-k_a}}\right]} \tag{6.18}$$

The term MEC in Equation 6.18 may be based on the EC_{50}, EC_{75}, or EC_{90} for phenotypic outcome in relevant human cell culture studies (see above), however, is more commonly based on the PPB-adjusted systemic exposure resulting in the desired level of efficacy (e.g., tumor regression) in animal models. The human values of F and k_a are predicted from allometric scaling and compartmental PK modeling, and k_e is determined as the ratio Cl/V_{ss}, with human Cl and V_{ss} also predicted from allometric scaling and compartmental PK modeling. An excellent review of dose predictions has been provided by Maurer et al. (2020) in which they provide equations, related to Equation 6.18, for basing dose projections on C_{max}, C_{min}, AUC, and other key PK parameters.

6.7 Preliminary Safety Assessment in Two Species

As part of the highly prescriptive studies required for an IND application (see Appendix 2), the applicant (referred to by FDA as the sponsor) must provide comprehensive dose-dependent toxicology data for the drug in one rodent and one non-rodent animal species, performed under good laboratory practices guidelines (GLP; 21 CFR Part 58, www.fda.gov). These are costly and time-consuming studies that are generally performed only on DC compounds that are expected to transition to clinical studies. Prior to DC declaration, it is common for drug discovery teams to perform preliminary safety studies in the same two species (one rodent, one non-rodent) that are expected to be subsequently used for GLP safety studies.

The preliminary, non-GLP safety studies are often performed in rats and dogs, although other species may be used, depending on the compound (e.g., ensuring that the target orthologue of the species is modulated by compound with reasonable affinity), and the therapeutic indication. Whereas GLP studies are typically performed for a period of 28 days or longer, the non-GLP studies are often run over shorter time courses of, for example, 7 or 14 days. A range of doses (typically, 3–5 doses) are chosen to bracket the predicted efficacious dose in each species, based on equations like Equation 6.18. These doses are then tested *in vivo* as single doses, in what is referred to as a dose range finding (DRF) study, to determine the range of doses that are tolerated by the organism(s) and to estimate the maximum tolerated dose (MTD). Generally, both males and females of each species are studied as separate groups to evaluate any potential gender-specific safety issues, or differences in drug metabolism. For example, it is well known that rats show a species-specific gender dimorphism in hepatic CYP450 expression that can result in significantly different drug exposures in male and female rats (Gerges and El-Kadi, 2023). Prior to initiating the in-life portion of such safety studies, medicinal chemists will explore different salt forms of the compound and different formulations to optimize compound exposure in the species to be tested. Additionally, formulation can play a significant role in tolerability to compound. Gad et al. (2016) provide a comprehensive review of the tolerability of different formulations in preclinical species used for safety studies. The reader is referred to this publication for further details.

Once the two species have been selected and compound salt form and formulation are finalized, the in-life portion of the studies can begin. Compound is dosed by a route (e.g., oral gavage) and schedule (e.g., QD, BID) that matches what one expects to use clinically, and a control group of each species and gender is dosed with vehicle alone. Animals are monitored daily for any gross signs of adverse effects; gross signs may include body temperature, food and water intake, lethargy, panting, changes in coat appearance, and general morbid and moribund behaviors. Because these studies are generally done in larger animals – such as rat and dog – serial bleeds can be taken over the course of the study to assess compound exposure, PD signals, PBMC composition and cell counts, and blood chemistry. In cases where the compound has some hERG inhibitory activity, one may decide to monitor cardiac function and rhythm in dogs by use of telemetry jackets, but this adds a great deal of cost to the study and so is only done in non-GLP studies when there is a clear risk of arrhythmia that must be discharged.

If, during the study period, animals are found to display severe morbidity, they will be euthanized. Such events and any spontaneous deaths are recorded, and the animals may be subjected to postmortem necropsy. At the end of the study period, all surviving animals are euthanized and subjected to necropsy. A battery of organ examinations and histology are performed to determine any adverse effects that may be related to the compound. The overarching goal of these studies is to determine the no adverse effect level (NOAEL; the highest dose at which no adverse effects are observed) and MTD (the highest dose at which the level of toxicity is acceptable) in both species, and from these the therapeutic indices of the compound for each species. In favorable cases, a sufficient margin (e.g., 5 to 10-fold) between the predicted efficacious dose and the MTD may be determined and used to justify progression of the compound to IND-enabling studies (Appendix 2). In certain cases (e.g., some oncology indications), a narrower therapeutic margin may be acceptable for monitorable adverse effects such as thrombocytopenia, neutropenia, and the like.

6.8 Summary

In this chapter, we saw that the continuum of compound optimization that started within the hit-to-lead stage of drug discovery (Chapter 5) expands significantly during lead optimization to include a battery of *in vivo* studies aimed at assessing the bioavailability, PK, PD, efficacy, and safety of compounds. Successful lead optimization results in a compound or compounds that meet all the criteria to be declared as clinical DCs. The declaration of a DC marks the formal end of a drug discovery campaign and the beginning of preclinical development. While the discovery campaign typically ends with the successful declaration of DC(s), the discovery project may continue to identify back-up and follow-on compounds as a risk mitigation strategy, in case the lead DC were to prove insufficient for clinical progression. The DC(s) must then go through an additional battery of highly prescriptive studies, referred to as IND-enabling studies, to fulfill the regulatory requirements for submitting an IND application to the FDA, or similar documents to regulatory agencies in other countries. The acceptance of an IND application is a key milestone that must be satisfied to proceed to clinical testing of a drug candidate in human subjects. IND-enabling studies are the topic of Appendix 2. The successful completion of IND-enabling studies and the submission of an IND application to the FDA mark the beginning of the clinical development of the IND.

References

Bowes, J., Brown, A.J., Hamon, J., Jarolimek, W., Sridhar, A., Waldron, G. and Whitebread, S. (2012) Reducing safety-related drug attrition: the use of *in vitro* pharmacology profiling. *Nature Rev. Drug Discov.* **11**, 909–922.

Brach, D., Johnston-Blackwell, D., Drew, A., Lingaraj, T., Motwani, V., Warholic, N.M., Feldman, I., Plescia, C., Smith, J.J., Copeland, R.A. and Keilhack, H. (2017) EZH2 inhibition by tazemetostat results in altered dependency on B-cell activation signaling in DLBCL. *Mol. Cancer Therapeut.* **16**, 2586–2597.

Castro, J.B., Daniels, M.H., Nayak, S., Laidlaw, M. et al. (2024) DHX9 inhibition as a novel therapeutic for cancers with loss-of-function mutations in DNA repair genes BRCA1 and BRCA2. *In*: *Proceedings of the 115th Annual Meeting of the American Association for Cancer Research*; 2024April 5–10, San Diego, CA. Philadelphia (PA): AACR, 2024. Abstract 3908.

Castro, J., Daniels, D.H., Brennan, D., Johnston, B., Gotur, D., Lee, Y.T., Knockenhauer, K.E., Lu, C., Wu, J., Nayak, S. and Collins, C. (2025) A potent, selective, small-molecule inhibitor of HDX9 abrogates proliferation of microsatellite instable cancers with deficient mismatch repair. *Cancer Res.* **85**, 758–776.

Chen, C., Zhu, H., Stauffer, F., Caravatti, G., Vollmer, S., Machauer, R., Holzer, P., H, M., Scheufler, C., Klumpp, M. and Tiedt, R. (2016) Discovery of novel DOT1L inhibitors through a structure-based fragmentation approach. *ACS Med. Chem. Lett.* **7**, 735–740.

Copeland, R.A. (2013) *Evaluation of Enzyme Inhibitors in Drug Discovery: A Guide for Medicinal Chemists and Pharmacologists*. 2nd Edition. Wiley, Hoboken, NJ.

Copeland, R.A. (2016) The drug-target residence time model: a 10-year retrospective. *Nature Rev. Drug Discov.* **15**, 87–95.

Copeland, R.A. (2021) Evolution of the drug-target residence time model. *Expert Opin. Drug Discov.* **16**, 1441–1451.

Copeland, R.A. (2023) *Enzymes: A Practical Introduction to Structure, Mechanism and Data Analysis.* 3rd Edition. Wiley, Hoboken, NJ.

Copeland, R.A. and Knockenhauer, K.E. (2023) The importance of binding kinetics and drug-target residence time in pharmacology. *Brit. J. Pharmacol.* **181**, 4103–4116.

Copeland, R.A., Castro, J., Daniels, M.H., Gotur, D. et al. (2024) Small molecule inhibitors of RNA modifying enzymes as precision cancer therapeutics. *In: Proceedings of the 115th Annual Meeting of the American Association for Cancer Research*; 2024 April 5–10, San Diego, CA. Philadelphia (PA): AACR, 2024. Abstract SY11-02.

Copeland, R.A., Pompliano, D.L. and Meek, T.D. (2006) Drug-target residence time and its implications for lead optimization. *Nature Rev. Drug Discov.* **5**, 7330–7739.

Daigle, S.R., Olhava, E.J., Therkelsen, C.A., Basavapathruni, A. et al. (2013) Potent inhibition of DOT1L as treatment of MLL-fusion leukemia. *Blood* **122**, 1017–1025.

Daniels, M.H., Castro, J., Lee, Y.-T., Gotur, D., Knockenhauer, K.E., Grigoriu, S., Lockbaum, G.J., Cheong, J.E., Lu, C., Brennan, D. and Buker, S.M. (2025) Discovery of ATX968: An orally available allosteric inhibitor of DHX9. *J. Med. Chem.* **68**, 9537–9554.

Deshaies, R.J. and Potts, P.R. (2025) Load and lock: an emerging class of therapeutics that influence macromolecular dissociation. *Science* **389**, eadx3595.

Drew, A., Motwani, V., Eichinger, L., Smith, J. and Raimondi, A. (2018) Mechanism of action of synergistic activity of EZH2 inhibition and IMiDs in preclinical multiple myeloma models. *Cancer Res.* **78** (13 Suppl), 807.

Euhus, D.M., Hudd, C., LaRegina, M.C. and Johnson, F.E. (1986) Tumor measurement in the nude mouse. *J. Surg, Oncol.* **31**, 229–234.

Franklin, M.R., Platero, S., Saini, K.S., Curigliano, G. and Anderson, S. (2022) Immuno-oncology trends: preclinical models, biomarkers, and clinical development. *J. Immunother. Cancer* **10**, e003231.

Gad, S.C., Spainhour, C.B., Shoemaker, C., Stackhouse, D.R., Stricker-Krongrad, A., Downing, P.A., Seals, R.E., Eagle, L.A., Polhamus, K. and Daly, J. (2016) Tolerable levels of nonclinical vehicles and formulations used in studies by multiple routes in multiple species with notes on methods to improve utility. *Int. J. Toxicol.* **35**, 95–178.

Gao, H., Korn, J.M., Ferretti, S., Monahan, J.E., Wang, Y., Singh, M., Zhang, C., Schnell, C., Yang, G., Zhang, Y. and Balbin, O.A. (2015) High-throughput screening using patient-derived tumor xenografts to predict clinical trial drug response. *Nature Med.* **21**, 1318–1325.

Gerges, S.H. and El-Kadi, A.O.S. (2023) Sexual dimorphism in the expression of cytochrome P450 enzymes in rat heart, liver, kidney, lung, brain and small intestine. *Drug Metabol.Dispos.* **51**, 81–94.

Graphodatsky, A.S., Trifonov, V.A. and Stanyon, R. (2011) The genome diversity and karyotype evolution of mammals. *Mol. Cytogenet.* **4**, 22.

Hammes, G.G., Chang, Y.-C. and Oas, T.G. (2009) Conformational selection or induced fit: a flux description of reaction mechanism. *Proc. Natl. Acad. Sci. USA* **106**, 13737–13741.

Ho, P.T.C. and Copeland, R.A. (2024) Discovery and development of tazemetostat. In *Epigenetic Drug Discovery* (H.M. Chan and C.H. Arrowsmith, Eds.). Royal Society of Chemistry, London, UK, pp. 563–595.

Huang, Q. and Riviere, J.E. (2014) The application of allometric scaling principles to predict pharmacokinetic parameters across species. *Expert Opin. Drug Metab. Tox.* **10**, 1241–1253.

Italiano, A., Soria, J.C., Toulmonde, M., Michot, J.M., Lucchesi, C., Varga, A., Coindre, J.M., Blakemore, S.J., Clawson, A., Suttle, B. and AA, M.D. (2018) Tazemetostat, an EZH2 inhibitor, in relapsed or refractory B-cell non-Hodgkin lymphoma and advanced solid tumours: a first-in-human, open-label, phase 1 study. *Lancet Oncol.* **19**, 649–659.

Jairam, R.K., Malluwar, S.R., Sulochana, S.P., Chandrasekhar, D.V., Todmal, U., Bhamidipati, R.K., Richter, W., Srinivas, N.R. and Mullangi, R. (2019) Prediction of human pharmacokinetics of fomepizole from preclinical species pharmacokinetics based on normalized time course profiles. *AAPS PharmSci. Tech.* **20**, 221.

Kerbel, R.S. (2003) Human tumor xenografts as predictive models for anticancer drug activity in humans. *Cancer Biol. Therapy* **2**, S134–S139.

Kim, M.P., Evans, D.B., Wang, H., Abbruzzese, J.L., Fleming, J.B. and Gallick, G.E. (2009) Generation of orthotopic and heterotopic human pancreatic cancer xenografts in immunodeficient mice. *Nature Protocols* **4**, 1670–1680.

Kleiber, M. (1932) Body size and metabolism. *Hilgardia* **6**, 315–353.

Knockenhauer, K.E. and Copeland, R.A. (2024) The importance of binding kinetics and drug-target residence time in pharmacology. *Brit. J. Pharmacol.* **181**, 4103–4116.

Knutson, S.K., Kawano, S., Minoshima, Y., Warholic, N.M., Huang, K.C., Xiao, Y., Kadowaki, T., Uesugi, M., Kuznetsov, G., Kumar, N. and Wigle, T.J. (2014) Selective inhibition of EZH2 by EPZ-6438 leads to potent antitumor activity in EZH2-mutant non-Hodgkin lymphoma. *Mol. Cancer Therapeut.* **13**, 842–854.

Knutson, S.K., Warholic, N.M., Wigle, T.J., Klaus, C.R., Allain, C.J., Raimondi, A., Porter Scott, M., Chesworth, R., Moyer, M.P., Copeland, R.A. and Richon, V.M. (2013) Durable tumor regression in genetically altered malignant rhabdoid tumors by inhibition of methyltransferase EZH2. *Proc. Natl. Acad. Sci. USA* **110**, 7922–7927.

Kuntz, K.W., Campbell, J.E., Keilhack, H., Pollock, R.M. Knutson SK, Porter-Scott M, Richon VM, Sneeringer CJ, Wigle TJ, Allain CJ, Majer CR (2016) The importance of being Me: magic methyls, methyltransferase inhibitors, and the discovery of tazemetostat. *J. Med. Chem.* **59**, 1556–1564.

Kwon, Y. (2001) *Handbook of Essential Pharmacokinetics. Pharmacodynamics and Drug Metabolism for Industrial Scientists*, Kluwer Academic, New York.

Lee, Y.T., Sickmier, E.A., Grigoriu, S., Castro, J. and Boriack-Sjodin, P.A. (2023) Crystal structures of the DExH-box RNA helicase DHX9. *Acta Crystallogr. D Struct. Biol.* **79**, 980–991.

Mao, J., Mohutsky, M.A., Harrelson, J.P., Wrighton, S.A. and Hall, S.D. (2011) Prediction of CYP3A-mediated drug–drug interactions using human hepatocytes suspended in human plasma. *Drug Metab. Dispos.* **39**, 591–602.

Maurer, T.S., Smith, D., Beaumont, K. and Di, L. (2020) Dose predictions for drug design. *J. Med. Chem.* **63**, 6423–6435.

Nguyen, V.T.M., Namba, H., Porter, H., Shlyueva, D., Lopez, E., Melcher, A., Béguelin, W., Melnick, A.M. and Helin, K. (2025) Synergistic antitumor effect of combined EZH2 and DOT1L inhibition in B-cell lymphoma. *Blood* **145**, 2873–2886.

Nishino, M. (2018) Tumor response assessment for precision cancer therapy: response evaluation criteria in solid tumors and beyond. *ASCO Educational Book*, 1019–1029. http://asco.org/ebook

Peterson, J.K. and Houghton, P.J. (2004) Integrating pharmacology and *in vivo* cancer models in preclinical and clinical drug development. *Eur. J. Cancer* **40**, 837–844.

Riviere, J.E. (1999) *Comparative Pharmacokinetics Principles*. Iowa State University Press, Ames, IA, Techniques and Applications.

Scott, C., Dodson, A., Saulnier, M., Snyder, K. and Racz, R. (2022) Analysis of secondary pharmacology assays received by the US Food and Drug Administration. *J. Pharmacol. Toxicol. Methods* **117** (107), 205.

Shargel, L. and Yu, A.B.C. (1999) *Applied Biopharmaceutics and Pharmacokinetics*. McGraw Hill, New York.

Smith, D.A., Beaumont, K., Maurer, T.S. and Di, L. (2015) Volume of distribution in drug design. *J. Med. Chem.* **58**, 5691–5698.

Smith, D.A., Beaumont, K., Maurer, T.S. and Di, L. (2019) Clearance in drug design. *J. Med. Chem.* **62**, 2245–2255.

Stein, E.M., Garcia-Manero, G., Rizzieri, D.A., Tibes, R., Berdeja, J.G., Savona, M.R., Jongen-Lavrenic, M., Altman, J.K., Thomson, B., Blakemore, S.J. and Daigle, S.R. (2018) The DOT1L inhibitor pinometostat reduces H3K79 methylation and has modest clinical activity in adult acute leukemia. *Blood* **131**, 2661–2669.

Swinney, D.C. (2004) Biochemical mechanisms of drug action: what does it take for success? *Nature Rev. Drug Discov.* **3**, 801–808.

Swinney, D.C. (2009) The role of binding kinetics in therapeutically useful drug action. *Curr. Opin. Drug Discov. Devel.* **12**, 31–39.

Teicher, B.A. (2006) Tumor models for efficacy determination. *Mol. Cancer Ther.* **5**, 2435–2443.

Tian, G. (1996) In vivo time-dependent inhibition of human steroid 5 alpha-reductase by finasteride. *J. Pharm. Sci.* **85**, 106–111.

Tummino, P.J. and Copeland, R.A. (2008) Residence time of receptor-ligand complexes and its effect on biological function. *Biochemistry* **47**, 5481–5492.

U.S. Food and Drug Administration (2020) Final guidance for industry: *In vitro* Drug Interaction Studies – Cytochrome P450 Enzymes- and Transporter-Mediated Drug Interactions Guidance for Industry. (http://www.fda.gov/media/135587).

Wienkers, L.C. and Heath, T.G. (2005) Predicting *in vivo* drug interactions from *in vitro* drug discovery data. *Nature Rev. Drug Discov.* **4**, 825–833.

Williams, J. (2018) Using PDX for preclinical cancer drug discovery: the evolving field. *J. Clin, Med.* **7**, 41.

Wilson, B.G., Wang, X., Shen, X., McKenna, E.S., Lemieux, M.E., Cho, Y.J., Koellhoffer, E.C., Pomeroy, S.L., Orkin, S.H. and Roberts, C.W. (2010) Epigenetic antagonism between polycomb and SWI/SNF complexes during oncogenic transformation. *Cancer Res.* **18**, 316–328.

Wood, E.R., Truesdale, A.T., McDonald, O.B., Yuan, D., Hassell, A., Dickerson, S.H., Ellis, B., Pennisi, C., Horne, E., Lackey, K. and Alligood, K.J. (2004) A unique structure for epidermal growth factor receptor bound to GW572016 (lapatinib): relationships among protein conformation, inhibitor off-rate, and receptor activity in tumor cells. *Cancer Res.* **64**, 6652–6659.

Appendix 1

Kinetic and Equilibrium Assessment of Receptor–Ligand Interactions

In Chapter 1, we reviewed the work of A. V. Hill that first provided a quantitative description of the concentration–response relationship between drug doses and pharmacological effect. With over a century of experimental validation of the receptor theory of drug action, we can today describe this concentration–response relationship in molecular terms, based on the binding of drug molecules (more generally ligands, and represented below as L) to specific target proteins (here referred to generically as receptors and represented below as R).

Here, we provide a more mathematical description of the binding and dissociation of ligands to receptors, as the system approaches equilibrium conditions, for the simplest case: a single ligand molecule binding to one molecule of receptor, for which the association of the two binding partners (R and L) occurs in a single kinetic step. Likewise, dissociation of the bimolecular complex (RL) also occurs in a single kinetic step. The reader should note that not all receptor–ligand interactions are well represented by this simplistic model. In some cases, stoichiometries other than 1-to-1 must be invoked to explain the experimental binding curves. In other cases, the assumption of single-step association and/or dissociation is not consistent with the experimental data (see Chapter 6). Finally, in the treatment below, we assume that the concentration of ligand is in great molar excess to that of the receptor. This assumption simplifies the mathematics considerably, but it is not always true in the case of very high-affinity binding interactions (again, see Chapter 6). Our goal here is to introduce the reader to the quantitative analysis of simple receptor–ligand interactions, which are applicable to many situations in pharmacology. The treatment below is taken from Copeland (2023) with some modifications. More advanced treatments of situations that deviate from this most simple model of receptor–ligand binding interactions can be found in the allied texts by this author (Copeland, 2013, 2023).

Molecular Targeted Drug Discovery: A Guide to How Modern Medicines are Created,
First Edition. Robert A. Copeland.

According to the receptor theory of drug action, the binding event resulting in formation of the RL complex is the predicate for all the resulting *in vitro*, cellular, and/or organismal biochemical and pharmacological activity associated with receptor function. Hence, the kinetic and equilibrium treatments to follow are equally applicable (with minor modifications) to cellular and *in vivo* pharmacodynamic (PD) and phenotypic effects that attend drug treatment, as we discuss throughout this text.

A1.1 The Equilibrium Dissociation Constant, K_d

We begin by defining the equilibrium between free (i.e., unoccupied) and ligand-bound receptor molecules. We will assume that the receptor has a single ligand-binding site that is conformationally fixed, so that any molecule of receptor is either free or ligand-bound. Likewise, any ligand molecule must be either free or bound to a receptor molecule. This assumption leads to the following pair of mass conservation equations:

$$[R] = [RL] + [R]_f \tag{A1.1}$$

$$[L] = [RL] + [L]_f \tag{A1.2}$$

where $[R]$ and $[L]$ are the *total* concentrations of receptor and ligand, respectively, $[R]_f$ and $[L]_f$ are the free concentrations of the two molecules, and $[RL]$ is the concentration of the binary receptor–ligand complex.

Under any specific set of solution conditions, an equilibrium will be established between the free and bound forms of the receptor. The position of this equilibrium is quantified in terms of the equilibrium dissociation constant, K_d, for the binary complex at equilibrium:

$$\mathrm{R + L \rightleftharpoons RL}$$

$$K_d = \frac{[R]_f\,[L]_f}{[RL]} \tag{A1.3}$$

The relative affinities (i.e., strength of binding) of different receptor–ligand complexes are inversely proportional to their K_d values; that is, the lower the value of K_d, the higher the affinity (i.e., the greater the strength) of the receptor–ligand binary complex. An easy way to remember this is by referring to the analogy of the required dose of a drug. The lower the required dose, the more effective it is in producing a pharmacological impact.

Equilibrium dissociation constants are thus used to compare affinities of different ligands for a particular receptor, and likewise to compare the affinities of

Table A1.1 Relationship between K_d and $\Delta G_{binding}$ for receptor–ligand complexes at 25 °C.

K_d (M)	$\Delta G_{binding}$ (kcal/mol)[a]
10^{-3} (mM)	−4.08
10^{-4}	−5.44
10^{-5}	−6.80
10^{-6} (μM)	−8.16
10^{-7}	−9.52
10^{-8}	−10.87
10^{-9} (nM)	−12.23
10^{-10}	−13.59
10^{-11}	−14.95
10^{-12} (pM)	−16.31

a) Calculated using equation A1.4. Note that every 10-fold change in K_d corresponds to a change in free energy of approximately 1.36 kcal/mol at 25 °C.

different receptors for a common ligand. The dissociation constant can be directly related to the Gibbs free energy of binding for the receptor–ligand complex (Table A1.1) as follows:

$$\Delta G_{binding} = RT\ln(K_d) \quad (A1.4)$$

At this point, the reader may wonder why drug discovery scientists choose to express affinity relationships in terms of equilibrium dissociation constants rather than the more conventional equilibrium association constants, as found in most chemistry and physics textbooks. The reason is that the dissociation constant has units of molarity and can thus be equated with a specific ligand concentration that leads to half-maximal saturation of the available receptor-binding sites.

A1.2 The Kinetics of Approach to Equilibrium

Suppose that the two molecules, R and L, bind reversibly to each other to form a binary complex, RL. The equilibrium between the free components, R and L, and the binary complex, RL, will be governed by the rate of complex formation (i.e., association of the complex) and by the rate of dissociation of the binary complex. Here, we will define the second-order rate constant for complex association as k_{on}

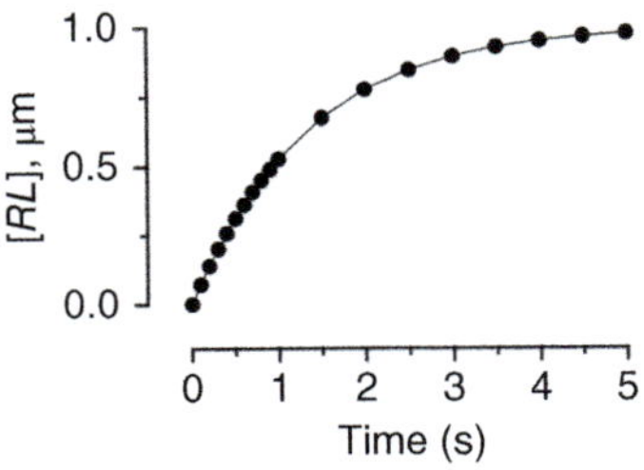

Figure A1.1 Time course for approach to equilibrium after mixing of a receptor and a ligand. The data are fit by nonlinear regression to equation A1.5, from which an estimate of the observed pseudo-first-order rate constant (k_{obs}) is obtained.

and the first-order rate constant for complex dissociation as k_{off}.

$$\mathrm{R} + \mathrm{L} \underset{k_{off}}{\overset{k_{on}}{\rightleftharpoons}} \mathrm{RL}$$

In many cases, the strength of interaction between the receptor and ligand is such that a large excess of ligand concentration over receptor concentration is required to effect significant binding to the receptor (typically, from 100- to >1000-fold excess of ligand over receptor). Hence, under most experimental conditions, association to form the binary complex proceeds without significant diminution of the concentration of free (i.e., unbound) ligand. In this case, the free concentration of ligand can reasonably be approximated by the total concentration of ligand, and we can therefore treat the rate of RL complex formation as a pseudo-first-order reaction:

$$[RL]_t = [RL]_{eq}\left[1 - e^{-k_{obs}t}\right] \tag{A1.5}$$

where $[RL]_t$ is the concentration of binary complex at time t, $[RL]_{eq}$ is the concentration of binary complex formed at equilibrium (at a specific ligand concentration), and k_{obs} is the experimentally determined value of the pseudo-first-order rate constant for approach to equilibrium. Figure A1.1 illustrates a typical kinetic progress curve for a receptor–ligand pair approaching equilibrium binding. The line drawn through the data in this figure is a nonlinear least squares best fit to equation A1.5, from which the researcher can obtain an estimate of k_{obs}.

For reversible binding, it can be shown that the value of k_{obs} is directly proportional to the concentration of ligand present, as follows:

$$k_{obs} = k_{off} + k_{on}[L] \tag{A1.6}$$

Hence, one can determine the value of k_{obs} at a series of ligand concentrations from experiments such as that illustrated in Figure A1.1. A replot of k_{obs} as a function of ligand concentration should then yield a linear fit with slope equal to k_{on} and y-intercept equal to k_{off} (Figure A1.2). From these values, the equilibrium dissociation constant can be determined by means of equation A1.7.

$$K_d = \frac{k_{off}}{k_{on}} \tag{A1.7}$$

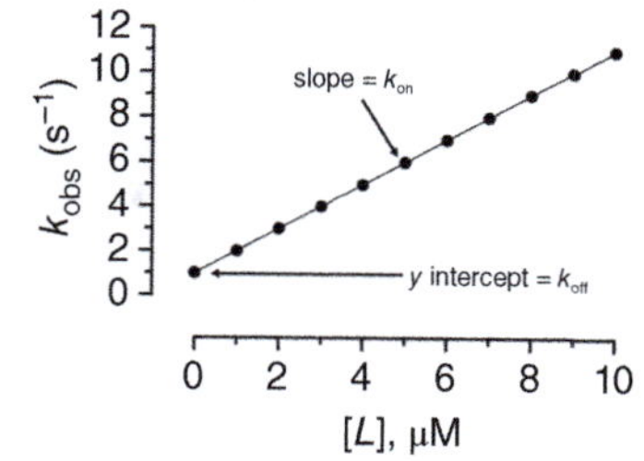

Figure A1.2 Plot of the observed pseudo–first-order rate constant (k_{obs}) as a function of ligand concentration as the reaction approaches equilibrium. As illustrated, the data are fit to a linear function (equation A1.6), from which the values of k_{on} and k_{off} are estimated from the values of the slope and *y*-intercept, respectively.

Pseudo–first-order and first-order rate constants describe the time dependence of many pharmacological processes, including *in vitro* CYP450, microsome, or hepatocyte-catalyzed drug metabolism, *in vivo* drug pharmacokinetics (PK), and the like. These rate constants have units of reciprocal time. In many cases, it is more useful to quantify the rate of a process in units of time to more easily relate this to other relevant time periods (e.g., dosing intervals and the like). If we simply take the reciprocal of the rate constant, we obtain the relaxation time, τ. Another useful metric of process rate is the half-life of reaction, $t_{1/2}$, which is the time interval required for the concentration of reactant to be reduced from its starting value to half of the starting value. The relationship between rate constant (k), τ, and $t_{1/2}$ is presented in equation A1.8.

$$t_{1/2} = 0.693 \times \tau = \frac{0.693}{k} \tag{A1.8}$$

where 0.693 is the value of the natural logarithm of 2 (ln (2)), and k is the first-order or pseudo–first-order rate constant. The mathematical derivation of these terms is presented further in Copeland (2013).

A1.3 Kinetics of Irreversible or Pseudo-Irreversible Processes

Copeland (2023) provides a thorough review of various kinetic processes that are relevant to biological and biochemical mechanisms. This includes situations of consecutive, irreversible (or pseudo-irreversible) reactions such as the following.

$$A \xrightarrow{k_1} B \xrightarrow{k_2} C$$

For this reaction, the concentration of intermediate species B at any specific time is given by the following equation.

$$[B]_t = \frac{[A]_0 \times k_1}{(k_2 - k_1)} \times \left(e^{-k_1 t} - e^{-k_2 t}\right) \tag{A1.9}$$

where $[A]_0$ is the starting concentration of the initial reactant, A.

This situation is analogous to that for a one-compartment model of the PK for an extravascularly administered drug, as described in Chapter 6. In this specific case, the constant $[A]_0$ in equation A1.9 is replaced by the fractional bioavailability of the drug (F), times the dose (D), divided by the volume of distribution (V_d). Additionally, the rate constants k_1 and k_2 of equation A1.9 are replaced by k_e (the rate constant for drug elimination) and k_a (the rate constant for drug absorption) to yield the following equation.

$$[C] = \frac{F \times D \times k_a}{V_d(k_a - k_e)} \times \left(\mathrm{e}^{-k_e t} - \mathrm{e}^{-k_a t}\right) \tag{A1.10}$$

Equation A1.10 was used in fitting the data shown in Figure 6.9B. Note, however, that this model is only valid for a one-compartment model that assumes instantaneous equilibration between drug concentration in systemic circulation and in tissue(s). When equilibration is not rapid, a more complex, two-compartment model must be invoked (see Chapter 6).

A1.4 Binding Measurements at Equilibrium

The equilibrium dissociation constant can be determined from kinetic studies, but in practice, researchers more commonly study receptor–ligand interactions after equilibrium has been established.

A1.4.1 Derivation of the Hill–Langmuir Isotherm

At equilibrium, the concentration of the RL complex is constant. Hence, the rates of complex association and dissociation must be equal. Thus

$$k_{\text{on}}[R]_f[L]_f = k_{\text{off}}[RL] \tag{A1.11}$$

or

$$[RL] = \frac{k_{\text{on}}}{k_{\text{off}}}[R]_f[L]_f \tag{A1.12}$$

Using the equality that $k_{\text{on}}/k_{\text{off}}$ is equivalent to the equilibrium association constant K_a, we obtain

$$[RL] = K_a[R]_f[L]_f \tag{A1.13}$$

Now, in most experimental situations, a good measure of the actual concentration of free ligand or protein is lacking. Therefore, we would prefer an equation in terms of the total concentrations of added protein and/or ligand – quantities that are readily controlled by the experimenter. If we rearrange the mass conservation equation A1.1, using the equality $K_d = 1/K_a$, and a little algebra, we obtain

$$[RL] = \frac{[R][L]_f}{K_d + [L]_f} \tag{A1.14}$$

Again, we consider that under most conditions the concentration of receptor is far less than that of the ligand. Hence, formation of the binary complex does not significantly diminish the concentration of free ligand. We can thus make the approximation that the free ligand concentration is about the same as the total ligand concentration added.

$$[L]_f \approx [L] \tag{A1.15}$$

With this approximation, equation A1.14 can be rewritten as follows:

$$[RL] = \frac{[R][L]}{K_d + [L]} = \frac{[R]}{1 + \frac{K_d}{[L]}} \tag{A1.16}$$

This equation describes a square hyperbola that is typical of saturable binding in a variety of chemical, physical, and biochemical situations, as already described in Chapter 1 and illustrated in Figure A1.3A and B.

It is not necessary, and often not possible, to know the concentration of total receptor in an experiment calling for the use of the Hill–Langmuir isotherm equation. If one has some experimentally measurable signal that is unique to the receptor–ligand complex, one can still construct a binding isotherm. Let us say that we have some signal, Y, that we can follow as a measure of [RL]. In terms of signal, the Hill–Langmuir isotherm equation can be recast as follows:

$$Y = \frac{Y_{\max}}{1 + \frac{K_d}{[L]}} \tag{A1.17}$$

The values of $Y_{\max}$ and K_d are then determined from fitting a plot of Y as a function of ligand concentration to equation A1.17. This form of the Hill–Langmuir equation is equivalent to that presented in Chapter 1 (equation 1.5), except that in the more generic form presented in Chapter 1, we replace Y and $Y_{\max}$ with E and $E_{\max}$ and replace K_d with EC_{50} (see below).

Note that the ratio $Y/Y_{\max}$ is equal to $[RL]/[R]$ for any concentration of ligand. The ratio $[RL]/[R]$ is referred to as the *fractional occupancy* of the receptor and is often represented by the symbol B (for *bound* receptor; binding isotherm shown in Figure A1.3C and D). Equations A1.16 and A1.17 can both be cast in terms of fractional occupancy by dividing both sides of the equations by $[R]$ and $Y_{\max}$, respectively:

$$\frac{[RL]}{[R]} = \frac{Y}{Y_{\max}} = B = \frac{1}{1 + \frac{K_d}{[L]}} \tag{A1.18}$$

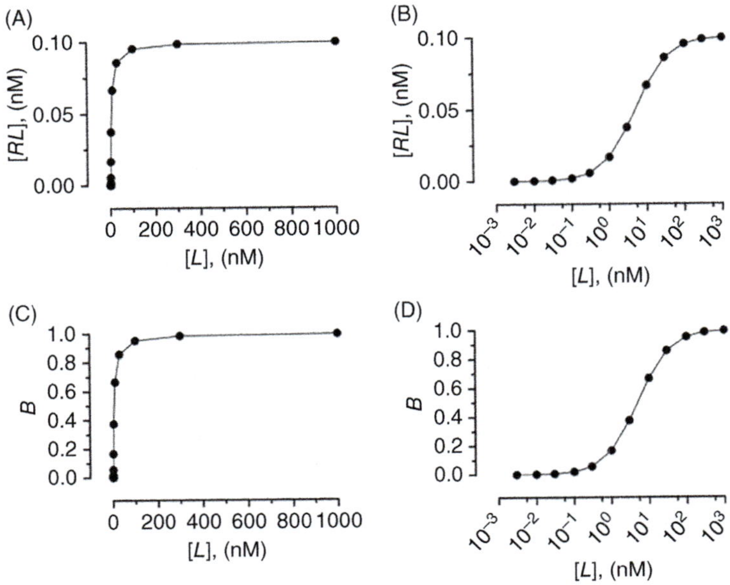

Figure A1.3 (A) Hill–Langmuir isotherm for formation of the binary RL complex as a function of ligand concentration. (B) As in (A) but plotted on a semilog scale. (C) The data from (A) expressed in terms of fractional receptor occupancy, *B*. (D) As in (C) but plotted on a semilog scale. The data in plots (A) and (B) and plots (C) and (D) are fit to equations A1.16 and A1.18, respectively.

This form of the Hill–Langmuir isotherm equation is useful in normalizing data from different receptor samples, which may differ slightly in their total receptor concentrations, for the purposes of comparing K_d values and overall binding isotherms.

A1.4.2 Competition Among Ligands for a Common Target Receptor

In many situations, one may wish to identify and/or optimize compounds that bind to a target in a fashion that is mutually exclusive with the binding of a known ligand, such as the natural substrate of an enzyme or the natural agonist of a G-protein–coupled receptor (i.e., competitive ligands as defined in Chapter 2). Let us designate the known ligand as L and a competing ligand as *X*. The equilibrium dissociation constant for these two ligands can be designated K_L and K_X, respectively.

If ligand L has some unique physicochemical property that can be quantitatively measured (e.g., a unique spectroscopic signal) or can be chemically labeled to impart a unique physicochemical property (e.g., fluorescence, radioactivity), then an assay may be configured to measure the amount of RL complex formed at a fixed concentration of total receptor (R) and specific concentrations of total ligand (L); from such data, the value of K_L can be determined as described by equation A1.18. Let us say that we fix the concentration of L and measure the amount of RL complex formed at varying concentrations of the competing ligand X. As we increase the concentration of X, the concentration of RL formed will decline from a maximum value at $[X] = 0$ to a value of zero (or some baseline value) at $[X] = \infty$, due to the competition between the two ligands for their common receptor binding site. It can be shown that the fractional occupancy of ligand L as a function of the concentration of ligand X can be described by the following equation.

$$\frac{[RL]}{[R]} = \frac{1}{1 + \frac{K_L}{[L]}\left(1 + \frac{[X]}{K_X}\right)} \tag{A1.19}$$

Alternatively, if we use some signal, Y, that is unique to the RL complex (see discussion above), we can measure the value of this signal at a fixed concentration of $[R]$ and $[L]$ and varying concentrations of the competing ligand X. Let Y_X represent the signal amplitude when the ligand X concentration is $[X]$ and Y_0 the signal amplitude when the ligand X concentration is zero. The ratio of these signal values can be related to the midpoint concentration of X that elicits half-maximal inhibition of RL complex formation, the EC_{50}, as follows.

$$\frac{Y_X}{Y_0} = \frac{1}{1 + \frac{[X]}{EC_{50}}} \tag{A1.20}$$

And it can be shown that when X is competitive with respect to L, the EC_{50} is given by

$$EC_{50} = K_X\left(1 + \frac{[L]}{K_L}\right) \tag{A1.21}$$

Equation A1.21 represents the Cheng-Prusoff equation for competitive inhibition of RL formation by X. This equation and that for other forms of inhibition have been presented elsewhere (e.g., Copeland, 2013; 2023) and are discussed further in Chapters 2 and 4 of this text.

A1.5 Extension to Other Pharmacological Situations

The concentration–response relationship captured in the Hill–Langmuir equation (equation A1.18) applies to many pharmacological situations beyond biochemical and biophysical observations of receptor–ligand binding. Indeed, in Chapter 1, we saw that Hill originally derived this equation in terms of the phenomenological observations of muscle contractions that attend treatment of muscle tissues with nicotine, and the inhibition of these contractions by curare. In such cases, rather

Table A1.2 Some examples of situational terms for the concentration of compound eliciting a half-maximal effect and descriptions thereof.

Term	Situation	Description
K_d	Receptor–ligand binding experiments	Thermodynamic equilibrium dissociation constant
EC_{50}	Measure of compound potency used generally in biochemical, biophysical, PD, and phenotypic cellular assays	Concentration of compound eliciting an effect equal to 50% of the maximal effect amplitude expected at infinite compound concentration
IC_{50}	Measure of inhibitor/antagonist potency in receptor–inhibitor binding, PD, and phenotypic cellular assays	Concentration of compound eliciting 50% inhibition of receptor function or downstream PD, relative to function observed in the absence of inhibitor
AC_{50}	Measure of activator/agonist potency in receptor–agonist, enzyme–activator and similar binding, and PD assays	Concentration of compound eliciting 50% of the activation of receptor function expected at infinite activator/agonist concentration
K_i	Inhibition constant, a measure of inhibitor potency	Thermodynamic equilibrium dissociation constant in the specific case of a receptor–inhibitor/antagonist complex
K_A	Activation constant, a measure of activator potency	Thermodynamic equilibrium dissociation constant in the specific case of a receptor–activator/agonist complex
K_S	Measure of enzyme–substrate affinity	Thermodynamic equilibrium dissociation constant in the specific case of an initial enzyme–substrate complex under equilibrium conditions
K_M	Michaelis constant, a measure of enzyme–substrate affinity	Substrate concentration eliciting 50% of the maximum velocity (V_{max}) of enzyme catalysis expected at infinite substrate concentration under steady-state conditions
K_I	Measure of potency for irreversible (e.g., covalent) inactivator/antagonists	Concentration of an irreversible receptor inactivator/antagonist that elicits a half-maximal rate of inactivation relative to that expected at infinite concentration of inactivator

than measuring RL complex formation *per se*, one is measuring some pharmacological effect resulting from ligand (i.e., drug) binding to its target *in vitro*, in cells, tissues, organs, or animals. Thus, here we are measuring the pharmacological consequence(s) of receptor–ligand complex formation (according to the receptor theory of drug action), which are generally referred to as pharmacodynamics (PD).

This is why in Chapter 1, we used the term E instead of $[RL]$ for the amplitude of the effect, and EC_{50} instead of K_d in our expression of the Hill–Langmuir equation. This is the most general form of the Hill–Langmuir concentration–response relationship and is universally applicable. However, it is important to note that the value of EC_{50} may not be quantitatively equivalent to the K_d of the receptor–ligand complex, depending on the experimental details at play. This point is discussed in greater detail in the case of enzyme–inhibitor interactions in Copeland (2013; 2023). Also, as pointed out in Chapter 1, the term EC_{50} is often replaced with alternative terms that relate to specific pharmacological situations, such as inhibition of target function, target activation, and the like. Table A1.2 summarizes some of the more common situational replacements for EC_{50} as the term used for the concentration of compound eliciting a half-maximal effect.

Appendix 2

Investigational New Drug (IND) Applications

As stated in Chapter 6, the declaration of a development candidate (DC) marks the culmination of drug discovery and the starting point for preclinical drug development. Preclinical drug development encompasses a plethora of activities, including biomarker assay refinement and validation (translational biology and translational medicine), exploration of additional clinical indications for a targeted drug, refining patient stratification strategies and the understanding of the underlying biology of selective essentiality, drug resistance mechanisms, drug combination potential, and chemistry efforts aimed at optimizing the salt form and formulation for the drug (see below). However, the most immediate and important role of preclinical drug development is preparing an investigational new drug (IND) application (or its equivalent outside the United States) as a prerequisite to beginning clinical testing of a drug in human subjects.

Under US federal law, a drug cannot be transported across state lines unless it is subject to an approved marketing application. The IND application provides an exception to this legal requirement, allowing the sponsor (e.g., the company, organization, or physician/researcher) to distribute an IND to clinical trial site locations across multiple states. The Food, Drug and Cosmetic Act (United States Code of Federal Regulations [CFR] Title 21, Chapter 1, Subchapter D, Part 312) codifies the definitions and requirements of the IND application. Once a sponsor files an IND application, thus articulating their desire to test a new drug for its potential use (i.e., diagnostic or therapeutic) in humans, the legal status of the compound changes; it now becomes a new drug and is subject to the specific requirements of the drug regulatory system, under the auspices of the FDA.

Hence, the IND application is a critical aspect of early drug development. It must be written and assembled according to specific rules, templates, and processes. This is best left to experts that specialize in medical writing, IND

Molecular Targeted Drug Discovery: A Guide to How Modern Medicines are Created,
First Edition. Robert A. Copeland.

applications, and other regulatory interactions; most companies either have devoted internal groups to oversee these processes or contract with specialists in this area. However, as described below, the IND application is rich in preclinical scientific data. Therefore, the groups that write and assemble the IND application typically work in collaboration with the research team that is responsible for data generation. From the viewpoint of the regulatory agency (e.g., FDA), the primary focus of the IND application is to assure that the data collected within is sufficient to establish that the drug will not expose human subjects to unreasonable risks in its use during early-stage clinical investigations.

A2.1 Types of IND Applications

There are two categories of IND applications: commercial and research (noncommercial). Commercial IND applications are filed by companies and similar organizations, with the goal of collecting sufficient clinical data to eventually support the filing of a new drug application (NDA) or biologics license application (BLA) for FDA approval and subsequent marketing of a therapeutic. Research IND applications are mainly filed by academic groups or not-for-profit organizations for the purposes of advancing scientific knowledge and exploring novel therapeutic approaches. Research IND applications are not intended to generate data in support of eventual product marketing.

Within these two broad categories, there are three basic types of IND applications that may be filed: Investigator, Emergency Use, and Treatment INDs. For companies and other organizations seeking to explore the clinical utility of a novel drug, an Investigator IND is what is usually filed. This must be submitted by a physician who initiates and conducts the investigation and under whose direction the drug is administered or dispensed. Investigator INDs usually study unapproved drugs but can also involve study of an approved drug for a new indication or in a new patient population.

Emergency Use INDs provide the FDA with the authority to allow the use of an experimental drug in life-threatening conditions for which the urgency of the need does not allow time for the submission of an Investigator IND and where the potential benefit outweighs the risk of the new investigational drug being used.

Treatment INDs are intended for experimental drugs that show promise in early clinical testing, for serious or immediate life-threatening situations, while the final clinical work is conducted and the review by FDA takes place (see FDA.gov for further information).

A2.2 Key Content of an IND Application

The content of an IND application is arranged into eight major sections that cover three key categories of information for review: safety assessment, clinical trial design, and manufacturing and quality control of the drug substance and drug product. The eight major sections are described below (Panchal 2024).

1) **General Investigational Plan:** An overview of the development plan for the drug. This includes the intended use of the drug, a rationale for development and a clinical trial plan.
2) **Investigator's Brochure:** A document that provides clinical investigators with a detailed description of the therapeutic thesis, preclinical data, clinical plans, potential risks, and potential benefits of the drug to be studied.
3) **Clinical Protocols:** A description of the proposed design of the clinical trial, including details of the objectives, endpoints, patient population, inclusion and exclusion criteria, dosing plan and schedule, safety monitoring, data collection, and data analysis protocols.
4) **Chemistry, Manufacturing, and Controls (CMC):** Details of the manufacturing process for the investigational drug, its composition, and quality control measures. This includes setting specifications for drug purity, potency and stability.
5) **Pharmacology, Pharmacokinetics, and Toxicity Studies:** A description of the preclinical studies performed to address the pharmacological effects, pharmacokinetics (PK), toxicity, and safety of the drug in animal models. Although the emphasis here is on *in vivo* studies, for current, molecular targeted drugs, this section also contains detailed information on *in vitro* studies of target affinity, selectivity, cellular effects, cellular selectivity (i.e., selective essentiality), structural biology, biophysics, and other supporting data that establish a causal link between specific target engagement and pharmacology.
6) **Previous Human Experience with the Drug** (if applicable): A summary of any previous studies of the investigational drug in human subjects that may have been conducted in different countries or in different clinical indications.
7) **Investigator Qualifications:** A summary of the qualifications and experience of the investigators that will conduct the clinical trial, to ensure the safety and integrity of the trial.
8) **Informed Consent and Institutional Review Board (IRB) Approvals:** Each clinical site must establish an independent institutional review board (IRB) that reviews the trial design and protocols to ensure that it conforms to

ethical standards and that the rights of participating subjects are protected. The IND application must document commitments to obtain informed consent from research subjects, to obtain review of the study by an IRB, and to adhere to the IND regulations.

A2.3 Review by FDA

Upon FDA receipt of the IND application, a 30-calendar-day review period begins. After an initial administrative review of the application, a scientific review of the data ensues, during which various FDA staff evaluate the application with emphasis on safety, clinical trial design, and CMC. The goal of the review by FDA is to ensure that the investigational drug is safe for human studies and that the clinical trial design is ethical and scientifically sound.

During this review period, the FDA may request additional information, clarification, or modifications to the IND. These requests are typically made in writing, and the sponsor is asked to respond promptly (within 30 days), also in writing. At the end of the 30-day period, the FDA will render a decision based on its review of the IND. According to Panchal (2024), there are three possible actions that the FDA can take at this point. These are as follows.

1) **Trial May Proceed (No Clinical Hold):** This indicates that the sponsor may move forward with the clinical studies. Nevertheless, the FDA may still provide recommendations for improving the trial, which the sponsor should consider implementing.
2) **Clinical Hold:** This occurs when the FDA review finds that the investigational drug represents an unacceptable risk to human subjects or when the application is deemed incomplete or deficient. Hence, the sponsor may not proceed with the clinical trial until the issues raised have been satisfactorily resolved. A clinical hold can be full or partial. A full clinical hold means that the clinical trial is suspended in its entirety; hence, the sponsor may not proceed with any aspect of the trial until the hold is rescinded. A partial clinical hold allows certain aspects of the trial to proceed whereas other aspects are on hold. Clinical holds may be issued for several reasons such as safety concerns, insufficient information of manufacturing, insufficient preclinical data, or protocol deficiencies (Panchal 2024). With any type of clinical hold, the FDA will provide the sponsor with details of why the hold was issued and the steps the sponsor must take to resolve these issues. The sponsor can then address the concerns raised by the FDA and submit a written response. The FDA will reevaluate the IND application considering the response(s) to decide whether to lift the hold.
3) **Recommendation for Changes:** Even without placing an IND on clinical hold, the FDA may offer recommendations for changes to the trial protocol

or other specific items within the applications. Examples of recommendations that the agency might make include adjustments to dose, changes to inclusion/exclusion criteria, additional safety monitoring during the trial, and the like. These recommendations are not mandatory, but sponsors would be wise to consider such recommendations with great care.

A2.4 IND-Enabling Studies

Two broad categories of data are generated to enable the filing of an IND application: CMC and animal pharmacology, PK, and toxicity studies. In aggregate, these studies are commonly referred to as IND-enabling studies. Pivotal toxicity studies included in support of an IND application must be performed in compliance with good laboratory practices (GLPs) for nonclinical studies, as detailed in 21 CFR, Part 58 (www.fda.gov). These standards provide a framework for ensuring the quality and integrity of nonclinical laboratory studies and give detailed requirements for the development of study protocols, data integrity, quality assurance, and data archiving. Likewise, the production of any drug materials for clinical use (drug substance, drug product, etc.) must be performed in compliance with current good manufacturing practices (GMPs), as detailed in 21 CFR, Parts 210 through 212.

A2.4.1 CMC Studies

With respect to therapeutic agents for use in human subjects, the FDA uses the terms *drug substance* or *active pharmaceutical ingredient* (API) to refer to the molecular component of the drug that exerts the intended pharmacological or therapeutic effect. Additionally, the FDA uses the term *drug product* to refer to the final formulation to be administered to human subjects. In addition to the API, the drug product may include inert/inactive components known as excipients (e.g., binders, fillers, disintegrants, and coatings) used to stabilize or deliver the drug.

CMC studies are performed and documented to provide detailed information on the manufacturing and formulation of the drug substance and drug product and on the stability and storage conditions for the drug product. The critical data that must be collected and reported within the IND application are as follows.

A2.4.1.1 Drug Substance

A detailed description of the drug substance, including its chemical structure, molecular formula, and physical and chemical properties relevant to its activity, must be provided. For small-molecule drugs, details of the synthesis, starting materials, intermediates, and reaction conditions must be detailed. For biologics, which may be produced in living organisms, a complete description of the biological processes used to produce the drug must be provided.

A2.4.1.2 Drug Product

The sponsor must provide a detailed description of the formulation of the drug product, including the concentration of drug substance, choice of excipients, and physical form of the drug to be delivered to patients (e.g., tablets, capsules, injections, topical creams). The sponsor must also provide details on the container/closure system to be used to package the drug product, as well as any necessary storage conditions to maintain the stability of the drug. Stability testing (see below) must provide data to demonstrate that the drug product remains stable and effective under the storage and handling conditions proposed by the sponsor.

A2.4.1.3 Manufacturing Process

The sponsor must provide a detailed description of the step-by-step process used to synthesize, purify, and formulate the drug product, including details of the equipment and facilities used in the manufacturing process. For small-molecule drugs, the chemical reactions, purification methods, and the quality control measures used must be documented. For biologics, details of the cell culture methods, fermentation, purification of proteins or nucleic acids, and viral inactivation or removal steps must be documented. In either case, the FDA requires the sponsor to demonstrate that the manufacturing process is well-controlled and capable of producing a consistent product within acceptable quality standards. This is often accomplished using process validation studies, aimed at demonstrating that the manufacturing process performs as intended over a range of conditions and at varying production scales.

A2.4.1.4 Specifications and Quality Control

Specifications are defined as predetermined criteria that the drug substance and drug product must meet to be considered acceptable for clinical use. Specifications of purity, potency, identity, and contamination level must be defined and documented. Quality control measures are put in place to ensure that each batch of drug substance and drug product produced meets the predetermined specifications. The sponsor must provide details of the analytical methods used to test the quality of the drug, including validated assays of potency, purity, and stability. The CMC section of the IND application must also provide data from batch release testing to show that the drug substance and drug product consistently meet all specifications for each batch to be used clinically.

A2.4.1.5 Stability Testing

Stability testing is performed under a variety of storage conditions (e.g., room temperature, refrigeration, freezing) and environmental conditions (e.g., different light levels, humidity) to provide data on how long the drug product remains chemically and physically stable and effective. The sponsor must provide results of stability testing to support the proposed shelf life of the drug product and the specific proposed conditions for storage and handling. These data also inform on the appropriate expiration date for the drug product.

A2.4.1.6 Environmental Impact Considerations

The potential environmental impact associated with the manufacturing process or disposal of drug product must be assessed by the sponsor. This may be particularly important for biologics for which the manufacturing process involves the use of genetically modified organisms (GMOs), or small-molecule drugs for which the manufacturing process requires hazardous chemicals (e.g., solvents). The sponsor may be required to submit an Environmental Assessment or request a Categorical Exclusion if the drug product does not pose a significant environmental risk.

A2.4.2 Pharmacology, Pharmacokinetics, and Toxicity Studies

Both *in vitro* and *in vivo* studies inform on different aspects of compound pharmacology and toxicity, particularly for molecular targeted drugs. Hence, both types of data must be collected and reported within the IND application. *In vivo* pharmacology and toxicity studies are typically referenced to a vehicle control group, as described earlier in this text. The *in vivo* PK properties of the drug are also reported within this section of the IND application.

A2.4.2.1 Pharmacology Studies

The required pharmacology studies can be divided into two broad categories: primary and secondary pharmacology. Primary pharmacology studies include *in vitro* receptor binding or enzyme inhibition data (as described earlier), cellular assays of target engagement, mechanism of action, and function, as well as animal models of efficacy that establish a dose–response relationship and help to inform on the therapeutic index of the drug by defining the minimum efficacious concentration (see Chapter 6).

Secondary pharmacology studies are focused on the interactions of the drug with nontarget receptors, enzymes, transporters, channels, and other proteins. Secondary pharmacology also includes *in vivo* studies of the effect of a compound on physiological systems not directly related to the disease to be treated; examples include effects on heart rate, heart rhythm, renal function, liver function, and the like (e.g., Safety Pharmacology).

A2.4.2.2 PK Studies

Drug PK must be reported in at least two species – one rodent and one non-rodent species. Key parameters to be reported include C_{max}, T_{max}, AUC, half-life, clearance rate, and volume of distribution at steady state. In addition to the systemic circulation (e.g., plasma), researchers may also study PK in specific tissue compartments and organs, such as tumor, liver, kidney, brain, and bone marrow, depending on the compound, the target, and the disease indication.

A2.4.2.3 General Toxicity Studies

The toxicity of the drug must be reported in at least two species (as above) at multiple doses tested in each species. The FDA requires the sponsor to report both acute and chronic toxicity studies. Acute toxicity refers to the effects of single-dose studies, whereas subchronic and chronic toxicity refers to the effects of repeat-dose exposure.

Acute Toxicity Studies Acute toxicity studies (also referred to as maximum tolerated dose or MTD studies) are intended to assess the effects of a single, high dose of drug. Typically, three levels of drug in each of the two animal species are tested in multiple groups of each species (both males and females of each), with the goal of defining the range of tolerated doses and the MTD (the highest dose of drug that does not cause significant adverse effects). The drug is administered, and the health of the animals is monitored for signs of toxicity over a period of (typically) 24–72 hours. Key endpoints for such dose studies include mortality rate, clinical signs of toxicity (e.g., changes in behavior, mobility, respiration, and the like), and changes in body weights. Acute toxicity studies also inform on the appropriate dose range to be used in dose-range-finding (DRF) toxicity studies.

Subchronic and Chronic Toxicity Studies Subchronic and chronic toxicity studies assess the effects on organisms of repeated or long-term exposure to a drug. These studies are designed to evaluate the cumulative toxicity of the drug. Again, these studies must be conducted at multiple dose levels of the drug (typically three dose levels plus a vehicle control group) in two species (one rodent and one non-rodent) and are usually conducted over a 7–14-day (DRF), 28-day (IND enabling), or longer period (chronic toxicology) to support a marketing application.

The animals are administered drug according to a predetermined route (e.g., oral) and dose schedule (e.g., QD, BID, etc.) and are monitored for signs of toxicity such as clinical signs, changes in body weight, lethargy, food/water intake, stool changes, and changes to physiological functions. At the end of the study, surviving animals are euthanized, organs are weighed, and tissue samples are collected for histopathology analysis to assess any damage to organs or tissues. Key endpoints include changes to clinical chemistry, coagulation, and hematology parameters (e.g., liver enzymes, blood cell count), and any histological changes to organs and tissues. The dose dependence of any observed signs of toxicity is also defined, with the goal of determining the no adverse effect level (NOAEL), defined as the highest dose at which no significant adverse effects are observed. The NOAEL for the IND-enabling toxicity study is in turn used, with an appropriate safety factor, to define the recommended starting dose for human trials.

In at least one pivotal toxicity study in rodent and non-rodent species, a subset of the animals from two (control and high dose) or more dose levels are not euthanized at the end of the dosing period but instead are kept alive for a defined period without further drug administration (the recovery period). These animals are referred to as recovery animals. At the end of the predefined recovery period, these animals are also euthanized and subjected to necropsy. The purpose of the recovery animal groups is to determine the reversibility of any signs of toxicity that are observed in the main group of animals.

A2.4.2.4 Specialized Toxicity Studies

Developmental and Reproductive Toxicity (DART) Studies For drugs intended for use in females of child-bearing potential, or drugs to be used during pregnancy, developmental and reproductive toxicity (DART) studies are required. A battery of tests is conducted in both genders for animals during mating, pregnancy, and lactation. For drugs intended to treat cancer indications, only embryofetal toxicity studies are required. Further details of these specialized studies are presented in Panchal (2024) and Gad (2008).

Genotoxicity Genotoxicity is studied to determine if the drug may cause genetic mutations or DNA damage. This is typically studied using the Ames test for bacterial mutation, an *in vitro* test for chromosomal aberrations, the *in vitro* micronuclei test in cells, and the *in vivo* micronucleus test for chromosomal damage in animals (Panchal 2024). Not all these studies are required in the case of drugs intended to treat cancer indications (see International Conference on Harmonisation of Technical Requirements for Registration of Pharmaceuticals for Human Use [ICH] S9; http://www.database.ich.org).

Carcinogenicity Studies Carcinogenicity studies are required for drugs intended for long-term use, especially for non-life-threatening conditions. The purpose of these studies is to assess the potential of the drug to cause cancer in animals over an extended period. This is typically conducted in rodents (rats or mice) over a 24-month dosing period in normal or transgenic animals. Key endpoints for such studies include incidence of tumors, histopathology examination of tissues for signs of abnormal cell growth or tumor formation, and assessment of the dose–response relationship for carcinogenicity.

References

Gad, S.G. (2008) *Preclinical Development Handbook: Toxicology*. Wiley, Hoboken, NJ.

Panchal, N. (2024) *Navigating Investigational New Drug (IND) Applications: A Comprehensive Guide*. DrMedHealth, Stafford, VA.

Index

Molecular Targeted Drug Discovery: A Guide to How Modern Medicines are Created, First Edition. Robert A. Copeland.

b

d

e

i

k

l

Printed and bound by CPI Group (UK) Ltd, Croydon, CR0 4YY

17/09/2026

14964233-0001